Combustion in Engines
Technology, Applications and the Environment

Conference Planning Panel

D H Tidmarsh, *BSc, PhD, CEng, FIMechE, MBIM* **(Chairman)**
Sheffield Hallam University
Sheffield

M Ahearne, *BSc, CEng*
British Gas plc
London

A Cole, *BSc, PhD, MSAE*
Engineering Research and Application Limited
Dunstable

P Hurd, *CEng, FIMechE*
European Gas Turbines Limited
Lincoln

A Mirren
Ford Motor Company Limited
Laindon

A Osborn, *FIMechE*
Perkins Technology Limited
Peterborough

B L Ruddy, *CEng, MIMechE*
AE Piston Products Limited
Bradford

W A Woods, *PhD, DEng, FEng, FIMechE*
Queen Mary and Westfield College
University of London

Proceedings of the Institution of Mechanical Engineers

I MECH E

Combustion in Engines
Technology, Applications and the Environment

International Conference

1–3 December 1992
Institution of Mechanical Engineers, Birdcage Walk, London

Sponsored by
Combustion Engines Group of the
Institution of Mechanical Engineers

In association with
Institute of Energy
Institute of Metals
SAE International
American Society of Mechanical Engineers
Japan Society of Mechanical Engineers
Société des Ingenieurs de l'Automobile
Associazone Tecnica dell'Automobile

IMechE 1992–10

 Published for IMechE by
Mechanical Engineering Publications Limited

First Published 1992
This publication is copyright under the Berne Convention and the
International Copyright Convention. Apart from any fair dealing for the
purpose of private study, research, criticism or review, as permitted
under the Copyright, Designs and Patents Act, 1988, no part may be
reproduced, stored in a retrieval system, or transmitted in any form or by
any means, electronic, electrical, chemical, mechanical, photocopying,
recording or otherwise, without the prior permission of the copyright
owners. Reprographic reproduction is permitted only in accordance with
the terms of licences issued by the Copyright Licensing Agency,
90 Tottenham Court Road, London W1P 9HE. *Unlicensed multiple
copying of the contents of this publication is illegal.* Inquiries should be
addressed to: The Managing Editor, Mechanical Engineering Publications
Limited, Northgate Avenue, Bury St. Edmunds, Suffolk, IP32 6BW.

Authorization to photocopy items for personal or internal use, is granted
by the Institution of Mechanical Engineers for libraries and other users
registered with the Copyright Clearance Center (CCC), provided that the
fee of $0.50 per page is paid direct to CCC, 21 Congress Street, Salem,
Ma 01970, USA. This authorization does not extend to other kinds of
copying such as copying for general distribution for advertising or
promotional purposes, for creating new collective works, or for resale,
085298 $0.00 + .50.

© The Institution of Mechanical Engineers 1992

ISBN 0 85298 803 6

A CIP catalogue record for this book is available from the British Library.

Contents

Diesel engine combustion chamber design with three-dimensional flow computations

A A BORETTI, MSc, PhD, P NEBULONI, MSc, M G LISBONA, MSc, P MILAZZO, MSc
Fiat Research Centre, Orbassano, Italy

SYNOPSIS

IDI Diesel engines for passenger cars are now very attractive for their interesting compromise between emissions and performances. The use of advanced computational techniques, properly supported by advanced and standard experimental techniques, considerably speeds up the design process of these engines. The paper presents an application of a computational tool based on the solution of the three dimensional, unsteady, turbulent, compressible, reacting, two-phase flow within the main chamber and the prechamber of an IDI Diesel engine. The tool is based on the SPEED_DC code, developed by Imperial College of London and the Joint Research Committee of European Car Manufacturers for Diesel engines.

1. INTRODUCTION

At present, the automotive industry is showing a great interest in the application of the indirect injection system to small high speed Diesel engines, due to its low emissions and fuel consumption accompanied by an adequate specific power output.

In IDI Diesel engines, the combustion chamber is made by two separated volumes, the cylinder volume and the prechamber one, connected by a passage. In a swirl-type prechamber, the flow is forced by the connecting passage to tangentially enter the prechamber. During the compression stroke, a strong swirling flow is generated in the prechamber by the motion of the piston in order to achieve the required fuel-air mixing rate. A glow plug, used as a cold-starting aid, is located in the prechamber just in front of the region where the flow enters the prechamber, strongly affecting the set up of the swirling motion. Fuel injection occurs at relatively (for Diesel engines) low pressure. The injector generally is a single-hole Pintle-type nozzle. The spray produced has an hollow cone shape, characterised by high initial velocities and small droplet diameters. Due to the small dimensions of the prechamber and the high momentum of the fuel spray, liquid fuel usually impinges the walls in the bottom region of the prechamber. Droplets coalescence, collisions, break-up and evaporation produce a fuel vapour that mixes with the air, and finally, under

proper temperature and pressure conditions, autoignites. The initial burning rate shows a relatively (for Diesel engines) gradual increase, because of the ignition delay is reduced by the higher compression ratio. After autoignition, evaporation rises and combustion is controlled by the mixing rate between fuel and air. The beginning of the combustion process occurs in the prechamber near TDC and the combustion then propagates into the main chamber while the piston is moving downward.

During the last few years, there have been many experimental and computational studies on IDI Diesel engines. The paper of Heywood et al. (13,14) provides a global thermodynamic analysis dealing with both performances and emissions and heat transfer as well. Other thermodynamic analysis are provided by Watson et al. (15). Detailed fluid dynamic and combustion experimental data can be found in Zimmerman (7), Duggal (8), Meintjes et al. (9) and Charlton et al. (5). Other combustion data are given by Toshiaki et al. (20). A few computational fluid dynamic and combustion analysis can be found in the literature, namely Pinchon (6), Zellat et al. (10), and Komatsu et al. (12). Only the first two studies, made by using the KIVA code, deal with realistic engine geometries. The papers of Charlton et al. (11) and Sitkej (16) basically deals with heat transfer analysis. Reference works for Pintle injection are finally those of Whitelaw et al. (17,18) and Zhang et al. (19).

Due to the large number of physical processes involved in the engine operation and their mutual influences, the full understanding of the relationships between prechamber and cylinder geometries, fuel injection parameters and engine operating parameters require the use of advanced computational tools. Once a proper pre- and post-processing environment has been defined, those computer codes now available reach a wide range of usability within an engine design environment. The solution of the three dimensional, unsteady, turbulent, compressible, reacting, two-phase flow within the cylinder proves indeed to be a quite powerful optimisation tool. The paper shows some examples of application to an IDI Diesel engine, presented as a sequence of instantaneous flashes within the cylinder of the physical processes evolution.

2. COMPUTATIONAL METHODOLOGY

Computational fluid dynamic has now a very wide range of applications in industry, with clearly and well defined benefits. CFD allows to reduce the number of prototypes in a design cycle, thus reducing design timing and costs. CFD simulates flow conditions too costly or impractical for experiments. CFD finally provides comprehensive data throughout all the flow domain, thus improving design capabilities. Despite of all these benefits, the use of CFD in design is not widespread in industry. This is due to the fact that CFD codes are not black boxes easy to use that perfectly reproduce physics. A CFD methodology is a result of the understanding of basic physical processes, their mathematical modelling, the solution of the mathematical model equations with a numerical method. The practical usability of a CFD methodology in a design environment depends mainly on the computational efficiency and accuracy of the CFD code, and on the efficiency of the pre- and post-processing devices for model generation and visualisation of results.

The code used here is the SPEED_DC code (1), developed by Imperial College of London and the Joint Research Committee of European Car Manufacturers for Diesel engines. The code solves the three dimensional, unsteady, turbulent, compressible, reacting, two-phase flow within the cylinder of DI/IDI Diesel engines. The SPEED_DC code is only briefly described here. Better details of both the mathematical model and the numerical solution algorithm can be found elsewhere (1).

For the gas phase, the partial differential transport equations for mass, momentum, energy, chemical species and turbulent fields are solved. Turbulence effects are computed on the basis of a two equation model, describing the turbulence velocity and length scales, while chemistry is considered on the basis of a two equation, global, irreversible combustion model, involving generic "fuel", "oxidiser" and "products" species. The reaction rate after autoignition is computed by using a modified version of the Magnussen model (21). The time scale for turbulent mixing is now evaluated from averaged values of turbulence kinetic energy and turbulence kinetic energy dissipation rate, in order to avoid unrealistic acceleration of the flame close to the wall. A modified version of the "Shell" model (3,4), involving a larger number still of global species, with reaction rates expressed according to the Arrhenius formulation, is used to detect autoignition, resulting in a sharp increase in temperature. These standard equations are discretised by using a finite volume, implicit scheme on an unstructured, moving grid.

For the liquid phase, Lagrangian conservation equations of mass, momentum and energy are solved for statistically relevant parcels of droplets, containing a number of droplets with homogeneous properties. Models are used for atomisation, coalescence, collisions, evaporation, gas-liquid interactions. An implicit scheme with subcycling is used to discretise these standard equations.

The pre- and post- processing for the SPEED_DC code is mainly done by using the PROSPEED code, specifically developed by Computational Dynamic Ltd. of London, except the generation of the computational grid is performed by using our own developed software. Details of the PROSPEED code are given in (1).

The computational domain is represented by the volume limited by those surfaces defining piston head, cylinder liner, cylinder head, valve head, valve curtain, main chamber-prechamber connecting passage, prechamber and glow plug. The computational grid is made by hexaedral volume elements for the inner volume, and quadrilateral area elements for the boundaries. In our automatic mesh generator, several topological regions with structured grid are considered and coupled together. These regions are: the region in between the inlet valve head and curtain and the piston head; the region in between the exhaust valve head and curtain and the piston head; the region in between the main chamber inlet of the connecting passage and the piston head; the remaining volume of the main chamber between the cylinder head and the piston head; the connecting passage region between the main chamber inlet and the prechamber outlet; the prechamber region between the prechamber contours and the connecting passage outlet and the glow plug. During the

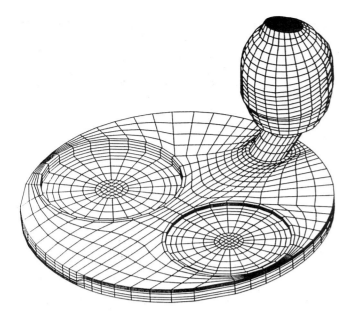

Figure 1 - Computational grid.

calculation the grid expands and contracts with the motions of the piston and of the valve according to a user defined algorithm. The automatic generation allows the user to easily fulfil the definition of the mesh movement algorithm.

3. SAMPLE COMPUTATIONAL RESULTS

Results of computations are presented here for a naturally aspirated IDI Diesel engine with the following characteristics:

bore	92. mm
stroke	93. mm
connecting rod	157. mm
compression ratio	23:1
prechamber volume	15.60 cm^3

The intake valve close at 228° c.a. after TDC, while the exhaust valve open at 492° c.a. after TDC. The calculations have been performed during the compression and expansion strokes. The flow case considered here refers to an engine speed 1600 Rpm and full load.

Initial air temperature and pressure at intake valve closure are supposed to be constant all over the computational domain and respectively equal to T=336 K, p=1.12*10^5 N/m^2, corresponding to a volumetric efficiency η_v=0.94. All the velocity components have been assumed equal to zero at the beginning of the computation. This assumption, also made by other authors [6,22], implies that the flow field, set up in the prechamber, is basically generated by the piston motion during the compression stroke, without any relevant influence given by the intake system. The initial turbulent kinetic energy and the turbulent kinetic dissipation rate have been set equal to

$\bar{K}$=1.2 m^2/s^2 and ϵ=72 m^2/s^3, from a prescribed inlet length scale of about one-third of the maximum valve lift, and a prescribed inlet turbulence level of about 6%.

During the compression and expansion strokes, all the valves are closed, and only the wall temperatures and those conditions describing fuel injection have to be specified. The wall temperatures have been defined as follows:

T$_{piston}$	527 K
T$_{liner}$	460 K
T$_{head}$	490 K
T$_{prechamber}$	510 K
T$_{glow plug}$	490 K
T$_{connecting passage}$	510 K

using available experimental and computational data on similar engines. The fuel injection conditions are finally those listed as follows:

mass fuel injected	36.8 mm^3/cycle
start of injection	10.75° c.a. BTDC
injection duration	13.6° c.a.
fuel temperature	350 K

The fuel injector is a Pintle type characterised by a conical needle with an asymmetrical flat surface inserted in a cylindrical hole of 1.000 mm. The annular gap varies between values of the order 0.01-0.04 mm. The angles of the cone varies between values 3-6°. The injection law has been defined by a sine-type shape covering the injection period.

The computational grid is shown in figure 1. The grid is made up to 17466 nodes, 15266 hexahedral volume elements and 4250 quadrilateral elements. The grid precisely reproduces the complex geometry of the IDI engine. The integration in time has been performed by using a time step of 1° c.a. up to the start of injection and 0.25° c.a. during the injection and combustion processes. Computations have been performed on a Silicon Graphics 4D/320 workstation, with 2 33 MHz processors and 128 Mbytes of main memory size. The memory requirement for the compression stroke is about 34.3 Mbytes with a resident set of 17.9 Mbytes and a CPU time of 590 s/time-step (27.53 hours for the entire compression stroke computation). For the injection and combustion phases the memory requirement is 37 Mbytes with a resident set of 19.7 Mbytes and a CPU time of 2460 s/time-step (164 hours for the entire expansion stroke computation).

The main result of the cold flow computation is the flow field generated in the prechamber at the beginning of the injection process. The results for both the mean and turbulent flow field are considered. Figure 2 and 3 present the velocity field during

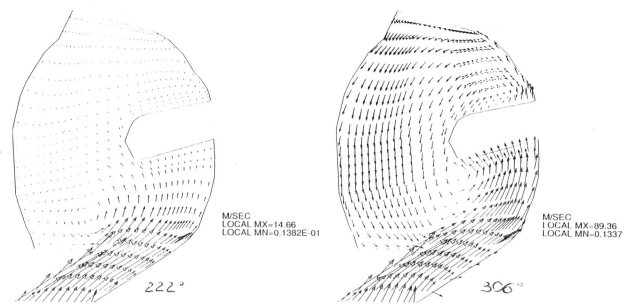

Figure 2 – Mean velocity within the prechamber before the start of injection.

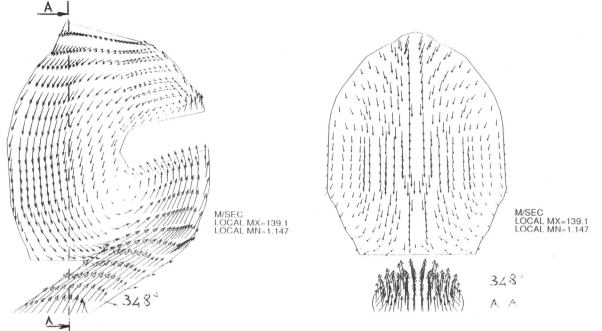

Figure 3 – Mean velocity within the prechamber at the start of injection.

compression. The piston movement produces a high-velocity flow through the connecting passage which then creates a high-momentum swirling flow in the prechamber. From the very beginning of the compression, the flow organises as a swirling flow but the modulus of the velocities rises considerably during the stroke. In the section, the vortex center moves from the bottom of the prechamber to the region under the glow plug. The glow plug behaves as an obstacle for the air stream coming from the cylinder. The flow impinges on the glow plug where it is deflected. As a consequence, in the region over the glow plug where the injection process will occur the velocity field is strongly modified by the presence of the glow plug.

At the time of injection (about 10° c.a. before TDC), Charlton et al. (5) measure on a very similar engine a swirl ratio (swirl angular velocity/engine angular velocity) of about 75 by using an hot wire anemometer, and about 20 by using a paddle wheel anemometer. Pinchon et al. (6) compute, on a reasonably similar engine, averaged values of about 20 with glow plug and about 47 without glow plug. The velocities computed here provide values of about 20-40, depending on the monitoring location, due to the structure of the swirl motion differing from the one of a solid body rotation.

In addition to the average flow field, the turbulent flow field at the start of injection is also of great interest. The lines of constant values

4

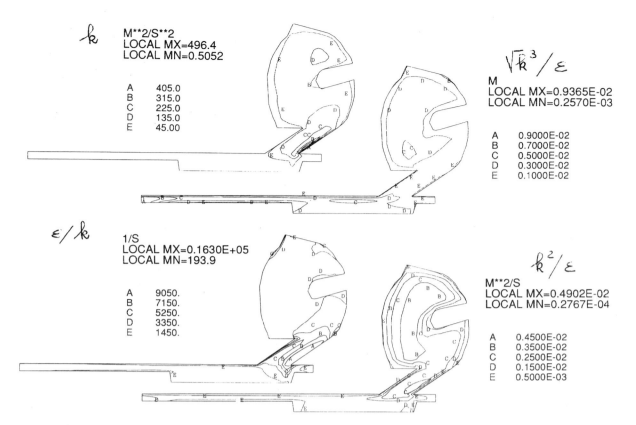

Figure 4 — Turbulent flow field at the time of injection.

for the turbulence kinetic energy K, the turbulent diffusivity K^2/ϵ, the inverse of the time scale of turbulent mixing ϵ/K and finally the turbulent length scale $K^{3/2}/\epsilon$ at 348° c.a. after TDC, just before the beginning of the injection process, are presented in figure 4. The turbulence kinetic energy K, representing the intensity of the turbulent fluctuations, is basically produced during the compression stroke by the velocity gradients, particularly intense at the inlet of the connecting passage. Production seems to be prevalent on transport, diffusion and dissipation, and the maximum values are located at the inlet of the connecting passage. Values within the prechamber are relatively high, of the order of 50-100 m^2/s^2. The distribution of the turbulence length scale, about 0.16 times the ratio $K^{3/2}/\epsilon$, shows relatively low values of the order of $5.0 \cdot 10^{-4}$ m. As a result, turbulent diffusivities are relatively high, while turbulent mixing times are relatively small. The turbulent diffusivity K^2/ϵ, representing a measure of the diffusion due to the turbulent mixing, shows maximum values in the centre of the prechamber of the order of 0.02-0.04 m/s^2. The inverse of the time scale of turbulent mixing ϵ/K, representing the velocity of the mixing process between separate eddies of fresh air-fuel mixture and burning products during combustion, is not precisely constant all over the prechamber. Turbulent time scales are of the order of $2.5-5.0 \cdot 10^{-4}$ s.

Cold flow computations have already been performed on different prechamber geometries, on the same engine and on different engines as well, at different operating conditions. Their results have been particularly helpful in the design of the prechamber and the positioning of the glow plug.

Results of the computation of the injection and combustion processes are also presented, but these results have to be considered with particular care, since the large number of processes involved in the computation makes the procedure particularly delicate. The spray and the auto-ignition models have been since now applied to low engine speed – partial load conditions. The definition of the injection law and injector geometry should be done very carefully, because global and local flow quantities have shown a considerable dipendence on these parameters. Furthermore, the auto-ignition model is very sensitive to the temperature distribution in the flow field and on the boundaries and requires the tuning of the constants involved. Figure 5,6,7 present the lines of constant temperature, vaporised unburned fuel and combustion products, while figure 8 finally presents the velocity field.

At 351° c.a. after TDC, the liquid jet shows no relevant deviation due to the prechamber flow field. The jet and the flow velocities have almost the same direction. During this first part of injection, droplets are bigger than the annular gap of the pintle injector. The vaporised unburned fuel shows that the jet is made by a liquid containing

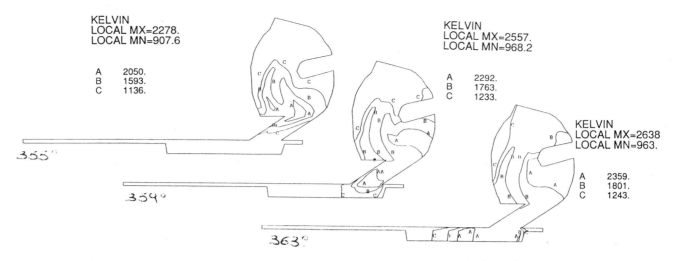

Figure 5 — Mean temperature after the start of injection.

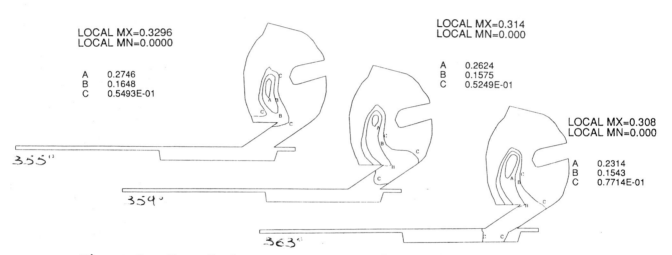

Figure 6 — Mean fuel vapour mass fraction after the start of injection.

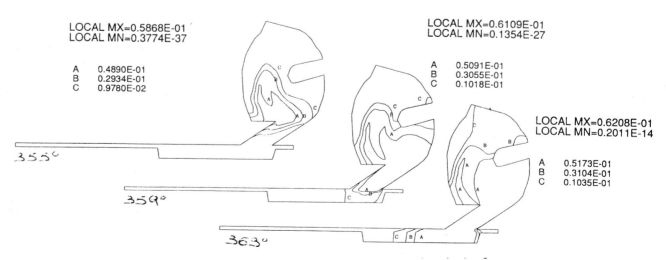

Figure 7 — Mean products mass fraction after the start of injection.

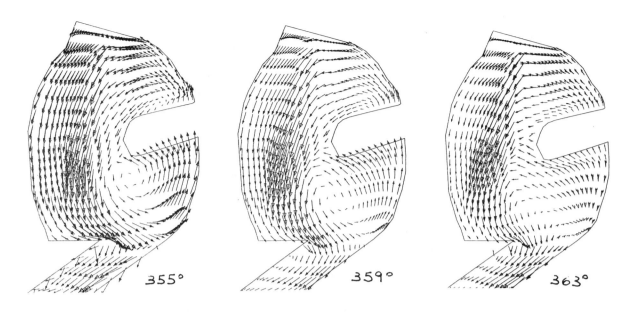

Figure 8 - Mean velocity within the prechamber after the start of injection.

core surrounded by fuel vapour. The temperature field is approximately constant over the computational domain, since combustion is delayed. The calculated ignition delay is about 3° of c.a., corresponding to $3 \cdot 10^{-4}$ s.

At 355° c.a. after TDC, the jet impinges on the wall in the bottom region of the prechamber with those droplets first exiting the injector. The velocity field is modified by the combustion process. Although the piston is still moving upward, the flow in the connecting passage has changed its direction and is moving from the prechamber towards the main chamber. The vaporised unburned fuel presents the highest values at the tip of the spray. The influence of the swirling motion on the fuel vapour is evident. After ignition, enhanced evaporation and mixing occur. The temperature distribution reveals that combustion has started in the bottom part of the prechamber and the hot combustion products are spread over the glow plug and towards the connecting passage. The top of the prechamber is not affected by the combustion process.

At 359° c.a. after TDC, the injection process continues. A considerable amount of fuel vapour is convected towards the main chamber. At this stage, combustion is controlled by the diffusion of fuel and air throughout the burning gases. Since the presence of oxygen in the bottom part of the prechamber is decreasing, because the first part of the combustion process occurred there, the vapour fuel has to reach the upper part of the prechamber or to move through the connecting passage where oxidiser is still available. The flame front is now divided into two parts; one transported by the swirling flow in the prechamber volume

in the region over the glow plug and the other towards the cylinder.

At 363° c.a. after TDC, the combustion products and the unburned fuel vapour have reached the main chamber. The temperature distribution shows that hot gases are spreading in all the prechamber and main chamber volumes. The highest temperatures are located close to the bottom part of the glow plug and in the middle part of the main chamber.

A limited number of fuel injection and combustion computations have been performed up to now, due to the increased difficulties. Therefore, the use in design of these results has been relatively limited.

4. CONCLUSIONS

Results of the application of a three dimensional code to the analysis and optimisation of an IDI Diesel engine has been presented. In order to make this kind of code an industrial tool for engine design, some problems of usability have been explored.

Usually a large part of the analysis is used in generating the computational mesh. The development of an automatic grid generator that allows the user to build up the grid in a short time makes the computational approach easier, thus widening the range of practical usability. Furthermore, the automatic mesh generation provides an accurate parametric representation of the engine geometry, thus introducing an efficient optimisation framework.

The analysis of the compression

stroke has pointed out how the geometry of the piston recess, the connecting passage, the positions and dimensions of the glow plug influences the characteristics of the setting up of the flow field in the prechamber. The three dimensional results provide detailed criteria in the optimisation of the local as well as global prechamber geometry.

The simulation of the injection and combustion processes has to be regarded much more carefully, considering the complexity of the phenomena to be described and the numerous simplification introduced in their modelling. Nevertheless, with this kind of tool, a first correlation among engine and injection parameters (volumetric efficiency, injection timing, mass of fuel injected,...) and spray and combustion evolutions can be introduced.

Improvements in the mathematical modelling and in the numerical solution algorithm as well, are currently being undertaken in order to produce a better modelling of those processes already included, like spray formation, autoignition and main combustion, and of those processes still to be included such as pollutants formation.

ACKNOWLEDGEMENTS

The development of the SPEED_DC code is founded by the Commission of the European Communities within the frame of the JOULE programme, by the National Swedish Board for Industrial and Technical Development (NUTEK) and by the Joint Research Committee of European automobile manufacturers (Fiat, PSA, Renault, Volkswagen and Volvo) within the IDEA Programme.

REFERENCES

(1) SPEED-DC USERS MANUAL, prepared for Joint Research Council by Computational Dynamics Limited, 1991.
(2) PROSPEED MANUAL, prepared for Joint Research Council by Computational Dynamics Limited, 1991.
(3) HALSTEAD, M.P., et al., The Autoignition of Hydrocarbon Fuels at High Temperatures and Pressures - Fitting of a Mathematical Model, Combustion and Flame, vol.30, 1977, pp.45-60.
(4) NATARAJAN B., BRACCO F.V.,On Multidimensional Modelling of Auto-Ignition in Spark-Ignition, Combustion and Flame, vol.57, 1984, pp.179-197.
(5) TAWFIG, M. E., CHARLTON, S. J. and PREST, P.H. An investigation of air and gas temperature in a motored indirect-injection Diesel Engine. IMECHE 1991, C433/002, 143-149.
(6) PINCHON, P., Three dimensional modelling of combustion in a Prechamber Diesel Engine. SAE, 1989, 890666.
(7) ZIMMERMAN, D.R., Laser anemometer measurements of the air motion in the Prechamber of an Automotive Diesel Engine. SAE, 1983, 830452.
(8) DUGGAL, V.K., PRIEDE T. and KHAN, I.M., A study of pollutant formation within the combustion space of a Diesel Engine. SAE, 1978, 780227.
(9) MEINTJES, K. and ALKIDAS A.C., En experimental and computational investigation of the flow in Diesel Prechambers. SAE, 1982, 820275.
(10) ZELLAT M., ROLLAND T. and POPLOW F., Three dimensional modelling of combustion and soot formation in an Indirect Injection Diesel Engine. SAE, 1990, 900254.
(11) CHARLTON S.J., CAMPBELL N.A.F., SHEPHARD W. J.J., COOK G. and WATTS M. J., An investigation of thermal insulation of IDI Diesel Engine swirl chambers. IMECHE 1991, D01191, 263-272.
(12) KOMATSU G., URAMACHI H., HOSOKAWA Y. and MONOSE K., Numerical simulation of air motion in the swirl chamber of a Diesel Engine. JSME, 1991, series 2, vol.34, 379-384.
(13) HEYWOOD J.B., HOSSEIN MANSOURI S., and RADHAKRISHNAM K., Divided-chamber Diesel Engine, Part 1: a cycle simulation which predicts performance and emissions. SAE, 1982, 820273.
(14) HEYWOOD J.B., RAJA T. KORT, HOSSEIN MANSOURI S., and AGOP EKCHIAN, Divided-chamber Diesel Engine, Part 2: experimental validation of a predictive cycle-simulation and heat release analysis. SAE, 1982, 820274.
(15) WATSON N., and KAMEL M., Thermodynamic Efficiency evaluation of an Indirect Injection Diesel Engine. SAE, 1979, 790039.
(16) SITKEI G. and RAMANAIAH G.V., A rational approach for calculation of heat transfer in Diesel ENgines. SAE, 1972, 720027.
(17) WHITELAW H.J., ARCOUMANIS C., COSSALL E., and PALL G., Transient Characteristics of single-hole Diesel Spray. SAE, 1989, 890314.
(18) WHITELAW H.J., HARDALUPAS Y., TAYLOR M.K.P., Unsteady Spray by a Pintle Injector. JSME, 1990, series 2, vol.33, no.2,
(19) ZHANG Y.S. and SHI S.X. Experimental and numerical study of the Diesel Fuel Spray from Pintle Nozzles. International Symposium Comodia 90, 1990, 287-292.
(20) TOSHIAKI T., KUNIHIKO S. and TAKAMASA U. Improvement of IDI Diesel ENgine Combustion through Dual-Throat Jet Swirl Chamber. SAE, 1986, 861184.
(21) MAGNUSSEN, B.F., and HJERTAGER, B.H., On Mathematical Modelling of Turbulent Combustion with Special Emphasis on Soot Formation and Combustion, Proceedings 16th Symposium on Combustion, Pittsburgh, 1976.
(22) SENDA J., OGAWA T., FUJIMOTO H., KUBOTA H., and KIMURA N.,Characteristics of Combustion in an IDI Diesel Engine with a Swirl Chamber Made of Ceramics, SAE, 1992, 920696.

C448/025

An investigation of the emission characteristics of the passenger car IDI diesel engine

S J CHARLTON, BSc, PhD, CEng, MIMechE, MSAE, A COX, BEng, B J SOMERVILLE, BEng
School of Mechanical Engineering, University of Bath
M J WATTS, BSc, CEng, MIMechE and R W HORROCKS, MSc
Ford Motor Company Limited, Research & Engineering Centre, Basildon, Essex

SYNOPSIS The paper describes an investigation of the emission and fuel economy characteristics of a number of alternative indirect-injection (IDI) Diesel engine swirl chamber concepts. The alternative chambers evaluated cover glow plug location and throat area, as well as novel dual throat and air cell configurations. The experimental work was carried out on a steady-state emissions test bed and on a rolling road equipped for full drive-cycle assessment of emissions, fuel economy and real-time particulates. The emissions considered are those subject to forthcoming legislation within Europe, namely: oxides of nitrogen (NOx), unburned hydrocarbons (HC) and particulates.
The experimental study has shown that particulate emissions may be reduced by adopting a tangential glow plug arrangement. Such a configuration, with the glow plug located downstream of the injector, also gives a significant reduction of visible smoke, especially in the part load regions important for vehicle applications. Real-time measurement of particulate emission during the FTP 75 / German drive cycle show considerable reductions during periods of vehicle acceleration.
The results of a computational fluid dynamics (CFD) simulation of the air motion within the swirl chamber, with 'cross-flow' and 'tangential' glow plug locations are presented.

1.0 INTRODUCTION

The IDI Diesel engine was first developed by Ricardo as a means of achieving rapid combustion in small automotive engines. The IDI combustion system consists of a swirl chamber in the cylinder head, which connects with the chamber above the piston via a passageway, or throat. During the latter part of the compression stroke air is transferred into the swirl chamber at high velocity, thereby forming a more or less orderly air motion, or swirl. Swirl is required to provide the mixing energy necessary for rapid break-up of the injected fuel, prior to rapid and clean combustion. The rate of heat transfer resulting from the additional surface area and high gas velocity is such that a glow plug has to be used to give reliable starting in cold conditions. It is a tribute to the effectiveness of this system that today it is used in the great majority of passenger car Diesel engines in a form virtually unchanged for thirty years, in a climate of acute concern for the environment.

The object of this research study is to evaluate the effects of a number of design changes to the basic Ricardo Comet chamber with special emphasis on exhaust emissions. The study has focussed on the location of the glow plug and the cross-sectional area of the throat between the two chambers. In addition, the dual throat system described by Tanaka [1] is evaluated, as is a novel air cell chamber. The research was carried out on an experimental version of the Ford 1.8 litre engine [2], in which the glow plug protrudes into the swirl chamber in a radial direction upstream of the fuel injector. The glow plug and throat exert a significant influence over the air motion in the swirl chamber, not only determining the swirl but also the intensity and scale of turbulence. The

relationship between swirl and turbulence for some of the geometries presented here is being studied at the University using the STAR-CD CFD package. Preliminary results for the datum 'cross-flow' glow

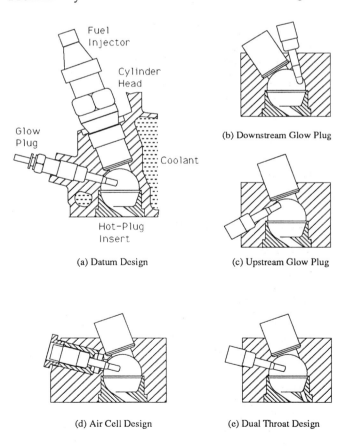

(a) Datum Design

(b) Downstream Glow Plug

(c) Upstream Glow Plug

(d) Air Cell Design

(e) Dual Throat Design

Figure 1   Details of the swirl chamber designs investigated

plug and 'tangential' glow plug arrangements are presented.

Growing concern for the environment has put pressure on manufacturers and governments to seek ways to reduce emissions from road vehicles. In Europe and North America new measures are being debated to limit further the emissions from passenger cars and trucks. In June 1991 the Council of the European Communities modified the test procedure for passenger vehicles by introducing a 400 second 'extra-urban' cycle to the existing urban cycle. The maximum speed during the test procedure was increased from 50 kph (31 mph) during the urban cycle to 120 kph (70 mph) in the extra-urban cycle. New limits for gaseous and particulate emissions were specified. This 'consolidated directive' is generally considered to be more severe than the 1987 US Federal limits for passenger cars. However, in the directive it was also stated that a further reduction in permissible levels will be proposed, to apply some time after 1st January 1996.

In the USA the Clean Air Act Amendments of 1990 introduce new Federal standards which will be phased-in over the period 1994 to 2003, with a 'Tier II' level proposed for 2004. From a Diesel perspective there is a NOx waiver at the current 1.0 g/mile (0.62 g/km) until 2003 and the particulate standard of 0.08 g/mile (0.05 g/km) is to be phased-in over the period 1994 to 2003 and remain at that level for 2004. The NOx proposal for 2004 is 0.2 g/mile, which is currently regarded as infeasible. The work reported here is a response to demands for a cleaner environment and tightening emission legislation.

2.0 SWIRL CHAMBER DESIGNS INVESTIGATED

The variations to the Ricardo Comet chamber investigated in the paper are as follows:

1. Datum design with conventional 'cross-flow' glow plug   *Figure 1a*
2. Glow plug in a position downstream of the injector and tangential in orientation *Figure 1b*
3. Glow plug in a position upstream of the injector and tangential in orientation   *Figure 1c*
4. Designs 1 to 3 but with enlarged throat area (5%, 10% and 20%)
5. Datum design with an air cell located around the glow plug in the head casting   *Figure 1d*
6. Datum arrangement with dual throat *Figure 1e*

The object of the study is to examine the effects of these chamber designs on fuel consumption and emissions under steady-state conditions and under transient (drive cycle) conditions on a rolling road emission facility.

These configurations were chosen principally for their effects on the orderly air motion (swirl) and on the turbulence levels in the chamber. The throat area, glow plug penetration and attitude are known to influence the level of swirl and the formation of

turbulence [3]. It has been assumed that the more 'streamlined' tangential configurations would be least detrimental to swirl. If a tangential glow plug is used with an enlarged throat the effect on swirl may be neutral, with the benefit of a reduction in the work needed to transfer the air into the swirl chamber, as expounded by Greeves et al [4].

The dual throat design is thought to enhance turbulence at the expense of swirl since some of the flow into the swirl chamber during compression is directed against the swirl. The secondary throat is aligned with the fuel spray, as in the research reported by Tanaka [1]. Tanaka suggested that such an arrangement would not only help to break up the fuel spray in the early part of combustion, but would allow some of the fuel to enter the main chamber at an early stage, thus reducing the inevitable overfuelling of the swirl chamber and the consequent particulate formation.

The air cell concept was inspired by the apparent benefits of conventional designs in which there is a clearance between the glow plug and the surrounding casting. It is believed that air is forced into such 'cells' during compression to be released later in the expansion process when combustion in the swirl chamber is starved of oxygen, thereby reducing soot formation in the otherwise fuel rich zones. In this research the concept was carried further by building in an enclosed volume around the glow plug, but separated from the swirl chamber by a small throat. The incorporation of the air cell reduced the compression ratio from 21.5 to 20.5.

3.0 STEADY-STEADY EXPERIMENTAL FACILITY

An engine research facility has been developed at Bath University for the investigation of exhaust emissions from passenger car Diesel engines under steady-state operation. The facility features electronic data acquisition for all experimental measurements and independent closed-loop control of engine speed and load. The engine 'environment' is carefully controlled to minimise the effects of variation in ambient conditions. Figure 2a shows the instrumentation system in outline. The heart of the data acquisition system is a PC computer with high-speed analogue to digital converter (ADC) and digital i/o cards. The ADC, which has a maximum acquisition rate of 1 MHz, has an integral multiplexer (MUX) which enables up to 16 channels of data to be read. Three of these channels are further multiplexed to increase the number of channels to 61. This configuration, under the control of software developed at the University, is capable of gathering all the data required for a thorough assessment of steady-state performance. The test procedure at a given operating point involves sampling each instrument a number of times, and taking an average for the calculation of performance and emissions. The data are corrected where relevant for ambient pressure, temperature and humidity,

© IMechE 1992 C448/025

according to the appropriate standards.

Figure 2b shows the control systems connected to the engine. These control torque, speed and the engine 'environment' in an attempt to obtain a high degree of repeatability, especially in emission measurements. To this end the controlled parameters are:

Intake air temperature
Fuel temperature
Lubricating oil temperature
Coolant temperature
Exhaust back pressure

In this paper the results are presented as maps of parameters of interest superimposed on load-speed axes. The maps were produced by collecting data at up to 80 load-speed points. This procedure takes up to three days to complete for each engine design, as the engine has to be allowed to stabilise thermally at each point and the calibration of exhaust analysers needs to be periodically checked.

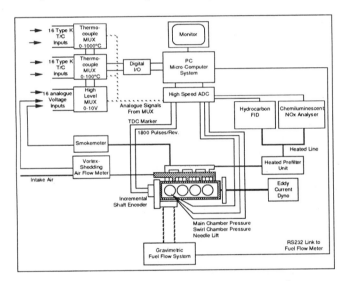

Figure 2a Data acquisition system and instrumentation

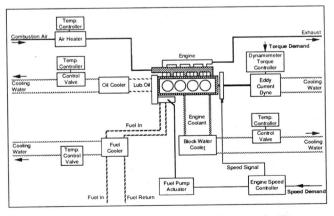

Figure 2b Control systems on the test facility

4.0 EXPERIMENTAL RESULTS

4.1 Steady-State Dynamometer Results
Glow Plug Position

Figure 3a compares the datum design with the downstream and upstream tangential glow plug designs on the basis of fuel economy, NOx, hydrocarbons (HC), smoke and calculated particulates. Each map is built up from data collected at speed intervals of 500 r/min and torque intervals of 10 Nm. The results for the datum design show a best fuel consumption of 260g/kWh at a bmep of 6.5 bar at a speed of 2500 r/min. As expected the NOx emission closely follows engine power. The highest levels occurring at maximum power and the lowest at idle. Maximum NOx has a value close to 150 g/h. The HC emission is very low at 4 to 10 g/h and is seen to increase with engine speed. The emission of visible smoke increases with load reaching values of 2.5 BSU at maximum bmep between 1500 and 4000 r/min. The particulate map has been derived from the HC and smoke results using the correlation of Greeves and Wang [5]. The calculated particulates are seen to increase with engine power from less than 5 g/h at idle to about 60 g/h at maximum power.

The results for the downstream glow plug design show a number of significant changes compared with the datum design. The best fuel consumption has improved to 250 g/kWh from 260 g/kWh (3.8%) and the consumption is generally improved over the entire map. The NOx emission has deteriorated somewhat, from 150 g/h to 170 g/h (13.3%) close to maximum power. The change in HC emission is quite small with some evidence of an increase at high loads and low speed. The most striking gain is in the emission of smoke, which has fallen by about 50% over the greater part of the low load, low speed region. As a consequence of reduced smoke, the particulates are also seen to be improved, at least over the greater part of the map.

The upstream glow plug design exhibits a similar characteristic to the downstream design. The fuel consumption is improved and the NOx emission is increased, although the upstream design has inferior NOx emission at light load. The upstream design has improved HC emission particularly at low speed. The smoke emission is improved over the datum but falls short of the improvement achieved by the downstream design. Particulates are improved over the datum, at least over the light load region, and the upstream design has a marginally better particulate emission.

Throat Cross-Sectional Area

Figure 3b shows the results of a study of the effects of swirl chamber throat cross-sectional area. Two cases were investigated : +10% and +20% cross-sectional area relative to the datum design. Results from a +5% area throat are presented later.
The trend with increased throat area is for the fuel consumption to increase slightly at higher loads with the global minimum almost unchanged,

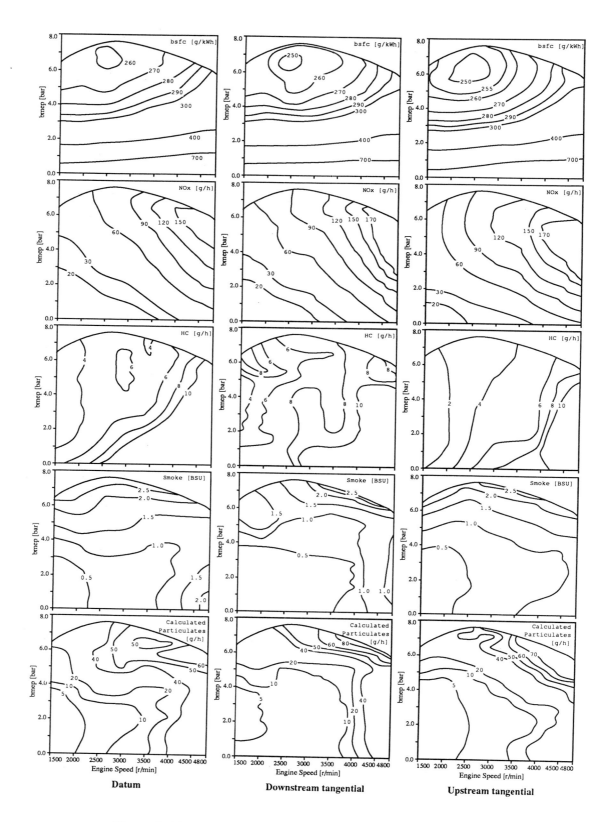

Figure 3a Experimental results for variation of glow plug location

whilst the NOx emission is decreased at maximum power. For both enlarged throat designs the NOx emission is seen to be reduced by about 25 g/h (16%) at maximum power. The HC emission is unchanged by throat area. The smoke emission is slightly increased at light load, but is significantly increased at loads above 50%. Smoke values of 3.5 BSU are seen, compared with 2.5 BSU for the datum design. Calculated particulates are slightly increased.

Tangential Glow Plug / Enlarged Throats
In Figures 3a and 3b it was shown that the enlarged throat swirl chambers produced lower NOx

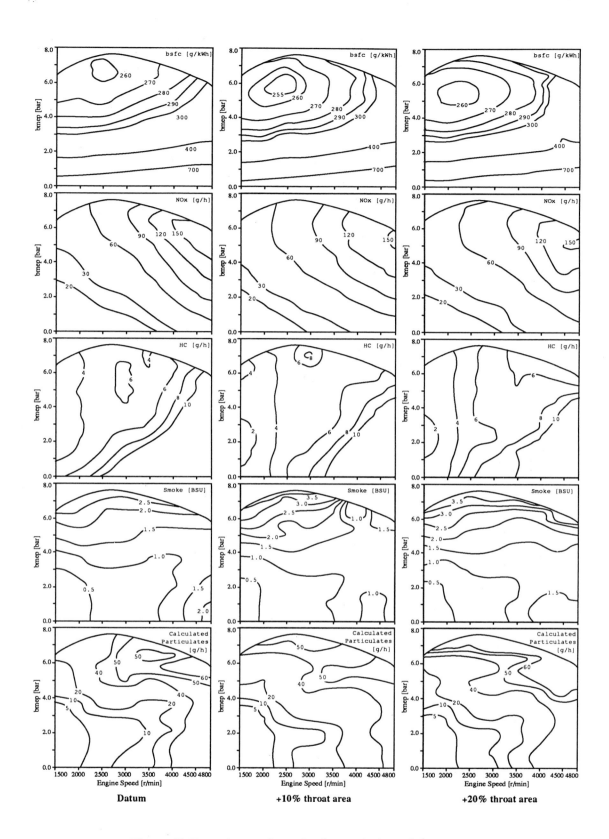

Figure 3b Experimental results for variation of throat area

emissions and inferior smoke and particulates, whilst the tangential glow plug arrangements produced the opposite effect - increased NOx and reduced smoke and particulates. Figure 3c shows experimental results from two combinations of tangential glow plug and enlarged throat compared with the datum chamber. The downstream

chamber is combined with a throat increased by 5% over the datum design. This produces very significant improvements in smoke and particulates at part load, but the NOx emission is even worse than for the downstream tangential design shown in Figure 3a. Also shown in Figure 3c are results from the upstream glow plug

Figure 3c Experimental results for combination of glow plug and throat area

arrangement with a throat having a 20% increase in cross-sectional area. The fuel consumption is slightly improved over the datum design, whilst smoke and particulates are marginally improved. The NOx emission is slightly worse than the datum design.

Air Cell and Dual Throat Designs

Figure 3d shows the response of the engine to the dual throat and air cell designs.

The dual throat design has a fuel consumption characteristic which has a best value comparable with the datum build, but elsewhere on the map is

14

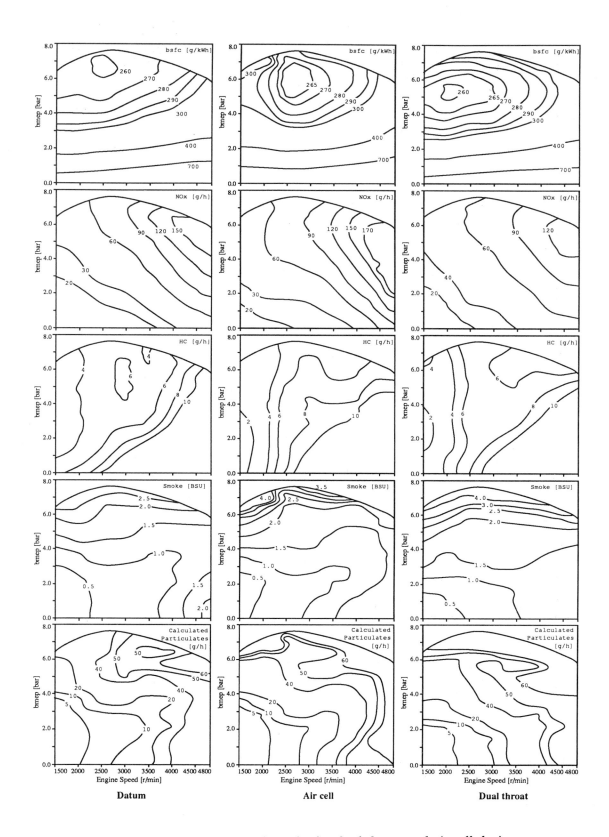

Figure 3d Experimental results for dual throat and air cell designs

considerably worse. The NOx emission is significantly reduced at high loads, from 150 g/h to 120 g/h (20%). The HC emission is little changed, however, the smoke characteristic is very much worse than the datum design. Smoke levels of more than 4 BSU are seen over most of the high load region. As a result of the higher visible smoke

emission, the calculated particulates are seen to be increased, by as much as 50%.

The air cell design also has a fuel consumption characteristic which is somewhat worse than the datum. The best fuel consumption island being slightly higher at 265 g/kWh. The air cell design has a higher NOx emission at high loads (150 g/h versus 170 g/h). The HC emission is little changed compared with the datum, whilst the visible smoke emission is significantly inferior at part loads and exhibits values in excess of 3.5 BSU at high bmep. Consequently, the calculated particulate emission is up to 50% higher than the datum.

4.2 Vehicle Testing

Having established that the downstream glow plug design gave the most encouraging results under steady-state dynamometer testing, this design was evaluated further by carrying out vehicle rolling road emission tests. Although a high degree of accuracy can be achieved from steady-state dynamometer testing by the close control of parameters such as water, oil and fuel temperature, there is an obvious difference with the transient nature of vehicle testing. This has become even more apparent recently as more and more emphasis is placed on exhaust emissions.

In order to obtain a direct comparison between the vehicle and dynamometer results, the fuel injection pump (FIP), cylinder head and injectors used on the dynamometer were transferred to the vehicle. Whilst on the dynamometer the FIP was set 3 degCA retarded, which earlier work had shown counteracted the increase in NOx seen with the downstream glow plug arrangement. This would most easily allow the advantage in the NOx-particulate trade-off to be seen. Vehicle testing was carried out on the rolling road facility at the Ford Research and Engineering Centre. The vehicle tested was a Ford Escort run with the chassis dynamometer set to simulate vehicle road load with vehicle inertia set to 1300 kg. Tests were run to the German emission cycle with the facility providing a second-by-second analysis of emissions throughout the drive cycle, including particulates, which were recorded in real-time using the Horiba TEOM analyser. Particulates were also measured in the traditional manner using a tunnel and weighed filter papers.

The vehicle was run with the datum cylinder head to establish a datum performance with which to compare the downstream glow plug design. The vehicle was finally returned to the datum build to enable the datum performance to be repeated in order to ensure the results were not subject to an underlying change in engine characteristics.

The vehicle emission and economy results are shown in Table 1, the datum results being an average of six tests, three run before the downstream design, three run after the downstream design. The results for the downstream design are the average of three tests. Figure 4a to c shows the real-time particulate measurement for the datum and the downstream designs, both being the average of three tests. The reduction of particulates predicted from the steady-state dynamometer testing is seen to have been confirmed in the vehicle although to a lesser

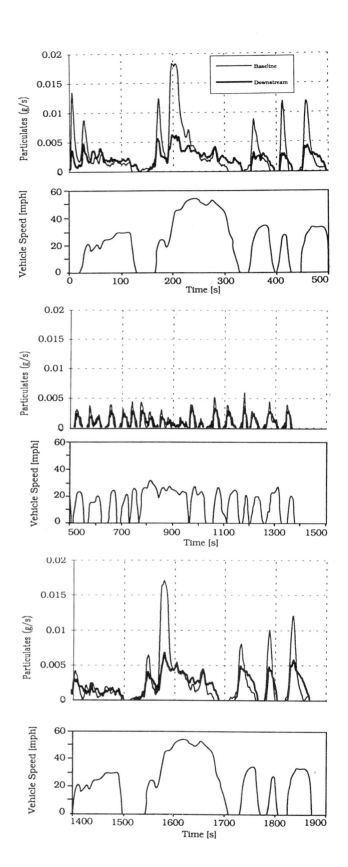

Figure 4   Real time particulate measurement during the FTP 75/German drive cycle.
Upper : Phase A (cold transient)
Centre : Phase C (stabilised)
Lower : Phase B (hot transient)

Table 1 Summary of results from the vehicle drive cycle tests - German Cycle

	HC g/km	CO g/km	CO2 g/km	NOx g/km	Partics[1] g/km	Partics[2] g/km	Economy 1/100km
Total							
Datum design	0.09	0.57	172	0.61	0.106	0.112	7.02
Downstream	0.18	0.61	174	0.53	0.077	0.083	7.14
Percentage gain	-100	-6.4	-1.5	+13	+27	+26	-1.7
Phase A (Cold 0-505 seconds Figure 4a)							
Datum design	0.14	0.76	183	0.67	0.174	0.222	
Downstream	0.27	0.74	183	0.57	0.106	0.134	
Percentage gain	-97	+1.8	+0.4	+15	+39	+40	
Phase C (Cruise 505-1372 seconds Figure 4b)							
Datum design	0.08	0.52	173	0.59	0.071	0.046	
Downstream	0.18	0.66	178	0.51	0.064	0.049	
Percentage gain	-124	-27	-2.8	+14	+9.8	-5.8	
Phase B (Hot 1372-1877 seconds Figure 4c)							
Datum design	0.08	0.53	159	0.60	0.120	0.154	
Downstream	0.12	0.42	161	0.56	0.079	0.110	
Percentage gain	-54.5	+21	-0.9	+7.7	+33.9	+28.6	

[1] Particulates measured by tunnel, [2] Particulates measured by TEOM real-time analyser.

extent, the vehicle averaging +27% against the test-bed +50%, compared with the datum. The most likely cause of the inferior particulates of the engine in the vehicle compared with the test-bed, is the increased hydrocarbon emissions seen in the vehicle results. This would appear to be a characteristic of the combustion system under transient conditions, particularly related to engine temperature. This latter point is borne out by the fact that the phase A hydrocarbons are much higher than the phase B hydrocarbons (0.14 g/km and 0.08 g/km), both phases being identical except that the engine is cold for phase A. This effect would be greatly reduced by the use of an oxidising catalyst, which would ameliorate the hydrocarbon contribution to the particulates, probably giving a gain over the datum design closer to the +50% seen on the steady-state test-bed. There was evidence of secondary injection occurring on this test which would have artificially increased the HC emission.

As with the particulate result the NOx result showed a different vehicle to test-bed relationship. Here the vehicle gave lower NOx levels than expected. The timing retard on the test-bed gave comparable results between the downstream and datum designs. The trend with engine temperature was, as expected, opposite to that of the hydrocarbons, the NOx being 13% less for phase A and 7.7% less for phase B. This reduction of NOx would indicate that a less retarded timing could be chosen if it was desirable to trade NOx for fuel economy.

Overall, the vehicle results confirmed the advantage of the downstream cylinder head design, showing the same improvement in the NOx-particulate trade-off as seen on the dynamometer. With the re-optimisation of injection timing for transient conditions and the addition of an oxidising catalyst it is predicted that a reduction of particulates of around 50% could be achieved with no increase in NOx or degradation of fuel economy.

5.0 CFD STUDY OF SWIRL AND TURBULENCE

CFD is now being used extensively by engine manufacturers in order to enhance understanding of mixing and combustion processes. Pinchon [9] reports work undertaken at the Institut Francais du Petrole on multi-dimensional modelling of combustion in an IDI Diesel engine. Zellat [10] reports on a similar investigation at Renault which included soot modelling. These studies used the KIVA code with the k-ε turbulence model, with other models added for ignition and combustion.

The CFD code STAR-CD was used to make a comparative study of the swirl and turbulence producing characteristics of the datum and downstream glow plug chambers. The program uses the finite volume numerical integration scheme to solve the Navier-Stokes equations for a wide range of flows. In this study the model requires a transient solution with compressibility, turbulence and a moving boundary to represent the piston crown. The simulation was run from a set of assumed initial conditions at the start of the compression stroke until the start of fuel injection close to top dead centre. The k-ε turbulence model was used in this study. This relates the Reynolds stresses and turbulent scalar fluxes to ensemble-averaged properties of the flow. k is the turbulent kinetic energy and ε is the dissipation rate. It is widely recognised that models of turbulence are inexact representations of the physical phenomena. The suitability of a particular model for a particular flow can only be established through experimental validation. Here, the CFD results are presented without validation and this should be borne in mind when considering the results presented. The empirical coefficients used in the k-ε model for this work may be found in reference [7]. The mathematical models, numerical techniques and other details may be found in references [6-8].

The meshes used for the study are shown in Figure 5. It will be seen that advantage was taken of the vertical plane of symmetry through the combustion chamber in order to reduce the computer solution time. Each mesh consists of about 40,000 cells, which has been found to be a realistic compromise between accuracy of solution and computer solution time. The meshes are seen to include the swirl chamber, glow-plug, throat, cylinder and piston crown recess. The cylindrical portion on top of the swirl chamber is the bore into which the injector is located. The model not only provides for movement of the piston crown, in order to drive the flow, but a process of cell layer removal from the cylinder has to be used to prevent cells from becoming excessively misshapen as the piston approaches top dead centre. At the crank angle shown in Figure 5, the number of cell layers in the cylinder has been reduced to four in order to retain cell aspect ratio within acceptable limits for numerical accuracy.

Figure 6 compares the velocity distributions in the swirl chamber of the datum and downstream tangential designs, at 14 deg btdc and 3000 r/min. The velocity distributions are shown as contours for clarity of presentation. At this late stage in the compression stroke the velocity at exit from the throat exceeds 150 m/s, however, the velocity distribution is far from uniform and the highest velocities are concentrated on the throat side of the chamber. The velocity gradient in the direction of flow is steep, falling from 150 m/s to less than 100 m/s in about 180 deg of rotation about the swirl chamber centre. The cross flow glow plug is seen to obstruct the progress of the incoming flow and to redirect it slightly towards the centre of the chamber. This effect is, however, restricted to the relatively small area occupied by the glow plug. In both cases the flow takes up an approximately solid-body rotation, with a clearly defined swirl centre.

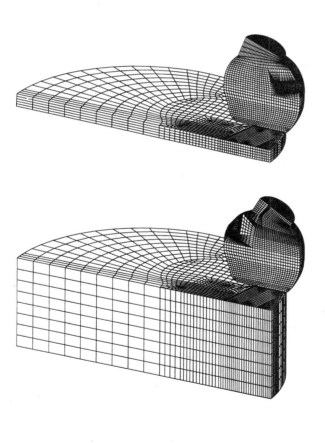

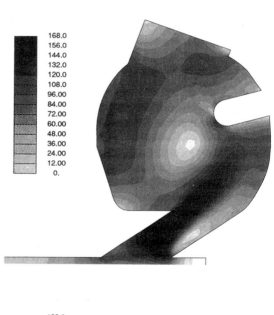

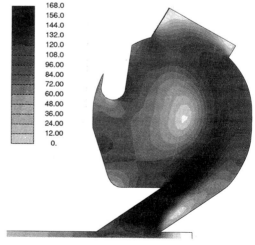

Figure 5   Computational meshes for the CFD study
of air motion
Upper: cross flow glow plug
Lower: downstream tangential

Figure 6   Calculated velocity distributions
Upper: cross flow glow plug
Lower: downstream tangential

Figure 7 shows contours of turbulent kinetic energy (k) calculated by the k-ε turbulence model. These show regions of high turbulence in the shear layers at the throat. The turbulence here will be convected into the chamber as the flow develops, as shown by the contours in the bowl. At this point in the compression stroke the turbulence distribution appears relatively unaffected by the glow plug location. There is some evidence that downstream of the cross flow glow plug the turbulence is of a slightly lower level, having been convected with a lower velocity due to the drag caused by the glow plug.

6.0 DISCUSSION

The experimental results show that the particulate emission may be reduced significantly by the adoption of a tangential glow plug arrangement, preferably in a location downstream of the injector, as shown in Figure 1c. The benefits observed during steady-state dynamometer tests were greater than the benefits seen during vehicle drive cycle tests. The steady-state results suggested particulate gains of up to 50%, whereas the actual drive cycle results showed a gain of only 27%. The differences were mainly due to the thermal lag of the engine, which ensures that much of the early part of the drive cycle is performed with chamber temperatures much lower than those obtained under steady-state tests. That this effect is real can be confirmed by comparing Figure 4a with Figure 4c. The former shows particulate emission from cold, whereas the latter shows particulate emission with the engine hot after more than 23 minutes of running. The integrated particulates for these parts of the drive cycle are 0.174 g/km and 0.120 g/km respectively for cold and hot phases, for the datum design, 0.106 g/km and 0.079 g/km for the downstream design. Particulate emissions would have been reduced even further but for the presence of secondary injection in the prototype FIE system.

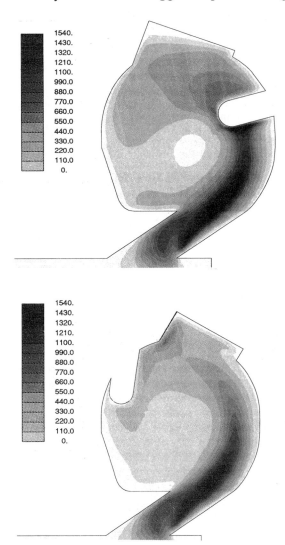

Figure 7   Turbulent kinetic energy distributions
Upper: cross flow glow plug
Lower: downstream tangential

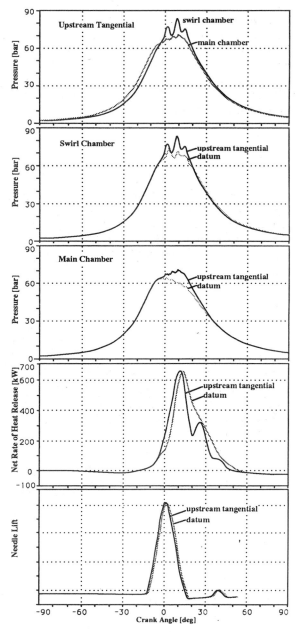

Figure 8   In-cylinder pressure, needle lift and heat release for the upstream tangential and baseline builds

It is evident from the marked change in emission performance between the datum and the tangential designs that combustion has been modified significantly. This is almost certainly due to an increase of swirl due to the less obstructive tangential design. Figure 8 shows pressure curves from the datum and upstream designs at a speed of 4000 r/min and a torque of 70 Nm (bmep = 5.0 bar). It may be seen that the swirl chamber and main chamber pressure curves are markedly different for the two cases. With the tangential glow plug the pressure in the swirl chamber rises both earlier and higher, indicating earlier, and possibly more rapid, combustion. This is confirmed by the rate of heat release, which was calculated by a single zone model based on swirl chamber pressure. The pressure development in the main chamber indicates that a higher proportion of combustion occurs in the main chamber with the tangential arrangement. This could explain the reduction of smoke and particulate emissions since the primary cause of those emissions is over-richness of the swirl chamber during combustion. It is possible that the early combustion seen in the swirl chamber, coupled with a changed swirl level, leads to the transfer of a larger proportion of the unburned fuel to the main chamber at an earlier stage. The assumed higher swirl level would be expected to provide better mixing in the swirl chamber, leading to a reduction of particulates and possibly earlier and faster combustion. Such a scenario is also consistent with the observed increase in NOx emission on two counts. Firstly, the change in combustion timing and rate would inevitably increase temperatures in the chamber. Secondly, the improved mixing characteristic would lead to less rich combustion giving higher flame temperatures.

The enlarged throat swirl chambers produced lower NOx emissions with increased particulates and smoke. From earlier work [3] it is clear that increasing the throat area will reduce swirl whilst also reducing pumping work.
In this case the reduced swirl appears to have made mixing worse, slowing down combustion and consequently reducing flame temperatures and NOx formation.

7.0 CONCLUSIONS

1. A number of alternative IDI Diesel engine swirl chamber concepts have been investigated in steady-state and transient experimental tests. The results show that particulate emissions may be significantly reduced by the adoption of a tangential glow plug, preferably downstream of the fuel injector. If injection timing is optimised for transient conditions and an oxidising catalyst is used, a reduction of particulates of around 50% could be achieved with no increase in NOx or degradation of fuel economy.

2. The downstream tangential glow plug arrangement appears to improve mixing, via its effect on swirl, such that combustion starts earlier. This leads to a more favourable distribution of combustion between the swirl chamber and the main chamber. Early transfer of fuel to the main chamber is thought to be important in reducing the formation of soot in the fuel rich regions of the swirl chamber.

3. The programme did not include evaluation of startability with the changed glow plug position, however, during experimental work the proposed arrangement did not appear to make starting any more difficult.

4. A CFD model has been used to study the generation of swirl and turbulence in the datum and downstream tangential swirl chambers. At 15 deg btdc, at an engine speed of 3000 r/min the differences between the two chambers are relatively small. On the central plane the cross flow glow plug deflects the incoming jet towards the centre of the chamber.

5. Novel dual throat and air cell chambers were investigated but gave relatively poor visible smoke emission, especially at higher loads. For the dual throat this was probably due to reduced swirl, for the air cell it may be due to the air in the cell being unavailable when needed for combustion.

ACKNOWLEDGEMENTS
The authors are grateful to the Ford Motor Company Limited for their support of this project and for permission to publish the research. The invaluable work of many others, in particular Ian Marsh and Martin Tanner at Bath University is acknowledged.

REFERENCES
(1) *Tanaka T, Sugihara K and Ueda T,* 'Improvement of IDI Diesel Engine Combustion Through Dual Throat Jet Swirl Chamber', SAE paper 861184, 1986.
(2) *Lawrence PJ, Knight D and Carnochan WA,* 'The Development of a 1.8 Litre Diesel Engine for Passenger Car Application' , IMechE Paper C382/070.
(3) *Tawfig M, Charlton SJ and Prest P,* 'An Investigation of Air Motion and Gas Temperature in a Motored Indirect-Injection Diesel Engine', Proc. IMechE, Conference on IC Engine Research in Universities, Polytechnics and Colleges, 30th-31st January 1991, pp143-150.
(4) *Greeves G, Wang C H T, Partridge I M and Black J R,* 'Improvements to Indirect Injection Diesel Combustion', IMechE Conference Publication C41/88, 1988.
(5) *Greeves G and Wang C H T,* 'Origins of Diesel Particulate Mass Emission', SAE Paper 810260, 1981.
(6) 'STAR-CD Version 2.1 User Guide', Computational Dynamics Limited, 1991.
(7) 'STAR-CD Version 2.1 Methodology - Part 1: Mathematical Modelling', Computational Dynamics Limited, 1991.
(8) 'STAR-CD Version 2.1 Methodology - Part 2: Numerical Solution Techniques', Computational Dynamics Limited, 1991.
(9) *Pinchon P,* 'Three Dimensional Modelling of Combustion in a Prechamber Diesel Engine', SAE Paper 890666, 1989.
(10) *Zellat M, Rolland Th and Poplow F,* ' Three dimensional Modelling of Combustion and Soot Formation in an Indirect Injection Diesel Engine', SAE Paper 900254, 1990.

APPENDIX - Basic Engine Data

Bore	82.5	mm
Stroke	82.0	mm
Swept volume	1.753	litres
Compression Ratio	21.5	
Firing order	1 - 3 - 4 - 2	
Nominal Datum Volume Ratio	0.50	
Nominal Throat Area / Piston Area	0.01	
Nominal Torque	110 Nm at 2500 r/min	
Nominal Power	44kW at 4800 r/min	
Injection timing	7 deg btdc at 1500 r/min	
	14 deg btdc at 4800 r/min	

C448/035

Development strategies for diesel combustion simulation using the SPEED code

A D GOSMAN, PhD, DSc, C KRALJ, Dipl-Ing, C J MAROONEY, BA, PhD
and P THEODOSSOPOULOS
Department of Mechanical Engineering, Imperial College of Science, Technology and Medicine,
London

SYNOPSIS   The stringency of recent engines emissions standards requires accurate three-dimensional modelling of the processes of fuel-spray evolution, autoignition, combustion and emissions formation, together with the evolution of the flowfield. This paper describes progress in Diesel combustion modelling for industrial design of engines, using the SPEED code. SPEED is a fully-implicit finite-volume code using an unstructured mesh permitting body-fitting and mesh adaptation with minimum distortion and memory usage. Models in the current version, SPEED-DC1, include the representation of spray as a stochastic ensemble of Lagrangian droplets, autoignition by a generic low-temperature chemical kinetics scheme, and combustion using the conventional Mangussen model. Future developments, including p.d.f. modelling to account for subgrid-scale inhomogeneous fuel distributions and for soot formation, are also outlined. Results from the code are discussed with reference to experimental comparisons with IDI engines.

1. Introduction

Demands on vehicle engine design for increased efficiency, either in the form of better performance or greater fuel economy, and for reduced exhaust emissions, have grown as a result of recent economic and legislative pressures. Up to a certain point, existing knowledge about the overall behaviour of the combustion process suffices to guide designers in meeting these demands; however, as the demands become stricter, more understanding is required of the details of this process. The complexity of interaction between the chamber geometry, the turbulent air flow and the combustion reaction means that it is necessary to be able to predict the combustion behaviour prior to building a prototype, as small changes in the design can have large effects on the performance.

This is perhaps true to a greater extent for Diesel engines than for spark-ignited engines, as Diesel combustion encompasses a greater range of physical processes. The details of liquid fuel injection, droplet breakup and the consequent distribution of the fuel vapour must also be predicted, together with the times and locations at which autoignition occurs. As Diesel engines are currently considered to have the best potential for meeting the above-mentioned performance criteria, it is the more important to have suitable computational tools for predicting how a given engine design meets the criteria. Computational fluid dynamics (CFD) codes such as KIVA (1) have now been available for some time to perform these predictions; and computer hardware now offers memory size and processing speed appropriate to industrial design schedules. It still remains true that given the number of interacting phenomena,

calculation times are still longer than desirable. The design of an efficient code for engine CFD is thus of great importance.

Two important areas for code design are the meshing and solution strategies. Earlier code have used structured meshes, where the mesh cell boundaries are defined by isosurfaces of curvilinear coordinates, can lead to areas of the mesh which are outside the flow boundaries, or which have too many or few mesh cells in order to fit the boundaries. Unstructured meshes of the finite-element type allow the mesh to be fitted to the boundary without overlap, and to be graded to control the spatial variation of mesh density according to the desired solution accuracy. The SPEED code (2,3) was created to give this meshing flexibility for engine design, but has applied the finite-volume discretization method to the unstructured mesh (4). This has a number of advantages, such as guaranteeing local conservation; it also achieves a greater physical transparency when coding the algorithm.

SPEED is now under development within the Integrated Diesel European Action (IDEA) project (5) , with the goal of producing a code, SPEED-DC, to predict the full range of Diesel phenomena, from air-charge preparation to emissions chemistry. Within this development, attention has concentrated both on the efficiency of the solution process, and on the addition of new submodels. Section 3 describes the solution techniques for the various models and their impact on the overall code efficiency. Section 2 provides an introduction to this by outlining the models currently in the first release code, SPEED-DC1, and those planned for the forthcoming SPEED-DC2. In section 4, the requirements for the overall coding strategy and their realization in SPEED are described, with particular reference to the addition of new models during their development. In section 5, the capabilities of the code are illustrated with reference to calculations for an indirect-injection Diesel engine. Section 6 concludes by looking at future developments for the code.

2. Diesel submodels.

2.1. Fuel injection

The liquid fuel injected into the chamber is characterized by a liquid jet, which may initially contain cavities as a result of effects in the nozzle, and which becomes progressively more multiply-connected as a result of dragging against the turbulent air-flow, finally breaking up in cascade into fine droplets, which subsequently evaporate to form the fuel vapour field. In SPEED, the liquid surface is not represented directly, but modelled by a changing Monte-Carlo population of sample droplets. This avoids the need to represent subgrid scales of the surface, and allows the spray position to be ensemble-averaged in line with the gas-phase calculations. In the currently-used version of the code, SPEED-DC1, the atomization process in the nozzle is modelled according to Huh (6), while breakup and coalescence of the resulting liquid core and droplets are modelled according to Reitz and Diwarkar (7) and O'Rourke (8) respectively. The droplets interact with the airflow through the drag force due to the mean flow and the turbulence, with the latter represented by a velocity of size proportional to the turbulence intensity, and with randomly chosen direction. Evaporation of the droplets cools the surrounding air and produces a fuel vapour. In the present modelling, droplets interact with wall surfaces by sticking, an ad hoc approach in the absence of more detailed modelling (but see section 6.) The droplets continue to evaporate as a result of their contact with the gaseous environment, which is in turn influenced by the wall temperature.

2.2. Autoignition and combustion

The presence of fuel vapour in the chamber is the signal for the chemical kinetics calculation to be turned on for autoignition. This is modelled in SPEED-DC1 by the Shell pseudo-reaction scheme (9) , modified by the Theobald acceleration parameter (10) for the overall reaction rate for Diesel fuel. Once the fuel vapour is ignited, combustion is represented by the one-stage turbulence-controlled reaction of the Magnussen model (11) . In contrast to the Shell model, which uses detailed rate constants and activation energies for the pseudo-reactions, this model subsumes the details of combustion into two global parameters sizing the reaction rate.

2.3. Emissions

The released code SPEED-DC1 currently has no emissions models in it. In current development work, models for NO_x and soot formation are being incorporated into SPEED-DC2. The NO_x work will use the Zel'dovitch mechanism (12) , while the soot modelling will use a p.d.f. formulation for the soot production rate coupled to an equivalent p.d.f representation of the combustion, as described in (5) and below in section 5. For emissions source terms representable by a chemical kinetics model, a generic kinetics solver is now used as described in section 3.5.

3. Meshing and solution strategies.

3.1. Mesh structure.

In the SPEED unstructured mesh, each control volume, or computational cell, is a hexahedron with quadrilateral sides (a topological cube,) The local connectivity of the mesh is uniform, with each face connecting just two cells; but beyond the first nearest neighbours there is no predefined structure. The lack of global structure serves the two purposes of boundary-fitting and selectable mesh-density referred to above. It also means that, provided the transport equations couple only nearest-neighbour cells, the local mesh structure simplifies the coding considerably.

3.2. Spray tracking

The use of a Monte-Carlo droplet representation allows each droplet to be treated separately. The number of droplets is usually small relative to the number of cells in the computational mesh, so that this calculation gives a relatively small contribution to the overall CPU usage, which is determined principally by the requirements of droplet tracking. In order to partition the interaction between the droplet and the gas in each cell it passes through correctly, the code must subdivide each computational timestep into subcycles, one for each cell the droplet traverses. In addition to the calculations for the process models, the code must also locate the droplet in an unstructured mesh. As there is no global coordinate description of such a mesh, special procedures are employed to find both the subcycle timestep size which retains the droplet within a cell, and the destination cell beyond.

3.3. Transport equations

The in-cylinder flow, heat transfer and mixing is represented by solution of the momentum, mass, species and energy transport equations discretized according to the unstructured-mesh finite-volume representation described in (4). To achieve the required turnround times, each equation is solved fully-implicitly in space. This ensures that the timestep size is not limited by Courant number stability conditions. The timescales for the interactions between the physical processes involved in Diesel combustion do however impose their own timestep limitations;

these can be quite severe in some cases.

The linearized and discretized transport equations yield implicit-solution matrices with the same lack of structure as the meshes on which they are defined. This limits the number of available matrix equation solvers that can be used: for serial computers, the Conjugate Gradient (CG) method, since it is defined solely in terms of matrix-vector products, and so does not require any particular matrix structure, has found favour. For efficient use in Diesel combustion problems, the matrix must first be preconditioned to cluster the eigenvalues: this significantly reduces the number of iterations required for solution. SPEED currently uses ICCG(0) preconditioning. This is well-known to inhibit code vectorization on Cray and other vector computers and research is currently in process for SPEED-DC to find alternatives. To date, a mesh-'colouring' technique, which ensures that each cell is surrounded by those of a different 'colour', offers promise. With this, each colour group vectorizes, provided the mesh is sufficiently large: for hexahedral meshes this size is of order 500, so that all time-consuming problems are vectorizable. The method of colouring affects the mesh bandwidth and hence the number of iterations to convergence, but vectorization factors of order 10 have been shown to be obtainable.

3.4. Chemical kinetics

Any chemical kinetics scheme comprises a set of coupled source terms which must be solved simultaneously, in species space as opposed to mesh space. This set is non-linear and in general stiff; and explicit methods, as used by other codes including earlier SPEED calculations for knock (13) needed small subcycling timesteps leading to very computationally intensive calculations. As both NO_x and autoignition can occur in principle at any location and for most of the combustion phase of the calculation, the kinetics calculation must remain activated throughout the mesh.

3.5. Equation coupling

Coupling between equations is represented by a predictor-corrector sequence. The pressure-velocity coupling uses a compressible pressure equation representation in the PISO method (14), embedded in the EPISO coupling sequence, in which the solution order is the causal order of the processes - spray (including evaporation), fuel species, energy, pressure/velocity - in each of the predictor/corrector steps. For Lewis numbers of the close order of unity, the energy and species equations can be decoupled by including the stored chemical energy with the thermal energy. The turbulence is solved for in a separate predictor-corrector sequence using the k-ε model.

3.6. Discretization.

Older versions of the SPEED code used a default spatial discretization practice of hybrid (central/upwind) differencing. A property of upwind differencing experienced by many codes including SPEED is that, in addition to the order of accuracy imposed by the scheme, advection of vector quantities in swirling flow leads to decay of the swirl strength. Central differencing considerably reduces this decay, but leads to negative matrix elements and consequent potential unboundedness of solution. If the method is filtered, however, so that upwind differencing is used at local solution extrema, then this effect is avoided. SPEED-DC1 now implements a self-filtered central differencing scheme of this kind, with improved results. Of the Diesel cases so far investigated, this effect is most marked with swirl flow in IDI prechambers; this is illustrated in section 5.

4. Code optimisation

4.1. Development strategies

It is important for a research code to have the flexibility to incorporate new models without substantial recoding; once this is so, maintenance of production codes is then also improved. While it is obviously not possible to anticipate all new modelling developments, SPEED has been designed to provide templates for a number of standard classes of model. The transport of species by the flow has been encapsulated into standard functions for matrix assembly, and a standard form of solution subroutine invoking them, which can be adapted for whatever particular form of source term is required. This has already been used for the pseudo-species of the Shell and soot models and the variances parametrizing combustion and soot p.d.f's. The chemical kinetics coupling has been generalised so that a reaction scheme of an arbitrary number of species and equations can be handled by a single common solver. Several solution techniques have been tested; for Diesel combustion time-explicit marching remains the most efficient method, while use of the same coding for knock calculations in spark ignition engines has shown greater efficiency with the fourth order Runge-Kutta scheme. Appropriate design of the liquid phase subroutines also has generic applications. The particle tracking algorithm described above can also be used to implement streaklines and monitor the environment of a gas 'particle'; it may also have ray-tracing applications.

4.2 Efficiency

Execution speed for the Diesel simulation reported in the next section was approximately 4 CPU minutes per time step during the combustion phase of the calculation (i.e. with all models in operation) when performed on a Silicon Graphics Power Series Iris 4D/340S with 4 R3000 processors and 128 MBytes memory. This compares with 1.5 CPU minutes per time step before start of injection, illustrating the additional demand during combustion, which is primarily due to the need to keep the autoignition kinetics on at all points in the domain. The mesh size for this run was 4300 cells, and a memory requirement of 180 64-bit words per cell. The code has also been run on various Crays: on an X/MP the standard ICCG(0) solver mentioned above yielded only serial speeds (order 5Mflops) while use of the mesh-colour reordering gave overall speeds of 50Mflops.

Aside from this improvement, the overall impact of the modelling is on accuracy of representation of the physical phenomena rather than on improvement of execution speed. The efficiency improvements to be expected from the SPEED meshing are rather due to the fact that the mesh requires no extra points outside the flow domain, and can vary the mesh density spatially as required wiithout the need for local refinement to be propagated into other regions to satisfy a global mesh topology.

5. Engine Calculations

To illustrate the nature of the results produced by SPEED-DC, we present calculations for a vertical glowplug IDI geometry. For this engine, of bore 86mm and stroke 89mm, the speed was 2500r/min. 20.5mg of fuel, characterized as n-dodecane, was injected between 347-377°CA through a pintle nozzle at temperature 350K. For this calculation the glowplug is not activated, i.e. normal running conditions are assumed. If a cold-start calculation were performed, the glowplug would be represented either by a constant temperature or a constant heat-flux surface.

Fig. 1 shows a sequence of results for the 25°CA period following start of fuel injection. For reasons of space, while all the field described are available at all the crank angles described, only those showing significant changes are included here. Fig 1(a) shows the flow field at start of injection, and 1(b) the corresponding temperature field. The flow is compressed by the piston into the prechamber forming a barrel swirl flow field, and the mixing of the hotter, more compressed gas from the cylinder with the cooler gas in the prechamber can be seen from the isotherms although the temperature differences are small. Fig. 1(c) shows the calculated droplet distribution, which indicates the injector position and the stochastic representation of the spray. Droplet sizes are represented by the circle diameters, while liquid temperature is represented by the (here grey-scale but originally colour) shading. The spray is shown at 367°CA, near maximum penetration, to give a reasonably-scaled figure, but its behaviour at this time is typical of earlier ones. Fig 1(d) shows the temperature field at 352°CA. The low (L) contour correlates with the spray position, where the gas is cooled by the high evaporation rate, and combustion has commenced in the region of the glowplug .Fig 1(e) shows the unburned fuel vapour distribution, which has a peak at the spray position. Figs 1(f) and 1(g) show the same fields 10° later at

362°CA. The main combustion zone has now been swept below the glow plug by the swirling motion, and the fuel vapour peak is now downstream of the spray position. This pattern now stabilizes over the next 10°. At the same time, the swirl motion in the chamber is sucked into the main chamber by the expansion stroke, as shown in fig. 1(h) for 367°CA..

As no publishable data are available for this engine geometry, comparison is made (fig. 2) with measured data for a similar engine, but with horizontal glowplug (16). The trend and peak are reasonably predicted but the ignition delay is seriously underestimated. In making this calculation, no 'tuning' of the parameters in the ignition or combustion models was made to obtain these results, and it is known that such tuning is often required to achieve good comparisons. As this calculation was made on a relatively coarse mesh (4300 cells,) good agreement is not expected; further calculations on meshes of order 100 000 cells will be needed to resolve the phenomena accurately.

In an IDI engine, the barrel swirl created in the prechamber by the compression provides a good example of the swirl decay caused by upwind differencing. Fig. 3 shows a comparison between the swirl flow at 10°CA ATDC calculated with hybrid upwind-differencing and with the SPEED self-filtered central differencing scheme under motored conditions, for a recessed-injector geometry with no glowplug. With the hybrid scheme, here dominated by upwinding, the swirl decays early with the swirl centre sucked into the main chamber by the piston motion. In contrast, the filtered central differencing scheme retains most of the original angular momentum. In firing conditions the liquid fuel is injected into this flow and its subsequent evolution will obviously be affected by the size of the swirl drag. As noted above, the use of this scheme does not improve the execution speed, but increases the accuracy of the calculation.

6. Future developments

From the objectives of the IDEA program for SPEED, we select here two particular features, relating to the code development paths already discussed: meshing and models. The meshing capabilities of SPEED have already been exploited in fitting irregular Diesel combustion chambers, and the major improvement to be expected here in the future will come from improved mesh generators rather than from within SPEED. Of more direct concern is the ability to adapt the mesh to fit the details of the flow. The unstructured mesh form allows the mesh to grade in size continuously from coarse to fine: application of a suitable criterion for a mesh size optimum to the solution, and a remeshing tool, then yield a new mesh. For cases where the remeshing time is expensive, it will also be possible to move the mesh without changing the connectivity to achieve the required size grading. SPEED already has the capability, now being exploited for spark-ignition pent-roof chambers, to move the interior of meshes with irregular moving boundaries by creating a displacement interpolation function throughout the interior. This is not needed for most simple Diesel chambers as the boundary motion is strictly one-dimensional, but it can also be used with a grading criterion to cluster the interior points at the required density.

The models already in SPEED-DC1, while standard, and allowing reasonable calculation times, do not represent the full range of phenomena, in particular by neglecting sub-grid-scale inhomogeneities in the species and temperature distributions, and the effect of turbulent mixing on them. The IDEA-project soot formation model of Borghi et al (5), which will be incorporated in a forthcoming version of SPEED-DC, includes these through a p.d.f. approach. In this, the statistical distribution of inhomogeneities is prescribed in terms of the mean values and variances of the reacting species, and yields overall formation rates for the emissions means and variances. The model requires that the combustion itself be described in these terms, and this will be implemented by the IDEA project pdf model for combustion of Bray and Rogg (5). P.d.f. models require in general that the source term be evaluated as the integral of the p.d.f. in all cells of the mesh; this again will give an increase in the solution time for combustion. We will be looking here for improvements in the p.d.f integration algorithms and where possible standardization of the p.d.f. representation to provide a framework for new modelling. The IDEA project will also be the source of more accurate spray-droplet/wall interaction models which include the effect of splashing as a function of wall temperature. Validation of these models will be part of the IDEA program.

SPEED-DC is already in use for engine evaluation by the JRC companies, and it is hoped that the various code development paths within the SPEED project of the IDEA program will converge to create a complete evaluation environment for Diesel engine design in the future.

Acknowledgements

Funding under the IDEA program from the Commission of the European Community and the JRC companies, Fiat, Peugeot, Renault, Volkswagen and Volvo, is acknowledged. We thank B N Adamson, C Hill, J A Jackson, M Sarantinos and H G Weller for their assistance.

References

1. Amsden A A, O'Rourke P J and Butler T D: "KIVA-II - A Computer Program for Chemically Reactive Flows with Sprays" Los Alamos Nat. Labs LA-11560-MS, 1989.

2. Gosman A D and Marooney C J: "Development and validation of computer models of in-cylinder flow and combustion in Diesel and Spark-ignition engines," IMechE C399/169, Autotech, Birmingham, 1989.

3. Adamson B N, Gosman A D, Hill C, Marooney C J, Nasseri B N, Sarantinos M, Theodoropoulos T and Weller H G: "Simulation of In-cylinder Flow and Combustion in Reciprocating Engines," 5th Intl. Conf. Science and Engineering on Supercomputers, London 1990.

4. B Adamson, A D Gosman, C J Marooney, B Nasseri and T Theodoropoulos, "A new unstructured-mesh method for flow prediction in internal combustion engines", International Symposium on Diagnostics and Modelling of Combustion in Internal Combustion Engines (Comodia), Kyoto, 1990

5. Schindler K-P, "Multidimensional Modelling of Diesel Engines", SAE 920591, 1992.

6. Huh K Y and Gosman A D, "A phenomenological model of Diesel spray atomization," Int. Conf. Multiphase Flows, Tsukuba, 1991.

7. Reitz R D and Diwarkar R, "Effect of Drop Breakup on Fuel Sprays," SAE 860469, 1986.

8. O'Rourke P J, "Collective drop effects on vaporizing liquid sprays, PhD thesis, Univ. Princeton, 1981.

9. Halstead MP, Kirsch L J and Quinn C P: Combustion and Flame 30, 45, 1977.

10. Theobald MA and Cheng WK, "A numerical study of Diesel Ignition", ASME 87-FE-2.

11. Magnussen B F and Hjertager B H: 16th Symp. (Intl.) on Combustion, p.1657, 1977.

12. Peters N, Comb. Sci. Tech. 19, 39, 1978.

13. Issa R I, J Comp Phys 62,40, 1986.

14. Gosman A D, Marooney C J and Weller H G: "Prediction of unburnt gas temperature in multidimensional engine combustion simulation," Comodia Intl. Symp. Diagnostics and Modelling of Combustion in IC Engines, Kyoto 1990.

15. Vafidis C, Vorropoulos G, Whitelaw J and Nino E: "In-cylinder flowfield in a motored engine with helical port and reentrant and square piston-bowl configurations," Imperial College Dept. Mechanical Engineering, Fluids Section Report FS/87/31.

16. Zellat M, Rolland Th and Poplow F: "Three-dimensional modelling of combustion and soot formation in an IDI Diesel engine," RNUR Report, 1989.

Figures

Fig 1. Calculation of combustion in an IDI prechamber:

(a) Flow field at 347°CA.

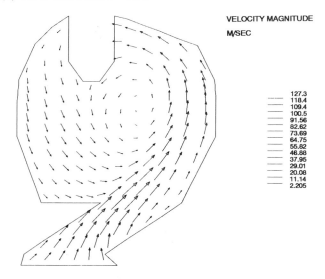

(d) Temperature field at 352°CA.

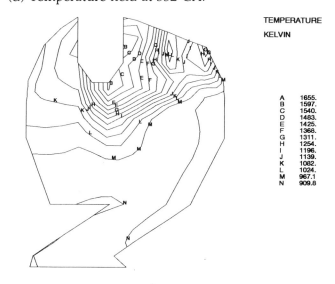

(b) Temperature field at 347°CA.

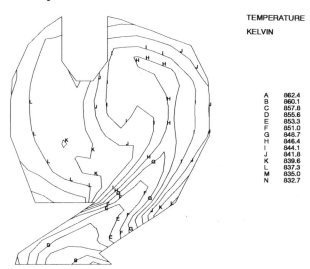

(e) Fuel vapour field at 352°CA.

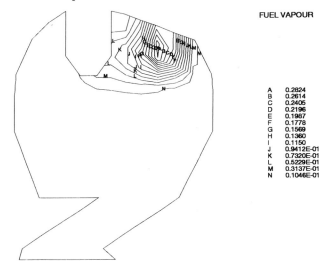

(c) Fuel spray distribution at 367°CA.

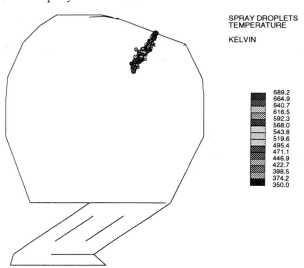

(f) Temperature field at 362°CA.

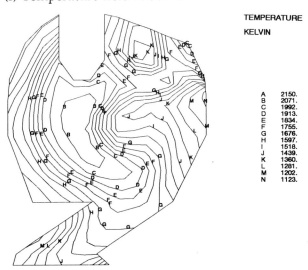

(g) Fuel vapour field at 362°CA.

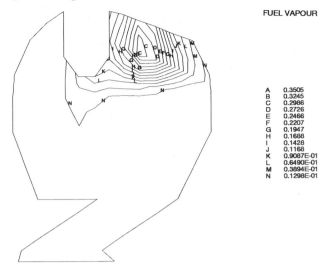

FUEL VAPOUR

A	0.3505
B	0.3245
C	0.2986
D	0.2726
E	0.2466
F	0.2207
G	0.1947
H	0.1688
I	0.1428
J	0.1168
K	0.9087E-01
L	0.6490E-01
M	0.3894E-01
N	0.1298E-01

(h) Flow field at 367°CA.

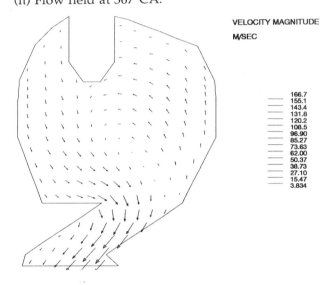

VELOCITY MAGNITUDE

M/SEC

166.7
155.1
143.4
131.8
120.2
108.5
96.90
85.27
73.63
62.00
50.37
38.73
27.10
15.47
3.834

Fig 3. Comparison between hybrid differencing and self-filtered central differencing for swirl in an IDI prechamber at 10° ATDC

Fig 2. Pressure history for an Indirect-Injection Diesel Engine. Comparison between calculated and measured data

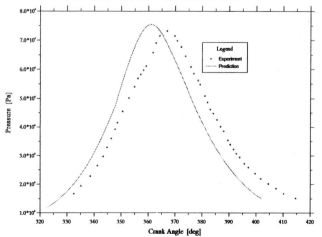

Application of CFD to the modelling of autoignition at high pressure in a cylinder

J F GRIFFITHS, BSc, PhD, DSc and D J ROSE, BSc, PhD
School of Chemistry, University of Leeds
M SCHREIBER, Dipl-Ing, J MEYER, Dipl-Ing and K F KNOCHE, Dr-Ing
RWTH, Aachen, Germany

A computational fluid dynamics code (KIVA II) has been modified to permit the simulation of autoignition based on detailed kinetics and heat release. The autoignition characteristics typical of alkanes were simulated for conditions representing combustion in a rapid compression machine. Results are reported from kinetics based on the "Shell" reduced scheme, which exhibits an overall negative temperature dependence of reaction rate and two-stage ignition within the reactant temperature range 700 - 780K. Autoignition 'centres' were predicted to occur at different spatial locations according to the compressed gas temperature and the nature of the prevailing reaction kinetics.

1 INTRODUCTION

There now appears to be a firm view that autoignition, resulting from the spontaneous combustion of hydrocarbon fuels in the "end-gas", is contributory to the onset of knock in spark ignition engines. There is also general agreement that the important chemical processes leading to spark ignition begin during the compression and heating of the charge in the firing cycle. The time interval from compression after the inlet valve has closed to late stages of propagation of the spark-ignited flame through the combustion chamber may be as long as 20 ms at low engine speeds and, more importantly, may encompass the development of chemistry in the end-gas itself over a wide temperature range (500 - 1000 K, say). There are marked changes in the kinetic interactions associated with this temperature range, which are illustrated most vividly in the manifestation of the two-stage, spontaneous ignition of hydrocarbons and other organic materials (1). The low temperature "cool flame" reactions, which constitute the first stage of the ignition, may also affect the temperature distribution in the vicinity of autoignition centres. This has been shown to have a dramatic effect on the subsequent knock severity in an engine (2). However, until now there has been a belief that end-gas autoignition in spark ignition (s.i.) engines originates in the "core gas", where the system is closest to adiabatic conditions (3). In this paper we show that the thermokinetic interactions that control hydrocarbon combustion in certain temperature ranges may cause the initiation of autoignition centres at different points and in a variety of different ways.

The broad principles of autoignition of hydrocarbon + air mixtures resulting from complex kinetic interactions, including chain branching coupled to self-heating, were put on a formal theoretical basis in 1969 by Gray and Yang (4). Today, the main theoretical approaches to the understanding of autoignition, especially with regard to knock in s.i. engines or ignition delay in Diesels, are through numerical simulation. The kinetic aspects of the simulations comprise either comprehensive models of the fuel + air system, in which there may be many thousands of reaction steps to represent the chemistry of combustion of a fuel through the temperature range 500 - 2500 K (5), or reduced kinetic models in which different classes of reactions are grouped to simplify the mechanistic structure in order to generalise the application and to ease the computational burden in the integration of the systems of stiff, non-linear equations (3,6,7). The latter approach is especially relevant to numerical analyses involving CFD. However, most thermokinetic modelling (and certainly all that which involves large kinetic schemes) is still represented in terms of spatially uniform temperature and concentrations in the combustion system. This is a useful description for testing developments in chemical interpretation, but it fails to capture the features that, in reality, relate to the spatial variations within combustion applications.

The purpose of the present paper is to demonstrate qualitative points of principle arising from the special kinetic characteristics, associated particularly with alkane oxidation, that cause the overall negative temperature coefficient (n.t.c.) of reaction rate and give rise to two-stage ignition. Thus to maintain simplicity and general familiarity the "Shell" reduced kinetic model was used to represent the overall kinetics. The kinetic and thermal parameters were chosen to represent the response of a 90 PRF fuel (0.9 i-octane + 0.1 n-heptane) at $\phi = 1$, as in previous work (6,8).

In order to take into account spatial temperature and concentration inhomogeneities in the combustion chamber, the main part of the numerical interpretation was based on CFD calculations simulating the autoignition of fuel + air mixtures in the combustion chamber of a rapid compression machine (RCM). The conditions were matched to the experimental system in use at Leeds University (9-13). When the piston of the RCM is brought to a halt at the end of compression of gaseous, premixed fuel and oxidant and there is no induction and exhaust of the charge, the constraints on the fluid mechanical calculations are not so stringent as in a full simulation of spontaneous ignition in a motored engine, or following spark ignition. Thus we were able to use a mesh of 14 x 10 during the combustion stage, and thereby admit the possibility of incorporating a quite complicated kinetic model. The CFD code used for this study was KIVA II. For the purpose of this work the program was adapted to accommodate reactive conditions described by a detailed thermokinetic model. The validation of the numerical procedures are outlined here. They are to be described in more detail elsewhere (14,15).

The most appropriate conditions to investigate the effect of spatial inhomogeneities on the onset of autoignition were determined from preliminary computations of the dependence of ignition delay on compressed gas temperature reached in the RCM. In the present numerical simulations, the ignition delay is defined as the interval from the end of the compression to the attainment of the maximum rate of pressure rise in ignition. Because of the computation time that is required for a full CFD calculation, these data were obtained on the basis of an assumed spatial uniformity of temperature and concentration in the chamber. The program used was an adapation of SPRINT (16), in which the LSODAR solver for the numerical integration of stiff, first order differential equations replaced the NAG D02QBF solver normally called within this program (17,18).

2 THE DEVELOPMENT OF KIVA TO INCORPORATE DETAILED CHEMICAL KINETICS AND HEAT RELEASE AND ITS APPLICATION TO RCM STUDIES

The experimental principles and procedures for operation of the RCM were followed in the CFD simulations, and are given in detail elsewhere (9-13). Briefly, these included the same combustion chamber dimensions, and the same piston speed, stroke and compression ratio (11.2 : 1). The premixed reactants were of such composition to give not only the required fuel to oxygen ratio, but also to include the appropriate proportions of carbon dioxide, nitrogen or argon as the inert components. These were required experimentally to change the ratio of the principal heat capacities of the initial reactant charge, and thereby give different compressed gas temperatures. In general the initial experimental pressures were 33 kPa, yielding compressed gas pressures of up to 1 MPa. However, in the present paper we focus on calculations from an initial pressure of 100 kPa to represent the autoignition characteristics at pressures that are more typical of those in normally aspirated engines.

KIVA II is generally familiar as a multi-dimensional, finite-difference code developed to solve the transient equations of motion of a turbulent, chemically reactive mixture of ideal gases in reciprocating, combustion engines (19-21). The spatial differences are formed on a mesh consisting of small hexahedrons. Curved surfaces, such as within a cylindrical combustion chamber, may thus be adequately described. A standard version of the k - ε model is used to represent the effects of turbulence, while the boundary layer drag and the surface heat transfer are calculated by a turbulent 'law of the wall'.

The geometric configuration of the combustion chamber of the rapid compression machine is fully axisymmetric, which permitted a two-dimensional treatment of spatial conditions on a plane representing piston displacement and half radius. Piston motion was simulated on a 14 x 50 mesh at the start of compression. The mesh was reduced to 14 x 10 cells as the piston reached "top-dead-centre", and this mesh was retained during the development of ignition in the constant volume combustion chamber. The resolution within the vicinity of the cylinder wall was enhanced by a non-equidistant grid structure. The thermophysical data base within the program was enlarged and modified to represent as far as possible the properties of the fuel and of the diluting gases representing typical hydrocarbon fuels or the compositions related to specific experiments.

The solving structure for the chemical source terms in the conservation equations was altered substantially to permit the computation of the low temperature oxidation processes leading to autoignition. A method was developed by an adaptation to the LSODAR program (22), in which there was an automatic switch between stiff and non-stiff integration routines. Whereas the corresponding post-induction equations are accuracy limited, the approximate equations describing the induction processes are stability limited. Computational efficiency was enhanced by choosing a "large" relative tolerance (10^{-3}). The performance of multipoint calculations in CFD codes is, at best, accurate to within a few percent. Thus exaggerated accuracy would be wasteful (23). During the course of iterative convergence of the equations the kinetic data were not re-evaluated, while the current temperature was within a local window (T + ΔT). ΔT was chosen to guarantee an error smaller than 1% in respective data. However, non-physical initial conditions

can lead to extremely small steplengths being taken. Algorithms have been developed to tackle this type of problem (23). We plan to adopt simular procedures.

The extent of detail incorporated in the kinetic mechanism was balanced against the available CPU time for each computation and the storage capacity that could be conveniently accessed. Taking 3 - 4 hrs of CPU and using up to 8 MBytes storage capacity on an IBM 3090 machine, it proved possible to compute the rapid compression stroke and subsequent thermokinetic development of a scheme comprising 30 species in 70 reactions over a real-time of 30 ms with the mesh adopted in these CFD calculations. In these circumstances a hybrid technique was adopted in which the explicit procedure was used during the compression stroke. A switch was then made to the the implicit solver (LSODAR) at the end of compression and throughout the post-compression interval, where the chemical evolution was most important.

Much shorter schemes (such as in the application of the "Shell" model discussed below) were far less demanding. However, kinetic mechanisms which contain up to 30 species are sufficiently large to permit the study of autoignition characteristics of "reduced" models incorporating significant mechanistic detail, as is required to investigate how mixtures of gasoline components of different molecular structures may interact. This size of kinetic model was also compatible with the initial target that we set for testing the CFD application.

The CFD calculation was made on an *ab initio* basis, starting with prescribed gas compositions at particular conditions corresponding to those in the RCM combustion chamber. Thus the first test was to be able to predict compressed gas pressures that corresponded to those observed experimentally. From this alone we could visualise the qualitative characteristics of gas motion generated by the piston and interpret typical temperature distributions achieved at the end of compression. The simulation of a stagnation zone associated with the roll-up vortex is clearly seen (Fig. 1). These features are governed by the nature of the CFD calculation itself, and are subject to modification as improved fluid dynamic models become available. A validation of the application under reactive conditions was based on experimental studies in the RCM of the decomposition of di-t-butyl peroxide in the presence of oxygen and other diluting gases (14,15). A kinetic model for the combustion of di-t-butyl peroxide in the presence of oxygen, comprising 27 species in 70 elementary reactions, was validated in earlier work against experimental, low pressure, spontaneous ignition studies (24).

3 THE "SHELL" REDUCED KINETIC MODEL AND PREDICTION OF THE N.T.C. REGION OF REACTION RATE

The reduced kinetic scheme and the rate data used to simulate autoignition in response to an n.t.c. of rate were derived from the well-established "Shell", four-variable thermokinetic model (6,8) in the form given in Table 1 (25). For kinetic simplicity, the chain propagation is established by a single type of free radical species R.

Table 1 Representation of the "Shell" model

initiation	$RH + O_2 \rightarrow 2R$	(1)
propagation	$R + (1 + l)(m^{-1}RH + pO_2)$	
	$\rightarrow qP + f_1B + f_2Q + R$	(2)
propagation	$R + Q \rightarrow B + R$	(3)
branching	$B \rightarrow 2R$	(4)
termination	$R + R \rightarrow P'$	(5)
termination	$R \rightarrow P'$	(6)

The modified form (25) adopted for the propagation reaction 2 is necessary because a mass balance of reactants and products must be retained in the CFD calculations. The fuel is regarded to be of composition C_nH_{2m}, so that if two H atoms are oxidised at each propagation cycle then the combustion of C_nH_{2m} is complete after m propagation cycles (6). Since the species P represents the final products from the propagation, the reaction stoichiometry

$$(1/m)C_nH_{2m} + pO_2 = n(CO_2 + CO) + mH_2O$$

leads to

$$q = (n/m) + 1.$$

The ratio of fuel to oxygen is fixed by the parameter p, which is governed by the relative yields of CO, CO_2 and H_2O. Thus

$$p = [n(2-\gamma) + mR]/2m,$$

where γ is the relative yield of CO and CO_2. These relative product yields are important also because they govern the overall heat release (6). The intermediate species B and Q are also products of the fuel oxidation. Thus the stoichiometric coefficient l is dependent upon the amounts of these products formed in the propagation cycle, and is given by

$$l = [f_1 M(B) + f_2 M(Q)]/[(1/m)M(RH) + pM(O_2)],$$

for which the molar masses M(i) and the coefficients f_1 and f_2 are defined as follows.

From reactions 3 and 4

$$M(B) = M(Q) \text{ and } M(B) = M(R)/2$$

respectively. The mass of R is determined from the initiation process 1,

$$M(R) = [M(RH) + M(O_2)]/2.$$

Although there is no realistic chemical identification of species R, B or Q that conform to these criteria, there is no constraint in the application of the very generalised "Shell" model that does not permit them. The mass of P is given by

$$qM(P) = (n\gamma/m)M(CO) + (n/m)(1-\gamma)M(CO_2) + M(H_2O).$$

The (minor) products of the termination reactions are deemed to be of the same mass as the inert gas components in the system. The coefficients f_1 and f_2 govern the rates of generation of the intermediates B and Q respectively, and must incorporate the temperature dependence of their rates of formation. The negative temperature dependence of rate is encapsulated in f_1. The temperature coefficient f_2 is chosen so that the formation of species Q is favoured only at higher temperatures. An alternative route to the branching agent B is then promoted, via 3.

3.1 Initial studies of the dependence of ignition delay on compressed gas pressure

The dependence of autoignition delay times on compressed gas temperature, derived from the application of the "Shell" model to combustion in the RCM in a simulation of a 90 PRF fuel, was evaluated on the basis of the integration of the differential equations for mass and energy assuming spatially uniform conditions. An averaged heat transfer coefficient was used to describe the heat loss to the walls. The total ignition delay computed over a range of compressed gas pressures at two different initial pressures are shown in Figure 2. As the pressure in the combustion chamber is raised there is a considerable reduction in ignition

delay, but the main qualitative features of the n.t.c. are retained. The numerical results in Figure 2 illustrate the relationship between the behaviour observed at the operating conditions in the RCM and the expected response at pressures compatible with normal induction and compression in an engine. The kinetic features of the "Shell" model are such that, as the pressure is raised, there is a shift in the range of temperature at which the n.t.c. is observed. There is recent supporting evidence for this from shock tube studies of the autoignition of n-heptane (25).

4 CFD CALCULATIONS INCLUDING THE "SHELL" MODEL AND THE IDENTIFICATION OF CENTRES OF INITIATION OF IGNITION

The results reported here were obtained at conditions simulating autoignition at a compressed gas pressure of about 2.6 MPa in the RCM at the three compressed gas temperatures $T_c = 670$ K, 750 K and 800 K (Figure 2). In order to facilitate the comparison of the simulated behaviour, the spatial and temporal variations of temperature and species densities in each of the three cases are represented on Figure 3. These results pertain to a radial coordinate at one plane of the combustion chamber, namely the central plane at 670 K and 800 K and at a plane close to the piston crown at 750 K. The latter was found to be the region of greatest activity at $T_c = 750$ K. The orientation of the axes to represent the reactant temperature, T, and density of the intermediate species Q differ from those that represent the densities of fuel, RH, and the intermediate species B. This is necessary in order for the main features of the changes that occur to be visualised satisfactorily.

The numerical results for the development of two-stage ignition at $T_c = 670$ K show that the temperature in the first stage is quite slow to develop after the end of compression (22 ms). Through the first stage of two-stage ignition there is an approximately uniform temperature across most of the chamber radius (Figure 3a). A temperature gradient exists close to the wall. When the hot stage of ignition begins to develop it appears first at the central axis of the chamber. This local development of ignition is mirrored by complete consumption of fuel being first achieved at the centre of the chamber (Figure 3b). There is an increase in the density of intermediate species Q also at the centre of the chamber. Q remains in lowest concentration close to the wall throughout the reaction period (Figure 3c). The density of the chain branching intermediate B exhibits a maximum in the first stage of two-stage ignition, and for part of the subsequent time during the ignition delay the density of B remains slightly higher in the vicinity of the combustion chamber wall than elsewhere (Figure 3d).

The numerical results from the CFD calculation at $T_c = 750$ K relate to behaviour at a radial coordinate close to the piston crown. At this cross section the behaviour is affected by the stagnation zone associated with the predicted roll-up vortex, and it is in that vicinity that hot ignition is first attained. As is shown in Figure 3e, the first stage develops very soon after the end of compression, but there is then a longer incubation period for the development of the second stage than that seen in Fig. 3a and a higher temperature is reached during this period. Moreover, the eventual development of the hot stage of two-stage ignition first appears at the edge of the chamber (which corresponds to the corner between the piston crown and the chamber wall). The fuel is consumed first there (Figure 3f). Undoubtedly, this is a consequence of the thermokinetic interactions that control the local densities of the intermediates B and Q, as discussed below. Their profiles are different from those predicted at 670 K insofar that, throughout the ignition delay, the density of B remains highest within the cooler regions of the chamber and, eventually, the density of Q is first to increase also in that vicinity (Figures 3g and h).

The calculation at a compressed gas temperature of 800 K leads to a single stage ignition. The time to ignition from the end of compression is still approximately 4 ms, and throughout most of this interval there is a virtually uniform, radial temperature distribution at the central plane of the combustion chamber (Figure 3i). There is a gradient close to the chamber wall. Ignition is predicted to occur simultaneously throughout the core gas and no preferred site at which complete consumption of the fuel first occurs can be distinguished (Figure 3j). Throughout the post compression period, the density of species B remains appreciably lower than that predicted at the lower compressed gas temperatures (Figure 3l). There is, however, a maximum in the density of B during the late stage of the compression. This is where the low temperature chemistry occurs. Thus there is already a reasonable free radical concentration at the start of the post-compression period, and the density of Q is already relatively high across the radial coordinate when piston motion stops (Figure 3k).

Contours for the predicted temperature variations in the plane of the piston motion, when temperatures in excess of 1000 K are first attained locally, are shown in Figure 4. The results pertain to the three compressed gas temperatures 670 K, 750 K and 800 K, as described above.

5 DISCUSSION

The effects on the fluid dynamics and heat transport in the fuel + "air" charge of the simulated compression stroke are qualitatively similar in all three applications of the "Shell" kinetic model. During the simulation of the piston motion the imposed field is too strong to allow the development of the roll-up vortex to be shown by the the $k - \varepsilon$ model. Nevertheless a roll-up vortex is expected to be formed by the shearing at the cylinder wall (27). Its development is seen in the post-compression period and this vortex creates a stagnation zone close to the corner between the piston and the chamber surface. The roll-up vortex survives throughout the ignition delay. Early in the post compression period, the highest gas velocities are associated with the charge close to the centre of the chamber, and correspond to the piston speed (12 m s^{-1}).

The prediction from KIVA II is that a virtually adiabatic temperature is attained within the core gases at the end of compression. The site at which autoignition is predicted to first occur is governed mainly by the interaction between the kinetics and the thermal environment, and how each modifies the other as a consequence of the associated heat release and heat loss. Mass transport may also contribute in regions where high gas velocities prevail.

Thus, during the first stage of two-stage ignition, following a compression to 670 K the steepest part of the temperature gradient exists at the centre of the combustion chamber at 24.8 ms (Figure 3a), which corresponds in time and space to the maximum density of B (Figure 3d). The local temperature is 755 K. By contrast, following a compression to 750 K the steepest part of the temperature gradient exists at the edge of the combustion chamber at 23.2 ms (Figure 3e), corresponding also in this case to the maximum density of B (Figure 3h). The maximum temperature gradient is appreciably lower than that in Figure 3a, commensurate with the somewhat lower density of B that is attained. The local temperature is 750 K. These features arise from the effect of the negative temperature dependence associated with the rate of formation of B (coefficient f_1) on the overall development of reaction. B accumulates at low temperatures because its decomposition rate, the chain branching step 4, is relatively slow. As the temperature is raised the decomposition rate increases, but the formation rate decreases, and so the local density (or concentration) passes through a maximum. The highest densities of B are attained at the centre of the vessel only when the compressed

gas temperature is low, and this is characterised by the range of compressed gas temperature within which there is an overall positive temperature dependence, as summarised in Figure 5.

At higher compressed gas temperatures, the criteria that favour the accumulation of B are optimised in the cooler regions at the edge of the combustion chamber. The consequences, which affect the development of the second stage, are that the negative temperature coefficient of reaction precludes the "low temperature" branching route, (f_1 in 2) + 4, in the "core" gas where the temperature is highest. There is a locally enhanced rate of production of R accompanied by heat release, and in due course there is also a preferential development of intermediate species Q (controlled by f_2) as the local temperature increases (Figure 3g). The full development of the hot stage of ignition occurs first where both the density of Q and the gas temperature are optimised, as a consequence of the second route to autocatalysis 3.

At the highest compressed gas temperatures studied, the reactants are forced by the compression stroke through the temperature range at which the low temperature chain-branching sequence is most active. Some reaction takes place during the compression stroke, as is shown by the accumulation of B prior to the end of compression (Figure 3l). There is greater activity associated with B at the edges of the chamber, although the densities that are attained are some orders of magnitude lower that those predicted at the lower compressed gas temperatures.

The development of ignition at a compressed gas temperature of 800 K is governed predominantly by the high temperature temperature mechanisms comprising (1), (4), (5) and the component of (2) involving the coefficient f_2, which gives rise to autoignition simultaneously throughout a large volume of the chamber (Figure 5).

6 CONCLUSIONS

From these numerical results we have shown qualitatively that the complexities of hydrocarbon oxidation can cause marked variations in the location at which ignition centres first occur in a combustion chamber according to the compression temperatures that are attained. The sites at which autoignition occur in a given system are also related to the gas motion and the associated heat transfer that takes place in the chamber.

The application of KIVA II to this study has had the considerable benefit of it being a well-developed fluid mechanical code. In due course we shall seek to apply CFD codes that improve quantitatively on the $k - \varepsilon$ treatment of turbulence, especially with regard to heat and mass transport close to the combustion chamber surface, since these properties are particularly important if acceptable quantitative simulations to the real behaviour are to be achieved. We shall also use kinetic models that give a quantitatively and kinetically more meaningful basis of the detailed interpretation of alkane oxidation and that of other major components of gasoline, than is possible with the "Shell" model.

ACKNOWLEDGEMENT

The authors wish to thank SERC for financial support

REFERENCES

(1) GRIFFITHS, J.F. and SCOTT, S.K., Thermokinetic interactions: fundamentals of spontaneous ignition and cool flames. *Progr. Energy Combust. Sci.* 1987, 13, 161

(2) KONIG, G. and SHEPPARD, C.G.W., End-gas autoignition and knock in a spark ignition engine. *SAE*, 1990, paper 902135

(3) HU, H. and KECK, J., Autoignition of adiabatically compressed combustible gas mixtures. *SAE*, 1987, paper 872110

(4) GRAY, B.F. and YANG, C.H., Unified theory of explosions, cool flames and two-stage ignitions. *Trans. Farad. Soc.*, 1969, 69, 1614

(5) AXELSSON, E.I., BREZINSKY, K., DRYER, F.L., PITZ, W.J. and WESTBROOK, C.K., Chemical kinetic modeling of the oxidation of large alkane fuels: n-octane and iso-octane. *Twenty-first International Symposium on Combustion*, The Combustion Institute, Pittsburgh, 1986, p 783

(6) HALSTEAD, M.P., KIRSCH, L.J. and QUINN, C.P., The autoignition of hydrocarbon fuels at high temperatures and pressures - fitting of a mathematical model. *Combust. Flame*, 1977, 30, 45

(7) COX, R.A. and COLE, J.A., Chemical aspects of autoignition of hydrocarbon - air mixtures. *Combust. Flame*, 1985, 60, 109

(8) NARAJAN, B. and BRACCO, F.V., On multi-dimensional modeling of autoignition in spark-ignition engines. *Combust. Flame*, 1985, 57, 179

(9) GRIFFITHS, J.F. and PERCHE, A., The spontaneous decomposition, oxidation and ignition of ethylene oxide under rapid compression. *Eighteenth Symposium (International) on Combustion*, The Combustion Institute, 1981, p893

(10) GRIFFITHS, J.F. and HASKO, S.M., Two-stage ignitions during rapid compression: spontaneous combustion in lean fuel-air mix- tures. *Proc. Roy. Soc. Lond.* 1984, A393, 371

(11) FRANCK, J., GRIFFITHS, J.F. and NIMMO, W., The control of spontaneous ignition under rapid compression. *Twenty-first International Symposium on Combustion*, The Combustion Institute, Pittsburgh, 1986, p 447

(12) INOMATA, T, GRIFFITHS, J.F. and PAPPIN, A.J., The role of sensitizers for the spontaneous ignition of hydrocarbons. *Twenty-third Symposium (International) on Combustion*, The Combustion Institute, 1991, p1759

(13) GRIFFITHS, J.F., PAPPIN, A.J., AL- RUBIAE, M.A. and SHEPPARD, C.G.W., Fundamental studies related to combustion in diesel engines. *I. Mech. E.* 1991, C433/018

(14) GRIFFITHS, J.F., JAIO, Q., SCHREIBER, M., MEYER, J. and KNOCHE, K.F., Development of thermokinetic models for autoignition in a CFD code: experimental validation and application of the results to rapid compression studies. *Twenty-fourth Symposium (International) on Combustion*, in press

(15) GRIFFITHS, J.F., JAIO, Q. KORDYLEWSKI, W., SCHREIBER, M. MEYER, J. and Knoche K.F., Experimental and numerical studies of ditertiary butyl peroxide combustion at high pressures in a rapid compression machine. Submitted to *Combust. Flame*

(16) BERZINS, M. and FURZELAND, R.M., A user's manual for SPRINT. *T.N.E.R. 85058, Shell Research Ltd.,* 1985

(17) BAULCH, D.L., GRIFFITHS, J.F., PAPPIN, A.J. and SYKES, A.F., Stationary-state and oscillatory combustion of hydrogen in a well-sirred flow reactor. *Combust. Flame*, 1988, 73, 163

(18) GRIFFITHS J.F. and SYKES A.F., Numerical studies of a thermokinetic model for oscillatory cool flames and complex ignition phenomena in ethanal oxidation under well-stirred flowing conditions. *Proc. R. Soc Lond.*, 1989, A422, 289

(19) AMSDEN, A.A., RAMSHAW, J.D., O'ROURKE, P.J. and KULOWICZ, J.K., KIVA: a computer program for two- and three- dimensional fluid flows with chemical reactions and sprays. Los Alamos Lab. Report, 1985, LA-10245-MS

(20) AMSDEN, A.A., O'ROURKE, P.J. and BUTLER, T.D., The KIVA II computer program for transient multidimensional chemically reactive flows with sprays. *SAE*, 1987, paper 872072

(21) AMSDEN, A.A., O'ROURKE, P.J. and BUTLER, T.D., KIVA II: a computer program for chemically reactive flows with sprays. Los Alamos Lab. Report , 1989, LA-11560-MS

(22) PETZFOLD, L.R. and HINDMARSH, A.C., LSODAR Livermore solver for ordinary differential equations with automatic method switching for stiff and nonstiff problems and with root-finding. LLNL report, 1987

(23) RHADAKRISHNAN, K. and PRATT, D., Fast algorithm for calculating chemical kinetics in turbulent flow. *Comb. Sci. Tech.*, 1988, 58, 155

(24) GRIFFITHS, J.F. and PHILLIPS, C.H., Experimental and numerical studies of the combustion of ditertiary butyl peroxide in the presence of oxygen at low pressures in a mechanically-stirred closed vessel. *Combust. Flame,* 1990, 82, 362

(25) SCHÄPERTÖNS, H., Ph.D. Thesis, 1984, Sinulation von Klopfvorgangen mit einem zweidimensionalen mathemattischen Modell. RWTH, Aachen

(26) CIEZKI, H. and ADAMEIT, C., Vergleich des Selbstzünddverhaltens von n-Heptan/Luft- und Benzol/Luft- Gemischen unter motorisch relevanten Bebingungen. *VDI Berichte*, 1991, 922, 495

(27) TABACZYNSKI, R.J., HOULT, D.P. and KECK, J.C., High Reynolds number in a moving corner. *J. Fluid Mech.*, 1970, 42, 249

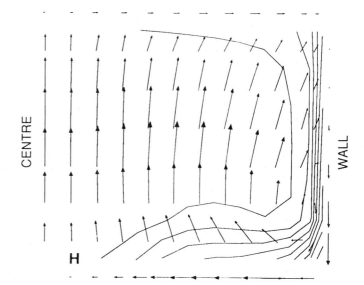

Fig 1 Calculated temperature and vectorial flow fields within the axisymmetric plane of the combustion chamber 0.1 ms after the end of compression to a temperature of 670 K. The adiabatically compressed, core gas lies within the contour marked **H**. The contour lines represent isotherms at approximately 15 K intervals. The separation between arrows shows the non-equidistant grid structure. The initial conditions simulate a 90 PRF fuel in air ($\phi = 1$) at 100 kPa in the combustion chamber of the Leeds RCM.

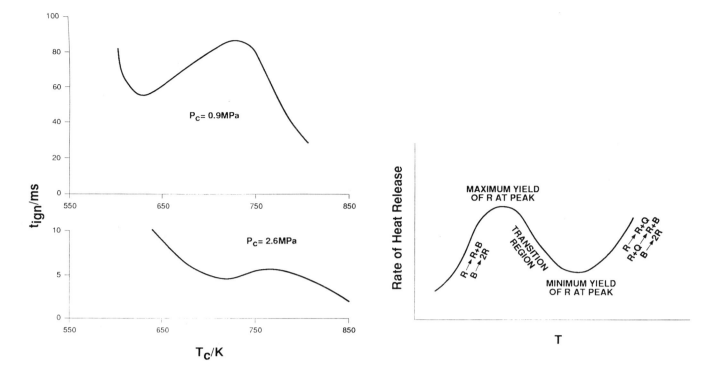

Fig 2   Ignition delays from the end of compression, as a function of compressed gas temperature, predicted on the basis of the "Shell" kinetic model for a 90 PRF fuel in the Leeds RCM. The results are given at two different initial pressures, yielding the compressed gas pressure averaged across the temperature range, as indicated.

Fig 5   Overall variation of heat release rate as a function of gas temperature and summary of the main kinetic or mechanistic features which control the ignition delay.

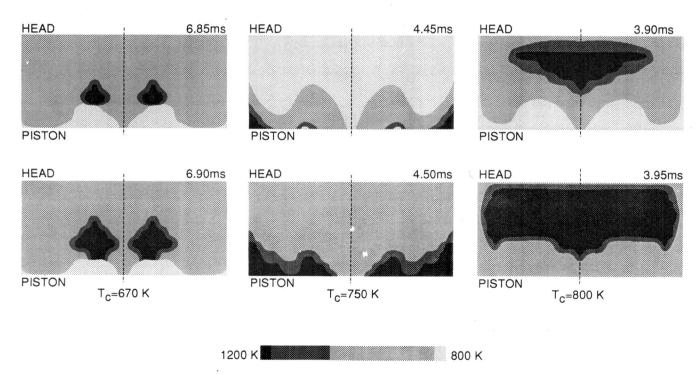

Fig 4   Predicted temperature distributions in the plane of the combustion chamber between the cylinder head and piston crown, at the onset of flame propagation. The times from the end of compression and the compressed gas temperatures (T_c) are given. The highest temperature is approximately 1200 K in each case. The lowest temperature is slightly less than 800 K for the case T_c = 670 K and greater than 800 K for the cases T_c = 750 and 800 K. The mirror image of the calculated profiles has been included to construct the full section of the combustion chamber.

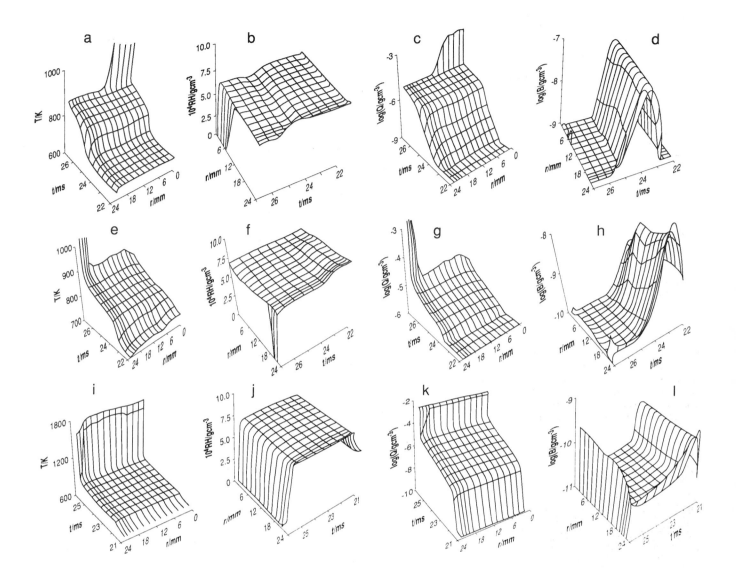

Fig 3   Simulated profiles from the "Shell" kinetic model for reaction of a 90 PRF fuel in air ($\phi = 1$) at an initial pressure of 100 kPa in the Leeds RCM.   The profiles show how species density and gas temperature evolve during two-stage ignition, at the compressed gas temperatures 670 K (a - d), 750 K (e - f) and 800 K (i - l).   The radial dimension signifies the distance from the centre to the chamber wall (24 mm).   For the results at 670 and 750 K, the time is taken from the moment that the piston stops (22 ms) and extends through the development of the second stage.   At 800 K, the initial time precedes the end of compression because reaction begins before the piston has stopped.   The directions of the co-ordinates are varied to emphasise the main features of each profile.   The densities of the intermediates, B and Q, are plotted logarithmically.

C448/057

Potential for engine control using neural networks

A BACON, BEng, PhD, AMIMechE and P J SHAYLER, MSc, PhD, CEng, MIMechE
Department of Mechanical Engineering, University of Nottingham
T MA, BSc, MSc, PhD, CEng, MIMechE
Ford Motor Company, Laindon, Essex

This paper discusses the viability of replacing elements of a modern electronic engine control system with trained neural networks. The basic technology required to implement a neural network solution is described. Also discussed are the potential benefits of using neural networks in an engine control application with particular emphasis on the areas of algorithm and strategy complexity, implementation issues and calibration effort.

NOTATION

ΔFUELPW	Correction to approximated fuel pulsewidth duration
ΔSAF	Correction to approximated spark advance value
ACT	Air Charge Temperature
ADC	Analogue to Digital Converter
BP	Back Propagation
CID	Cylinder Identification signal
CPS	Crankshaft Position Sensor
ECT	Engine Coolant Temperature
EEC	Electronic Engine Control
Epoch	One complete pass through a set of patterns during training
FUELPW	Calculated fuel injector pulsewidth duration
HEGO	Heated Exhaust Gas Oxygen Sensor
LOAD	Proportion of maximum possible air mass flowrate induced by engine
MAF	Mass Air Flowrate to engine
N	Engine Speed
PC	Personal Computer
PIP	Profile Ignition Pickup (cycle marker signal)
SAF	Calculated Spark Advance
TFC	Transient Fuelling Compensation
TPOS	Throttle Position sensor output
TSS	Total Sum of Squared errors
VMAF	Voltage signal from Mass Air Flowrate sensor
VSS	Vechicle Speed Sensor

1. BACKGROUND

The introduction of electronic engine control (EEC) systems has enabled engine performance improvements to be made in many areas. The success of EEC systems stems from the improved accuracy of calibration and the greater facility they provide to monitor and control various engine performance features compared to mechanical systems. The development of calibration and strategy software for new engine applications is, however, a complex task requiring a substantial investment of resources. EEC systems also require manufacturers to retain expertise to maintain and update the EEC strategies throughout the product cycle. Although many EEC tasks are generic to all applications, details vary. The development of truly generic EEC systems would offer advantages in reducing product development time, documentation, and the overhead of maintaining expertise for particular applications. The subject of the work described in this paper is an assessment of how neural networks might be exploited to achieve this.

Neural networks are processing algorithms which can be used as associative memories. They have been used as pattern classification tools [1, 2, 3, 4, 5] for image and drawing analysis [6, 7] and in speech detection [8]. For use as associative memories in control applications, neural networks can be trained to map a series of input (sensor) signals from a multi variable control system to a set of corresponding output (actuator) signals. This mapping processes is the basis of many of the functions carried out by EEC systems which employs 2 or 3 dimensional lookup tables, with interpolation between table entries. For each control output, several tables may be used with interim calculations also being made. Forming these lookup tables and the associated algorithms from test bed mapping data is a complex procedure.

In principle, neural networks can be used as operators relating input signals to desired outputs which can be substituted for look-up tables of calibration data or elements of control strategy. Once formulated for a generic task, such a network can be retrained for new applications of the same type if suitable training data is made available. Updating and modifications to encompass new data can be effected by retraining without a detailed knowledge of the network details.

The authors have so far been concerned with validating the accuracy of neural networks when used to solve the mapping problems encountered in EECs. The approach taken was to train neural networks to emulate elements of a current production engine control strategy. Data recorded during rolling road tests was used to train the networks and also provided a standard against which the accuracy of the trained network could be measured. The development of a neural network to give the required spark advance and fuel injector control outputs is covered in this paper. Three alternative solutions to this problem are compared. In the light of these results, the potential for the application of neural networks to other control tasks within engine management systems is discussed.

2. NEURAL NETWORKS

A concise description of neural network principles is given here. Readers requiring a more detailed treatment of neural network theory and background are directed to texts by Rumelhart and McClelland [9, 10, 11].

A network is constructed of a series of interconnected units or nodes as shown schematically in Figure 1. In this example the units are arranged in three groups: an input layer, a hidden layer and an output layer. There may be any number of hidden layers (including none) although, in practice, most problems may be solved using one or two hidden layers [9]. The units are linked by a series of connections. Units in the input layer are connected to those in the hidden layer and units in the hidden layer are connected to those in the output layer. There are no direct connections between input and output units (unless there are no hidden units). The networks used by the authors are fully interconnected (all units in the input layer are connected to all units in the hidden layer and, similarly, all units in the hidden layer are connected to all units in the output layer). Restricting the connections between layers of units is currently a topic for research and is not covered in this paper. Each connection has a strength or weight value associated with it. This determines the proportion of the incoming data value which is transmitted between the units.

When a set of inputs is applied to a network an activation value is calculated for each unit in the network. The activation value for a unit in the input layer is simply equal to the input applied to it. For units in the hidden and output layers, activations are calculated using the current values of activations and weights in connected layers. For a unit in the hidden layer, the activation is calculated by summing the products of the activation and weight strength of each unit in the input layer to which the unit in the hidden layer is connected. The activations of units in the output layer are calculated in the same way using the hidden layer activation values and the hidden to output weight values.

Networks are trained by adjusting the strengths of the connections between units so as to correlate the input and output patterns. This is an iterative procedure requiring each of a set of training patterns to be presented to the network several times. Each time an input pattern is presented to the network, the current set of weights associated with the network are used to calculate a set of outputs. These outputs are compared with the desired outputs for the input pattern and the errors are used to modify the values of the weights in the network. This procedure is then repeated for the other patterns in the training set. Training continues with as many complete passes (epochs) through the training set as are necessary for the weights to converge to a set of stable values.

Several procedures are available for modifying the weights during training. The most common is the back propagation (BP) method which is a gradient descent technique. During training the BP method adjusts those weights in the network which contribute most to the errors observed in the outputs. Sufficiently small modifications must be made to the weights on each presentation of a pattern to allow the weights to converge. The size of the adjustments made to the weights are determined globally by a learning rate constant which sets the proportion of the calculated correction to a weight value which is actually used.

The state of the network during training is monitored using the sum of the squared errors in each of the outputs summed over all patterns in the training dataset. This parameter, called total sum of squared errors (tss) reduces as training proceeds. An example of this is given in Figure 2. Since the network will rarely converge to a set of weights which give precise outputs for all training patterns, the network is deemed to have converged when the tss reaches a suitably small threshold value, corresponding to an epoch number of 600 in Figure 2. At this stage the weight values represent a solution to the mapping problem for the patterns in the training set.

The trained network can be used to generate output values for input patterns not included in the training set. When such an input pattern is presented to the trained network, the output activation values are calculated as during training, using the final weight values generated by the training process. When used in this way, no adjustments are made to the weight values. A trained network is thus able to generalise through a process which may be viewed as interpolation in multi-dimensional space. The degree to which this generalisation takes place is affected by the degree of similarity between the input pattern and the patterns used in training. Accurate output values will not normally be obtained for an input pattern which is completely different from all examples used in training, and the spacial separation of the training patterns must not be too large. In an engine mapping problem, the network must be trained on data taken at, for example, suitably spaced speed and load conditions. Also, if there is noise on, or other variation of, the inputs, this effect must be included by obtaining example patterns over a period of time for each training condition to ensure that the training data is representative. There are, however, no rules for ensuring that the correct training point spacing and sampling periods are chosen. These must be determined experimentally and are affected by the problem to be solved.

3. THE CONTROL PROBLEM

A current production EECIV control system capable of meeting the 1983 US emissions standard was selected as the basis for this work. Figure 3 is a schematic diagram of the system showing the important interfaces with the engine. The system has distributorless ignition with multi-point, sequential fuel injection. The control modules contain proprietary algorithms based on functions embedded in lookup tables. The algorithms are grouped together according to the mode of engine operation to which they relate. Examples of operating modes include cranking, idle, part load and wide open throttle. The control strategies for each mode are tuned to give optimum conditions at each operating regime.

The primary objective of the authors' work was to examine the accuracy with which neural networks could emulate particular segments of the control strategy, and attention has been focused on the control of steady state, part load operation with open loop fuelling control (i.e. where the fuelling requirement cannot be determined from the heated exhaust gas oxygen (HEGO) sensor fitted to the engine.) Spark timing and fuel injector pulsewidth were chosen as the two control outputs to be generated, since these require the greatest mapping and calibration effort in the present EEC system. It was envisaged that the accuracy of neural networks in carrying out this control task would be indicative of their performance on similar tasks in other modes of EEC IV operation.

The relevant input and output signals for spark and open loop fuel control under steady-state, part load conditions were identified through an analysis of the EEC IV strategy. The key signals are listed in Table 1.

engine controlled by a 1983 US emission level EEC IV strategy. The EEC input and output values listed in Table 1 were logged on an IBM-PC compatible computer via an RS-232 data link. Complete sets of input and output values were obtained via this link at approximately 300 ms intervals.

Two distinct groups of test datasets were acquired under open loop control. Firstly, data was acquired at the steady state speed and load conditions in Table 2.

Table 2: Test Conditions for Training Data Acquisition

Manifold Vacuum [N/m²]	Engine Speed [rpm]			
	1000	2000	3000	4000
65	X	X	X	X
50	X	X	X	X
40	X	X	X	X
35	X	X	X	X
30		X	X	X

Table 3: Test Conditions for Verification Data Acquisition

Test	Manifold Vacuum [N/m²]	Engine Speed [rpm]
I	60	1500
II	35	1500
III	45	2500
IV	25	2500
V	45	3500

Table 1:

Inputs		Outputs	
Air Charge Temperature (ACT)	[°F]	Spark Advance (SAF)	[°CA]
Mass Air Flow (MAF)	[lb/s]	Fuel Injector Pulsewidth (FUELPW)	[ticks]
Engine Speed (N)	[rpm]		
Load (LOAD)	[-]		

Clearly, the simplest network layout which could be used to solve this problem would have four inputs (one for each input signal) and two outputs (one for each output signal). This simple four input, two output network has been adopted for most of the work. The network layout has the advantage of requiring only straightforward scaling of the sensor inputs to convert them from engineering units to values in the range 0 to 1 which can be handled by the network. Similar scaling is carried out for the outputs, and the calculations carried out by the EEC IV system involving the use of several lookup tables are replaced by a single neural network.

4. DATA ACQUISITION

Data was logged in rolling road tests of a mule vehicle. The vehicle was fitted with a 1.8l high output Zeta

This matrix of data was used for training the neural networks. A second group of steady state tests were carried out at the conditions shown in Table 3. These conditions were chosen to lie between those in Table 2 and were used only to test the ability of the trained networks to interpolate. The ECE 15 test cycle formed the basis for the selection of the test points. The car was driven on the rolling road for a period of a few minutes at each test condition. After this stabilising period, data was acquired over a period of approximately 10 seconds. This yielded approximately 30 input/output sets of data at each test point which are assumed to be representative of normal variations in the inputs due to noise, normal movements of the throttle by the driver etc. at each of the test conditions.

The logged data was stored as engineering unit values, with the units shown in Table 1. Before being used to train or test a neural network, this data was scaled so that data for each input or output unit lay in the range 0 to 1. Since this scaling procedure is dependent upon the form of network solution chosen, the scaling factors used are discussed below along with the test results for each of three network formats.

5.   RESULTS

Three types of network implementations have been examined. Details of the network layout, training procedure and results for each solution are given separately. All three approaches carry out the same basic, four input, two output mapping.

Absolute Value Network

This network arrangement is the simplest solution to the four input, two output mapping problem. The four input values are simply scaled and applied to the network in Figure 4. The fuel injector pulsewidth and spark advance outputs generated by the neural network are directly proportional to the engineering unit values of these two outputs. Thus, calculation of spark advance in degrees crank angle, and fuel injector pulsewidth in EEC clock ticks outputs is simply a matter of re-scaling the network output values. Scaling ranges for the input and output values were chosen to cover only the conditions encountered in the training and verification datasets. Expansion of these ranges to cover the whole engine operating envelope has been carried out without any major reduction in the accuracy of the neural network outputs.

The 4:4:2 network shown in Figure 4 was trained on the data acquired at the test conditions in Table 2. The scaling factors in Table 4 were used to convert between engineering and network units.

Table 4: Scaling Factors for Absolute Value Network

Parameter	Abbreviation	Scale Factor
Air Charge Temperature	ACT	(x-90)/40
Mass Air Flow	VMAF	(x-0.9)/2.1
Load	LOAD	(x-0.18)/0.42
Engine Speed	N	(x-100)/3500
Spark Advance	SAF	x/30
Fuel Injector Pulsewidth	FUELPW	(x-1000)/2400

The patterns in the training set were presented to the network in random order and a small value of learning rate was used. The network weight values converged after approximately 1000 epochs (complete passes through the training set). The input data obtained at the test conditions in Table 3 and scaled according to Table 4 were then applied to the trained network and the outputs were compared with those obtained from EEC IV under the same conditions.

Figures 5a and 5b compare the spark advance and fuel injector pulsewidth results obtained from EEC IV and the neural network at the five test conditions. The spark advance values given by the network are retarded relative to those generated by EEC IV at each condition. The errors range from 0.7°CA to 5.7°CA yielding an overall RMS error of 34.1 per cent. The injector pulsewidth values are more accurate, having an overall RMS error of 5.4 per cent. The network has lower accuracy at condition IV where the pulsewidth value given is 11.3 per cent lower than that given by EEC IV. This suggests that the network is more accurate at interpolation between training conditions (required at conditions I, II, III and V) than at extrapolation which is required for test IV. (Compare tables 2 and 3 which detail the load/speed combinations used to train and test the network).

The substantial errors in spark timing at conditions requiring interpolation between training points are due to the complexity of the spark advance map used in the EEC system, and the wide separation in load/speed space of the data used to train the network. Increasing the density of the training points covering the load and speed ranges increases the interpolation accuracy of the network for the spark advance output. The fuelling output values are more accurate where interpolation is required because the fuelling map employed by the EECIV system is less complex in shape than that employed for spark advance and is, therefore, more accurately represented by the sparse matrix of training data used for these tests.

It should be noted that training and testing of all the networks described in this paper was carried out off-line on an IBM-PC using data stored during the test procedures described in the Data Acquisition section above. Comparative engine performance or emissions results could not, therefore, be obtained. Work is in hand to develop a neural network based control system to allow the emissions and performance implications of neural control to be evaluated. This system is described in more detail in the Discussion section below.

Additional Correction Network

The additional correction network approach was designed to address the problem of low accuracy of the spark advance results and to improve the extrapolation ability of the network. A schematic of the additional correction approach is shown in Figure 6. Simple mathematical functions are used to produce approximate spark advance and injector pulsewidth values. These are summed with correction values made by a neural network to give final output values. This approach may be suitable for the development of more generic control strategies using standard approximation functions, with the strategy being tuned to particular engine types using the neural network which can be easily trained to give appropriate correction values.

A network having the same layout and input data as that in Figure 4 was used. The output values generated by this network were corrections to the approximate spark advance and injector pulsewidth values. Several approximation functions were tested. The pair of functions yielding the best network accuracy were used to generate the results shown here. The 4:4:2 network was trained using the errors in the approximation functions when applied to the data obtained at the training conditions. Training was carried out in the same manner as for the absolute value network. When the verification data were presented to the network the total output (approximation function + network correction) values shown in Figure 7a and 7b were obtained. As for the absolute value network, the injector pulsewidth values are more accurate than the spark advance values. The RMS error in the spark advance values has reduced to 17.9 per cent with a corresponding slight reduction in the range of errors. Unlike the absolute value network results, both positive and negative errors in the additional correction outputs were obtained. An improvement in fuel pulsewidth accuracy was also obtained from this network. This was largely due to the effect of the approximation function which improved the extrapolation ability of the network reducing the error at test point IV. The overall RMS error for injector pulsewidth reduced to 2.3 per cent.

Proportional Correction Network

Although more accurate than the absolute value network, errors in the spark advance generated by the additional correction network were deemed unsatisfactory and attributed to excessive corrections being made by the neural network. The proportional correct network arrangement uses approximation functions in the same way, but makes corrections in proportion to the approximated values. This prevents large corrections being made to small output values. The maximum proportion by which the approximate values can be corrected can be adjusted for each output. Maximum correction limits of 50 per cent and 20 per cent have been found to be suitable for spark timing and injector pulsewidth respectively, for the same approximation functions used by the additional correction network.

When trained and tested on the same basic data used for the previous network layouts, the results shown in Figures 8a and 8b were obtained. The RMS errors in both spark advance and fuel injector pulsewidth are smaller than for the earlier two techniques. Although the system does not emulate the EEC IV control precisely, the errors in spark advance have reduced to a level which would allow an engine to be controlled by the network.

Clearly, the proportional correction network gives the best accuracy on both spark advance and fuel injector pulsewidth at the steady state conditions tested so far. This arrangement may be suitable for use in a generic strategy where a common set of approximation functions are used for all engine types and are tuned by training appropriate correction networks. Work is now

in hand to examine the accuracy of this network arrangement under a wider range of operating conditions and to further reduce the errors in the spark advance output.

6. DISCUSSION

The results presented indicate that a neural network is capable of interpolating between training conditions with reasonable accuracy. Good levels of accuracy, particularly for fuel control, can be obtained for both interpolation and extrapolation from training condition data if it is used to correct approximate values given by simple mapping functions. Such a network can be trained automatically to generate the equivalent of strategy and calibration lookup table arrangements currently used in engine control systems with significantly less development effort and expertise than normally required. The neural networks described were trained from real test bed data in under 30 minutes on an IBM-PC computer, and re-calibration to account for additional inputs/outputs or for a new engine configuration is easily accomplished. Development of the equivalent spark and fuel control algorithms and lookup tables would require many man hours of work.

Work in progress is concentrating on proving the viability of neural networks as an alternative to conventional engine control strategies. A neural network based control system is being developed, as shown schematically in Figure 9, which will both acquire training data and serve as a trial implementation of engine control. The system is able to carry out both spark and sequential fuel injection control for engine speeds up to 6500 rpm, and will allow the effects on performance of control errors to be assessed. It is envisaged that the interpolation and extrapolation accuracies for both spark and fuel control will improve when on line training and testing is carried out using a finer mesh of speed/load training points.

More work is required to apply the technique to other areas of control. The possibility of adapting the control strategy for a vehicle to account for changes in, for example, sensor calibration through the life of the vehicle warrants investigation. Here, the integrity of the resulting strategies would be paramount and a high degree of confidence in the re-training system would be required. However, if accurate example patterns can be generated for the re-training, it should be possible to adapt the strategy at intervals during the life of the vehicle.

The accuracy of control which can be achieved remains the most important open issue. In the context of spark and fuelling control, the latter is the most critical to constraining emissions and, fortuitously, has proved to be the more amenable to an approach based on a neural network system. The data presented shows that open-loop calibrations are maintained to within 2 per cent of target at present, for steady state conditions. It therefore appears that the effects of coolant temperature, for example, on fuelling calibrations during warm-up can be accounted for with confidence. As important, however, will be the accuracy of control during transient operating states which follow changes in throttle demand. These require special attention. During such transients, the induced mixture ratio deviates from steady-state target values unless some form of fuel compensation is employed to eliminate the effects of fuel accumulation and depletion in the intake port. The incorporation of transient fuelling compensation (TFC) into the neural network system will entail an increase in the number of inputs, and an extension of the types of training example required. Preliminary work suggests that two networks operating in parallel should be used to separate steady state and transient elements of fuelling, for simplicity of function and to enable the TFC component to continue to operate after steady state control has switched to closed-loop mode. During transients, it is necessary to contain peak mixture ratio excursions to within approximately two air/fuel ratio units of target to avoid unacceptably high contributions to drive-cycle emissions. Our experience of neural network performance for steady state control suggests this is a feasible target.

7.   CONCLUSIONS

o   Elements of current engine control strategies can be emulated by trained neural networks.

o   The interpolation and extrapolation accuracy of neural networks, particularly for fuelling control, can be high.

o   The use of neural networks in combination with simple analytical functions provides a means of generating generic control strategy elements.

o   Significant reductions in calibration expertise, time and effort can be achieved through the use of neural networks to generate control strategies directly from test bed calibration data.

o   Further work on the wider application of neural networks and to validate their accuracy and integrity is required before any commercial application could be considered.

8.   ACKNOWLEDGEMENT

The authors wish to acknowledge and thank the Ford Motor Company for financial support and permission to publish this work.

9.   REFERENCES

[1]   Marko K. A., James J., Dosdall J and Murphy J., Personal Communication, 1990.

[2]   Ebron S., Lubkeman D. L., White M., "A Neural Network Approach to the Detection of Incipient Faults on Power Distribution Feeders", IEEE Transactions on Power Delivery, Vol. 5, No. 2, April 1990.

[3]   Lippman R. P., "Pattern Classification using Neural Networks", IEEE Communications Magazine, pp 47 - 64, November 1989.

[4]   Gray J. O., Sanderson M. L., "Parallel Processing for Pattern Recognition Systems", IEE Colloquium on Vision Systems in Robotic and Industrial Control (Digest No. 24), pp 3/1 - 3, 1985.

[5]   Sigillito V. G., Wing S. P., Hutton L. V. Baker K. B., "Classification of Radar Returns from the Ionosphere using Neural Networks", John Hopkins APL Technical Digest, Vol. 10, No. 3, 1989.

[6]   Wan C. L., Dickinson K. W., Rowke A., Bell G. H., Xu Z., Hoose N., "Low-Cost Image Analysis for Transport Applications", IEE Colloquium on Image Analysis for Transport Applications, February 1990.

[7]   Banks R. N., Elliman D. G., "A Net Incorporating Feedback, for Recognising Poorly Drawn Characters", New Developments in Neural Computing, Proceedings of a Meeting on Neural Computing, pp 87 - 94, 1989.

[8]   Newman W. C., "Detecting Speech with Adaptive Neural Network", Electronic Design, pp 79 - 90, March 1990.

[9]   Rumelhart D. E., McClelland J. L., "Explorations in Parallel Distributed Processing", MIT Press, 1988.

[10]   Rumelhart D. E., McClelland J. L., "Parallel Distributed Processing Volume 1", MIT Press, 1986.

[11]   Rumelhart D. E., McClelland J. L., "Parallel Distributed Processing Volume 2", MIT Press, 1986.

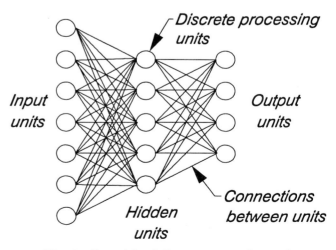

Fig 1   Graphical Representation of a Typical Neural Network

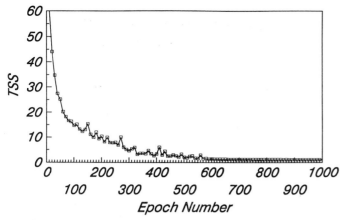

Fig 2 Typical Variation of TSS value
During Training

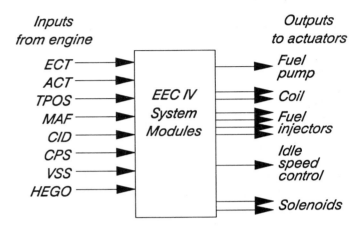

Key

ECT Engine Coolant Temperature
ACT Air Charge Temperature
TPOS Throttle Position
MAF Mass Air Flowrate
CID Cylinder Identification Marker
CPS Crank Position Sensor
VSS Vehicle Speed Sensor
HEGO Heated Exhaust Gas Oxygen Sensor

Fig 3 Schematic of the Ford EEC IV
Engine Control System

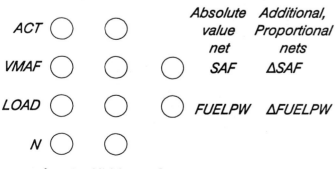

Each layer fully interconnected with adjacent layer

Fig 4 4:4:2 Spark and Fuel Network

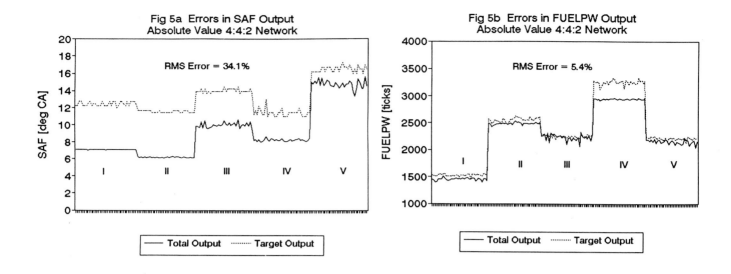

Fig 5a Errors in SAF Output
Absolute Value 4:4:2 Network

Fig 5b Errors in FUELPW Output
Absolute Value 4:4:2 Network

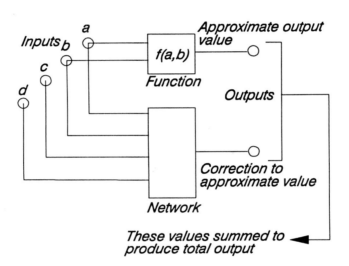

Fig 6 Approximation + Net Correction Method

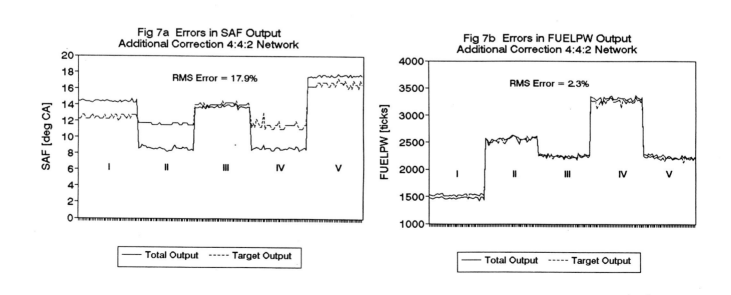

Fig 7a Errors in SAF Output
Additional Correction 4:4:2 Network

Fig 7b Errors in FUELPW Output
Additional Correction 4:4:2 Network

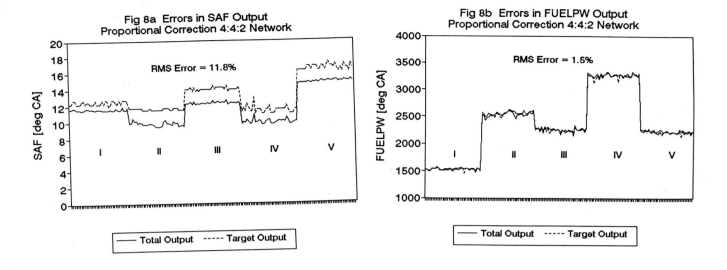

Fig 8a Errors in SAF Output
Proportional Correction 4:4:2 Network

RMS Error = 11.8%

Fig 8b Errors in FUELPW Output
Proportional Correction 4:4:2 Network

RMS Error = 1.5%

——— Total Output   ----- Target Output

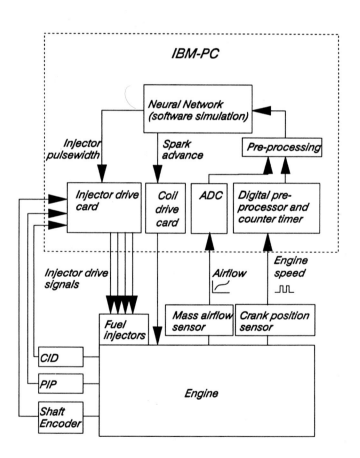

Fig 9 Neural AFR/Spark Controller

C448/013

A turbulent combustion model used to give insights into cycle-by-cycle variations in spark ignition engine combustion

C R STONE, MA, DPhil, CEng, MIMechE, MSAE and **A G BROWN**, BSc, PhD
Department of Mechanical & Engineering Science, Brunel University, Uxbridge, Middlesex
P BECKWITH, BSc
B P Research Centre, Sunbury-on-Thames, Middlesex

SYNOPSIS A phenomenological model of turbulent combustion has been developed and validated against data from wide ranging test conditions. The parameters varied included: the fuel, air fuel ratio, ignition timing, compression ratio and load. The cycle–by–cycle variations were characterised by the coefficients of variation in the: i.m.e.p., maximum cylinder pressure, maximum rate of pressure rise, flame travel times and burn rate. The combustion model will be described and evidence will be given of its validation.

The phenomenological model will then be used to study the effects of flame kernel displacement during the early stages of combustion on the subsequent combustion process. It will be shown that these flame kernel displacements predict the correct trends in cycle–by–cycle variations as the engine operating conditions are changed. Results will be presented from different fuels, showing comparisons between measurements and predictions for the mean values and coefficients of variation of a range of combustion performance parameters.

Attention will also be drawn to the shortcomings of using the maximum cylinder pressure, (as opposed to the i.m.e.p.), as a means of defining the cycle–by–cycle variations in combustion. The significance of the minimum cycle–by–cycle variations occurring at MBT ignition timing will also be discussed.

1. INTRODUCTION

The cycle–by–cycle variations in combustion that occur with spark ignition engines have an important influence on engine performance. Firstly, the differences in burn rate mean that cycles that burn faster than the mean are being ignited too early, and conversely, the slow burning cycles are being ignited too late. Clearly only a few cycles will be burning with the optimum ignition timing, and in consequence the cycle–by–cycle variations lead to a penalty in terms of both the power output and efficiency. Soltau (1) suggested that if cycle–by–cycle variation could be eliminated, there would be a 10 per cent increase in the power output for the same fuel consumption with weak mixtures.

Secondly, the cycle–by–cycle variations in combustion lead to differing amounts of work being produced in each cycle. This leads to fluctuations in the engine speed or its torque output; high levels of variation can be detected as poor driveability. Thirdly, at full throttle, the faster burning cycles are likely to knock. These cycles are the ones that will require the use of a higher octane rating fuel or an engine with a lower compression ratio. Perhaps surprisingly, the total elimination of cyclic dispersion may not be desirable, because of engine management systems that retard the ignition when knock

is detected. If there was no cyclic dispersion, then either none or all of the cycles would knock. It would be acceptable to the engine and driver for only a few cycles to knock, and the ignition control system can then introduce the necessary ignition retard. If all the cycles were to knock it is likely to lead to runaway knock, in which case retarding the ignition would have no effect.

Fourthly, when the engine is operating at a condition with inherently slow combustion, then the slowest burning cycles may not have completely burned by the time the exhaust valve opens; this is known as partial burn. Partial burn leads to high levels of unburned hydrocarbons, and it can define the weak mixture limit for highly diluted combustion.

A comprehensive review and discussion of the cause and effects of cycle–by–cycle variations has been prepared by Young (2). Cyclic dispersion is thought to occur because the turbulence within the cylinder varies from cycle to cycle, the air fuel mixture is not homogeneous (there may even be droplets of fuel present) and the exhaust gas residuals will not be fully mixed with the unburned charge. It is widely accepted that the early flame development can have a profound effect on the subsequent combustion. Firstly, the formation of the flame kernel will depend on: the local air fuel ratio, the mixture motion, and the exhaust gas residuals, in the spark plug gap at the time of ignition.

An investigation by Hamai et al (3) took a sample through the central electrode of the spark plug, and concluded that there was a correlation between the gas composition immediately prior to ignition, and the cycle-by-cycle variations in combustion. Hamai et al also investigated the effect of gas flow on the spark and flame kernel distortion. An apparently contradictory result has been reported more recently by Williams and Gover (4), who use the CARS technique to deduce the AFR in a small control volume, 1 degree crank angle before the spark. However, in this work the exact location of the control volume is unspecified, and the cycle-by-cycle variations in combustion were characterised by the peak pressure variations.

In contrast, the investigation here concerns the role of early flame displacement on the subsequent combustion process. Evidence for the flame kernel displacement comes from several sources. Beretta et al. (5) and Gatowski et al. (6) present photographs of flame propagation in a spark ignition engine. Photographs at a fixed angle after ignition demonstrate that the flame can be convected in different directions in different cycles. Beretta et al (5) use the concept of a best fit circle to the photographs of the flame front, in order to evaluate the flame front radius and apparent centre displacement. Limited data are presented, but these show a flame centre displacement of up to 10 per cent of the bore diameter away from the spark plug. Beretta et al (5) also show how displacing the flame towards the cylinder walls leads to a smaller flame area for a given enflamed volume; they infer that this will then lead to significant variations in the burn rate. Witze et al. (7) and Kerstein and Witze (8) developed and used a spark plug fitted with 8 optical fibres to study the early flame development. In effect they measured the flame arrival times at 8 equally spaced locations 5.35 mm away from the spark plug. In addition, laser schlieren photographs were taken of the flame front growth in a direction normal to the spark plug axis. This work provides data on the flame kernel displacement for successive cycles in low swirl and high swirl cases. However, the Sandia research engine used in these studies has an unconventional geometry with the valves in the cylinder wall; in consequence these data are not directly applicable to more conventional engines. None the less, there is clear evidence of the apparently random displacement of the flame kernel away from the spark plug on successive cycles. When the engine was tested with low swirl at 600 r/min, the flame kernel was displaced with an average velocity in the range of 4 to 5 m/s.

It can be hypothesised that in the absence of swirl these displacements will be random, and attributable to the cycle-by-cycle variations in the mean flow. Once the flame kernel is beyond a critical size then the effect of the turbulence and the mean motion will be to stretch the flame front rather than to displace the flame kernel. Thus after a certain time, when the flame has become a fully developed turbulent flame front, the apparent origin of the flame front will be stationary. The data presented by Beretta et al (5) show that when the flame kernel has a radius of less than 10 mm, then it can be displaced, but above this size the flame kernel centre stabilises. Thus in the investigation here, the effect of flame kernel displacement on cycle-by-cycle variations is to be investigated by assuming that the flame kernel centre is displaced to a sequence of fixed locations in different cycles.

Before the results of this flame kernel displacement investigation are reviewed, it is necessary to describe the experimental facility used, and the development and validation of the turbulent combustion model.

2. EXPERIMENTAL SYSTEM

The experimental results were obtained from a Ricardo E6 variable compression ratio spark ignition engine, fitted with all the usual instrumentation for quasi-steady measurements. The Ricardo E6 has a bore of 76 mm and a stroke of 111 mm; there is a quiescent disc-shaped combustion chamber. The cylinder pressure was measured with a Kistler 6121 piezo-electric transducer that was flush-mounted in the cylinder head. The transducer was used with a Kistler 5001 charge amplifier. The cylinder pressure data were logged with 1° crank angle resolution by a 12 bit resolution ADC in a HP1000 computer for 300 successive cycles. The pressure datum was defined by previous measurements relating engine speed and the inlet manifold pressure, to the minimum cylinder pressure measured by a transducer fitted with a clipper adaptor. The cylinder pressure data were used to deduce the burn rate. A laser schlieren system was also used for measuring flame speeds, and for providing additional data to validate the phenomenological model of turbulent combustion. Space precludes the inclusion of these results here, but full details can be found in Brown (9).

The tests were conducted with three pure fuels of widely differing structure (that can occur in gasoline). Iso-octane (2,2,4-trimethylpentane) was selected as it is a primary reference fuel, toluene was selected as a representative aromatic, and methanol was chosen as a representative alcohol. Table 1 summarises the fuel properties.

The combustion performance of the different pure fuel components was evaluated by measuring: indicated mean effective pressure (i.m.e.p.); maximum cylinder pressure (p_{max}) and the angle of its occurrence; the maximum rate of pressure rise ($\dot{p}_{max}$) and the angle of its occurrence; and the burn rate. These cycle-by-cycle performance parameters were evaluated in terms of their mean,

Table 1 Properties of the Pure Component Fuels

Fuel	Structure	Boiling Point (°)	Research Octane Number
iso-octane	C_8H_{18}	99	100
methanol	CH_3OH	65	105
toluene	$C_6H_5.CH_3$	111	120

standard deviation and coefficient of variation values over a wide range of operating conditions that encompassed variations in:

a) ignition timing   (2–44° b.t.d.c.)
b) equivalence ratio   (0.9, 1.0 and 1.1)
c) compression ratio   (7.5, 8.5)
d) throttle setting

The ignition timing was varied so as to encompass the MBT (minimum advance for best torque) timing, to investigate how the parameters for defining cycle-by-cycle variations in combustion varied, and to provide data from combustion cycles with knock present (at the compression ratio of 8.5).

The stoichiometric operating condition was chosen to reflect the operation of engines with 3 way catalysts, and the rich mixture represents an air fuel ratio that gives the maximum output. (It should be noted that even catalyst equipped engines can operate with rich mixtures at full throttle, in order to increase the output and reduce the heat load on the cooling system and catalyst). The weak mixture was chosen to represent diluted combustion, as might occur when exhaust gas recirculation (EGR) is applied to an engine operating at stoichiometric. The equivalence ratio was measured by a Micro Oxivision MO-1000 Air Fuel Ratio Meter, with an accuracy of $\pm$ 0.02 Ø.

3. TURBULENT COMBUSTION MODEL

3.1 Introduction

The combustion in a homogeneous charge spark ignition engine is commonly divided into three parts:

a) an initial laminar burn, before the flame kernel is large enough to be influenced by turbulence; this can be considered as corresponding to the first few per cent mass fraction burned.

b) Turbulent burning, with a comparatively wide flame front, and pockets of unburned mixture entrained behind the flame front.

c) A final burn period, ("termination period" or "burn-up"), in which the mixture within the thermal boundary layer is burned at a slow rate, due to a reduced fluid motion and a lower temperature.

Many different workers have published phenomenological combustion models, including: Keck (10); Beretta, Rashidi and Keck (5); Keck, Heywood and Noske (11); Tabaczynski (12); Tabaczynski, Ferguson and Radhakrishnan (13); Tabaczynski, Trinker and Shannon (14); Borgnakke (15); Tomita and Hamamoto (16); and James (17).

3.2 Geometric Considerations

For the majority of cases, the combustion chamber will have a complex geometric shape, in which it will be necessary to define: the enflamed volume, the flame front area, and the area wetted by the enflamed volume. Two different approaches to this problem are provided by Poulos and Heywood (18) and Cuttler, Girgis and Wright (19). Poulos and Heywood divide the combustion chamber surface into a series of triangular elements, and then check for interception by vectors, of random direction, radiating from the spark plug. Cuttler et. al. fill the combustion with tetrahedra, and employ vector algebra. However, for the purpose of this model, these complications are avoided, since the Ricardo E6 combustion chamber can be treated as a simple disc. Although there are small cavities associated with the: spark plug, pressure transducer, piston top land and optical windows, these represent only about 2½% of the swept volume. Since the standard assumption of a spherical flame front is subject to question (James (17)), it is contended here that modelling the combustion chamber with a disc is not unreasonable. The analytical expressions that define the intersection of a sphere with a disc can be integrated numerically, to give: the enflamed volume, the flame front area, and the wetted area.

3.3 Combustion Modelling

The turbulent combustion models used by Keck and co-workers and Tabaczynski and co-workers both seem capable of giving good predictions of turbulent combustion. The model here adapts ideas from both of these groups.

The flame is assumed to spread by a turbulent entrainment process, with burning occurring in the entrained region at a rate controlled by turbulence parameters. The flame front is assumed to entrain the mixture (i.e. spread into the unburned mixture) at a rate that is governed by the turbulence intensity and the local laminar burn speed. The rate at which the mass is entrained into the flame front is given by:

$$\frac{dm_e}{dt} = \rho_u A_f (u' + S_L) \qquad (1$$

where m_e = mass entrained into the flame front.

ρ_u = density of the unburned charge

A_f = flame front area (excluding area in contact with surfaces).

u' = turbulence intensity

S_L = laminar burning velocity

The turbulence intensity is assumed to be proportional to the mean piston speed. The rate at which the mixture is burned, is assumed to be proportional to the mass of unburned mixture behind the flame front:

$$\frac{dm_b}{dt} = \frac{m_e - m_b}{\tau} \qquad (2$$

where m_b = mass burned behind the flame front

τ = characteristic burn time for an eddy of size ℓ_m

ℓ_m = the Taylor microscale, that characterises the eddy spacing

Since the eddies are assumed to be burned up by laminar flame propagation:

$$\tau = \frac{\ell_i}{S_L} \qquad (3$$

This model assumes isotropic and homogeneous turbulence throughout the combustion chamber at the time of ignition, with the integral length scale (ℓ_i) at this point assumed to be proportional to the combustion chamber height, Ferguson (20).

$$\ell_i = \frac{0.5 C_i b^2 h}{b^2 + 2bh} \qquad (4$$

where h = combustion chamber height

b = combustion chamber bore

C_i = constant specific to the engine

and the turbulence intensity (u'_i) calculated from an empirical calculation derived from results obtained by Lancaster (21).

After ignition the integral length scale (ℓ_i) and the turbulence intensity (u') are assumed to be governed by the conservation of momentum for coherent eddies.

Thus

$$\ell_i = (\ell_i)_o \left(\frac{\rho_{ui}}{\rho_u}\right)^{\frac{1}{3}} \qquad (5$$

and

$$u' = u'_o \left(\frac{\rho_u}{\rho_{ui}}\right)^{\frac{1}{3}} \qquad (6$$

with the suffix "o" referring to the values at ignition.

For isotropic turbulence:

$$\ell_m = \sqrt{\frac{15 v \ell_i}{u'}} \qquad (7$$

where v = dynamic viscosity.

By using equations 4, 5, 6 and 7, the eddy burn-up term can be calculated at each time step. A full description of this model and its implementation can be found in Brown (9), along with the data and correlations used for determining the laminar burn speeds.

4. EXPERIMENTAL AND NUMERICAL RESULTS

4.1 Mean Performance Parameters

Figure 1 shows comparisons between the experimental and predicted values of: the mean i.m.e.p., mean maximum cylinder pressure and the mean maximum rate of pressure rise, for iso-octane with weak, stoichiometric and rich mixtures over a range of ignition timings. As the air fuel ratio varies from weak (lambda = 1.1) to rich (lambda = 0.9) with a fixed ignition timing, then the experimental i.m.e.p., maximum cylinder pressure and maximum rate of pressure rise, all increase in the expected way. As the ignition timing is advanced the maximum cylinder pressure and the maximum rate of pressure rise both increase. The i.m.e.p. values all rise to a maxima and then fall as the ignition timing is advanced beyond the MBT value. The i.m.e.p. is under-predicted at over-advanced ignition timings and it seems probable that the effect of heat transfer has been over-allowed for. As would be expected, the slowest burning mixture was the weak mixture (lambda = 1.1) and this has the highest value of MBT ignition timing. Conversely, the rich mixture (lambda = 0.9) was the fastest burning and required the lowest value of MBT ignition timing.

The numerical predictions have the same trends as the experimental data with the only notable difference occurring in the i.m.e.p. data. These discrepancies are greatest at the over-advanced values of ignition timing, as full allowance was not made for the effects of dissociation

50

ignition could be sufficiently advanced such that combustion was always complete before the maximum cylinder pressure was attained; a consequence of this would be negligible variations in the maximum cylinder pressure. Another reason for the reduction in cycle-by-cycle variations of the maximum cylinder pressure as the ignition timing is advanced, is that the mean value of the maximum cylinder pressure is increasing, so that any variations about this mean level will lead to lower values of the cycle-by-cycle variation. The only exception to these trends is the weak mixture with the most retarded ignition timing. The slow combustion with this operating point means that the maximum pressure detected is that due to piston motion, which occurs just prior to top dead centre.

Figure 5 shows a comparison between the measured coefficients of variation and the predicted combustion variations in the i.m.e.p., for: iso-octane, methanol and toluene, with a range of ignition timings and mixture strengths. The results for iso-octane have already been discussed in the content of Fig. 4, and Fig. 5 shows that methanol and toluene follow the same trends. In particular, the CoV i.m.e.p. is a minimum in the region of MBT ignition timing, and the faster burning richer mixtures have lower values of the CoV. The differences between the fuels are most clear with weak mixtures (lambda = 1.1) and retarded ignition, and it can be seen that the slowest burning fuel (iso-octane) has the highest CoV of i.m.e.p., whilst the fastest burning fuel has the lowest CoV of i.m.e.p.. However, in the region of MBT ignition timing the differences in the CoVs of i.m.e.p. are small. In general, the differences between fuels are smaller than the differences that would occur for a unity change in the air fuel ratio.

The predicted combustion variations in i.m.e.p. are in all cases smaller than the experimental CoVs of the i.m.e.p.; this is most evident for the weak mixtures with retarded ignition timings. This under-prediction of cycle-by-cycle variation with weak mixtures might be attributable to an underestimate of the flame kernel displacements. With weak mixtures (and part throttle operation) the reduced laminar burning velocity, will mean that it will take longer for the flame kernel to grow to a size at which it is no longer corrected by the mean flow. In consequence the possible displacements of the flame kernel from the spark plug will be larger, and in turn the cycle-by-cycle variations in combustion will be higher. In the work here a series of fixed displacements was used as there were insufficient experimental data on flame kernel displacement. Furthermore, it might be expected that local variations in the air fuel ratio and residuals, in the spark plug gap at the time of ignition, will lead to larger variations in the laminar burn speed of weak mixtures than rich mixtures. This in turn could influence the extent of the flame kernel displacements. It has also been argued (9) that the differences with changing equivalence ratio, are underestimated because of insufficient laminar burning data, and possible

uncertainties in the correlations for estimating the laminar burning velocities at high temperatures and pressures. There is also a lack of data on the laminar burning velocity of toluene under ambient conditions.

The predicted results in Figure 6 show consistently lower values than the experimental data for the cycle-by-cycle variations in burn rate. However, the trends are predicted correctly, and the discrepancy is in part due to errors in predicting the mean values of the burn times.

Although it is simple (and quite widely adopted) to characterise the cycle-by-cycle variations in combustion, in terms of the CoV of the maximum cylinder pressure, it gives a less valid measure of the cycle-by-cycle variations in combustion than by measuring the CoV of i.m.e.p. Furthermore, it is the cycle-by-cycle changes in the i.m.e.p. that lead to the slight speed variations in the engine that can ultimately be detected as engine roughness or poor driveability. Also, it should be remembered that most engines are operated with MBT ignition timing or a slightly retarded timing (either to reduce NOx emissions and or to avoid combustion knock). At such operating points the CoV of i.m.e.p. is insensitive to the ignition timing, whilst the CoV of the maximum pressure will show an increase as the ignition timing is retarded.

5.   CONCLUSIONS

Experimental data from wide ranging engine operating conditions have been used to validate a model of turbulent combustion. This turbulent combustion model has then been used to elucidate the origins of cycle-by-cycle variations in combustion.

*   When characterising cycle-by-cycle variations in combustion, the CoV of i.m.e.p. should be used in preference to the CoV of the maximum cylinder pressure. The CoV of maximum cylinder pressure is sensitive to the ignition timing in a way that could lead to misleading results.

*   The lowest cycle-by-cycle variations in combustion occur in the region of the MBT ignition timing, and this coincides with the shortest early burn period.

*   It has been demonstrated that displacement of the flame kernel can be a major source of the experimentally observed cycle-by-cycle variations in combustion.

*   There is a need for additional laminar burn data for pure fuel components under both ambient and high pressure and temperature conditions.

*   Cycle-by-cycle variations are reduced when the in-cylinder conditions at ignition lead to a fast flame kernel development.

– an effect that will be most significant at over-advanced ignition timings.

Figure 2 shows comparisons between the experimental and predicted values of the 10 per cent, 50 per cent and 90 per cent burn times for iso-octane with weak, stoichiometric and rich mixtures, over a range of ignition timings. There is good agreement between the experimental and predicted values of the burn rate, with essentially the same trends obtained from the tests using methanol and toluene, except that methanol burnt a little faster and toluene was a little slower than iso-octane.

The experimental 0–10 per cent burn time data (Fig. 2) all show a minima in the region of MBT ignition timing. The presence of a minima in the 0–10 per cent burn time can be explained by the interaction of two effects with opposing characteristics. Firstly, the later the ignition timing, the higher the cylinder temperature (and pressure) so the higher the laminar flame speed. Secondly, (and conversely), the later the ignition timing, the more the turbulence will have decayed during the compression stroke, and the lower the enhancement of the combustion by turbulence. As the 0–10 per cent burn time includes the early (laminar) burn period and the transition to fully turbulent combustion, then it is subject to both of the above influences. The 0–50 per cent and 0–90 per cent burn times also show a minimum, but for slightly more advanced ignition timings. This is presumably a consequence of earlier ignition leading to higher levels of turbulence during the main part of combustion (the 10–90 per cent burn).

4.2 Cycle-by-Cycle Variations

A hypothesis has been made here, that a major source of cycle-by-cycle variations in combustion is attributable to displacement of the flame kernel. It has been argued that this changes the effective location of the spark, and this has been represented by the 8 locations shown in Figure 3. It has then been assumed that combustion modelled from the spark plug position will generate the 'mean' value of any combustion parameter. The modulus of the difference between the 'mean' value of the combustion parameter and the value corresponding to each of the 8 locations was then averaged and divided by the mean value, to calculate the percentage difference or combustion variation, in other words:

$$combustion\ variation = \sum_{i=1,8} \frac{|X_o - X_i|}{8X_o} \times 100\% \quad (8)$$

where: X is any combustion parameter
suffix o denotes combustion originating at the spark plug
suffix i denotes combustion originating at any of the 8 locations defined in Fig. 3.

Ideally data are needed for this engine on either

a)   the cycle-by-cycle variations in the mean flow at the spark plug

or

b)   the cycle-by-cycle position of the effective centre of combustion.

If such data were available, then the combustion model could simulate combustion initiated from each centre of combustion. If this was done for a statistically valid number of cycles, then the standard deviation and the mean of the combustion parameters could be used to calculate the coefficient of variation. Although this would significantly increase the computation time it would be a more rigorous approach. However, it is being argued here that the current approach is sufficient to investigate the effect of flame kernel displacement, and to see if the engine operating conditions affect the computed combustion variation in the same way as the measured coefficient of variation (CoV).

Figure 4 shows the cycle-by-cycle variations in the: i.m.e.p., maximum cylinder pressure, and the maximum rate of pressure rise for iso-octane with weak, stoichiometric and rich mixtures over a range of ignition timings. For all mixture strengths at all ignition timings, the weak mixture shows the highest levels of cycle-by-cycle variations, and the rich mixture shows the lowest levels of cycle-by-cycle variations. This can be explained by the weak mixture having the lowest laminar flame speed and thus the slowest early burn period. It is well established that a slow early burn period leads to high levels of cycle-by-cycle variations (2).

In all three cases the numerical predictions show the same response to the ignition timing as the experimental data, although the predicted differences in changing the equivalence ratio are in all cases smaller than the experimental differences. This can in part be explained by no allowance being made for the weaker mixtures having lower laminar flame speeds, and thus taking longer to reach a size at which the flame kernel is no longer displaced by the mean flow. None the less, the trends predicted by cycle-by-cycle variation model are correct.

The cycle-by-cycle variations in i.m.e.p. reduce to a minimum in the region of MBT ignition timing. Figure 2 also showed that the 0–10% burn time (encompassing the early burn period) is also a minimum in the region of MBT ignition timing. This supports the observation that the cycle-by-cycle variations in combustion are reduced for shorter early burn periods. The cycle-by-cycle variations in the maximum cylinder pressure and the maximum rate of pressure rise (Figure 4) both show reductions as the ignition timing is advanced. The reduction in cycle-by-cycle variations of the maximum pressure are a consequence of the rise in pressure being partly due to piston motion. In the limiting case, the

ACKNOWLEDGEMENTS

Support is grateful acknowledged from the SERC and the BP Research Centre.

REFERENCES

1   Soltau, J.P. Cylinder pressure variations in petrol engines, pp. 99–116, July 1960. (*Instn Mech. Engrs Conf. Proc.*).

2.   Young, M.B. Cyclic dispersion in the homogeneous–charge spark–ignition engine – a literature survey. *SAE paper 810020, 1981.*

3.   Hamai, K., Ishizuka, T. and Nakai, M. Combustion Fluctuation Mechanisms Involving Cycle–to–Cycle Spark Ignition Variation Due to Gas Flow Motion in SI Engines. *21st International Symposium on Combustion*, The Combustion Institute, 1986.

4.   Williams, D.R. and Gover, M.P. The Effect of Variations in Local Air Fuel Ratio on Cyclic Variability in a Firing Production Engine. I.Mech.E. Seminar Experimental Method sin Engine Research and Development '91, pp. 143–150, MEP London, 1991.

5.   Beretta, G.P., Rashidi, M., and Keck, J.C. Turbulent flame propagation and combustion in spark ignition engines. *Combust. Flame*, vol. 52, pp. 217–245, 1983.

6.   Gatowski, J.A., Heywood, J.B., and Deleplace, C. Flame photographs in a spark–ignition engine. *Combust. Flame*, vol. 56, pp. 71–81, 1984.

7.   Witze, P.O., Hall, M.J., and Wallace, J.S. Fiber–optic instrumented spark plug for measuring early flame development in spark ignition engines. *SAE paper 881638, 1988.*

8.   Kerstein, A.R., and Witze, P.O. Flame–kernel model for analysis of fiber–optic instrumented spark plug data. *SAE paper 900022, 1990.*

9.   Brown, A.G. Measurement and modelling of combustion in a spark ignition engine. Brunel University PhD Thesis, 1991.

10.   Keck, J.C. Turbulent Flame Structure and Speed in Spark Ignition Engines. *Proceedings of the 19th International Symposium on Combustion*, The Combustion Institute pp 1451–1466, (1982).

11.   Keck, J.C., Heywood, J.B. and Noske, G. Early Flame Development and Burning Rates in Spark–Ignition Engines. *SAE Paper 870104, 1987.*

12.   Tabaczynski, R.J. Turbulent Combustion in Spark–Ignition Engines. *Prog. Energy and Combustion Sciences* Vol 2 pp 143–165, 1976.

13.   Tabaczynski, R.J., Ferguson, C.R. and Radhakrishnan, K. A Turbulent Entrainment Model for Spark–Ignition Engine Combustion. SAE paper 770647, *SAE Trans.* Vol 86, 1977.

14.   Tabaczynski, R.J., Trinker F.H. and Shannon, B.A.S. Further Refinement and Validation of a Turbulent Flame Propagation Model for Spark Ignition Engines. *Combustion and Flame* Vol 39 pp 111–121, 1980.

15.   Borgnakke, C. Flame Propagation and Heat–Transfer Effects in Spark–Ignition Engines, in *Fuel Economy of Road Vehicles Powered by Spark Ignition Engines*, eds J.C. Hilliard and G.S. Springer, Plenum Press, 1984.

16.   Tomita and Hamamoto. The Effect of Turbulence on Combustion in the Cylinder of a Spark Ignition Engine – Evaluation of an Entrainment Model. *SAE paper 880128, 1988.*

17.   James, E.H. Further Aspects of Combustion Modelling in Spark Ignition Engines. *SAE Paper 900684.*

18.   Poulos, S.G. and Heywood, J.B. The Effect of Chamber Geometry on Spark Ignition Engine Combustion. *SAE paper 830334, 1983.*

19.   Cuttler, D.H., Girgis, N.S. and Wright, C.C. Reduction and Analysis of Combustion Data Using Linear and Non–Linear Regression Techniques. *I.Mech.E. Conf. Proc.* Paper C17187, 1987.

20.   Ferguson, C.R. *Internal Combustion Engines.* Wiley, New York, 1986.

21.   Lancaster, D.R., Krieger, R.B., Sorenson, S.C. and Hull, W.L. Effects of Turbulence on Spark Ignition Engine Combustion. *SAE Paper 760160, 1976.*

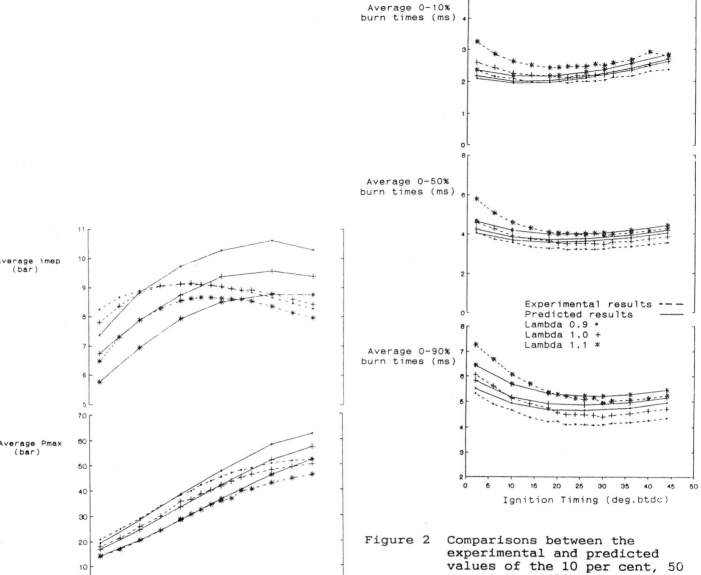

Figure 1 Experimental and predicted values of: the mean i.m.e.p., mean maximum cylinder pressure and the mean maximum rate of pressure rise, for iso-octane with weak, stoichiometric and rich mixtures over a range of ignition timings at full throttle.

Figure 2 Comparisons between the experimental and predicted values of the 10 per cent, 50 per cent and 90 per cent burn times for iso-octane with weak, stoichiometric and rich mixtures, over a range of ignition timings at full throttle.

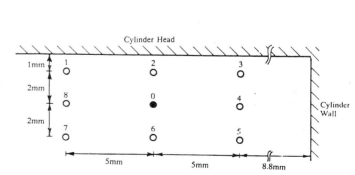

Figure 3 Assumed distribution of ignition sites.

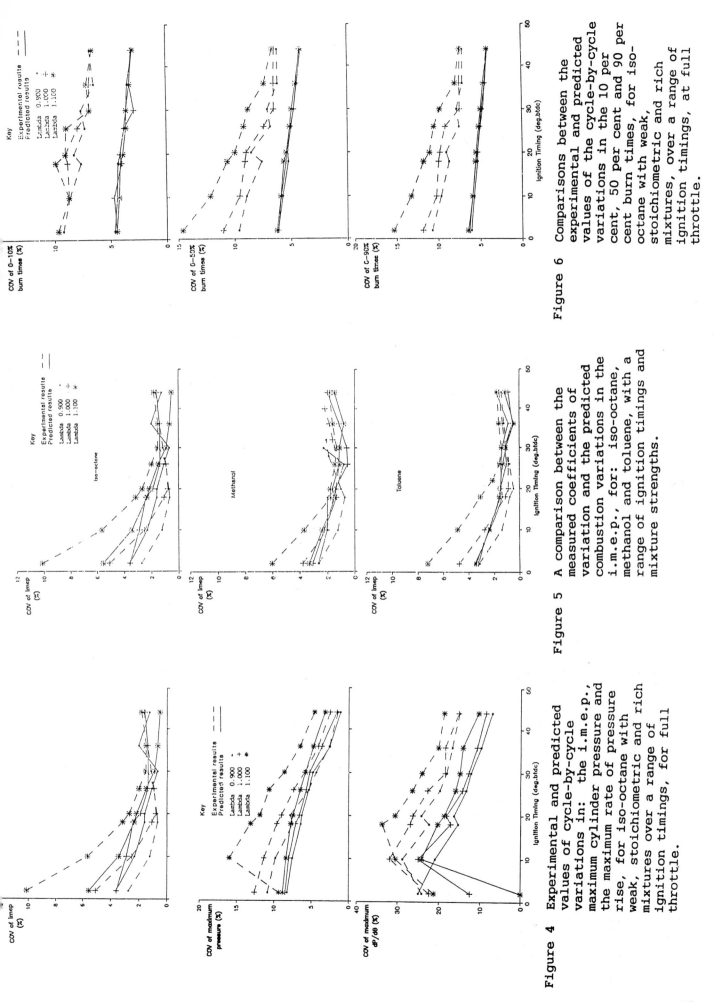

Figure 4 Experimental and predicted values of cycle-by-cycle variations in: the i.m.e.p., maximum cylinder pressure and the maximum rate of pressure rise, for iso-octane with weak, stoichiometric and rich mixtures over a range of ignition timings, for full throttle.

Figure 5 A comparison between the measured coefficients of variation and the predicted combustion variations in the i.m.e.p., for: iso-octane, methanol and toluene, with a range of ignition timings and mixture strengths.

Figure 6 Comparisons between the experimental and predicted values of the cycle-by-cycle variations in the 10 per cent, 50 per cent and 90 per cent burn times, for iso-octane with weak, stoichiometric and rich mixtures, over a range of ignition timings, at full throttle.

C448/064

The sources of unburnt hydrocarbon emissions from spark ignition engines during cold starts and warm-up

D J BOAM, BSc, National Engineering Laboratory, Glasgow,
I C FINLAY, BSc, PhD, CEng, FIMechE, MSAE, Thermal Systems Research, Glasgow,
T W BIDDULPH and T MA, BSc, MSc, PhD, Ford Motor Company Limited, Basildon, Essex
R LEE, BSc, and S H RICHARDSON, BSc, MSc, Jaguar Cars Limited, Coventry
J BLOOMFIELD, Lotus Engineering
J A GREEN, MIEE and S WALLACE, CEng, MIMechE, Rover Group Limited, Lighthorne, Warwick
W A WOODS, PhD, DEng, FEng, FIMechE, FRS and P BROWN, BSc, MSc, PhD, Queen Mary and Westfield College, London

SYNOPSIS The results of a three-year collaborative research study into the sources of unburnt hydrocarbon (uHC) emissions are reported. The study sought to extend existing knowledge of the sources in an engine to the crucial period following a cold start and before the exhaust catalyst becomes fully effective. The study, carried out on a range of engines but centred on the Rover M16 4-valve engine, identified a number of sources, all of which are equally important in the warm-up period. The paper concludes with some recommendations for the control of uHC emissions.

NOTATION

A   Arrhenius factor
E   Activation energy
P   Pressure
R   Universal gas constant
T   Temperature
W   Rate of disappearance of fuel
Y_f, Y_o Mass fractions of fuel and oxygen

1 INTRODUCTION

Legislative controls on motor vehicle exhaust emissions have been tightening progressively for over twenty years and this has made it increasingly difficult to find ways to meet the new demands. Recognition of this problem and the fact that benefits might flow from the pooling of knowledge and resources led to the formation, in 1988, of the United Kingdom Engine Emissions Consortium (UKEEC), a joint venture between the UK Department of Trade and Industry (DTI) and British industry.

The Consortium drew together some twelve industrial companies including diesel and petrol engine manufacturers, component suppliers and oil and chemical companies. Research organizations and UK universities were also involved as subcontractors. The DTI provided some £2.3 million of the total budget of over £5.5 million over a three-year period. The Consortium addressed eleven projects at the leading edge of emissions technology.

The work reported here sought to gain a deeper understanding of the mechanisms which contribute to the formation of unburnt hydrocarbon emissions (uHC). Considerable effort has already been devoted to identifying the sources in a warmed-up engine Fig 1) and these studies will be discussed later (Section 3). However the slow

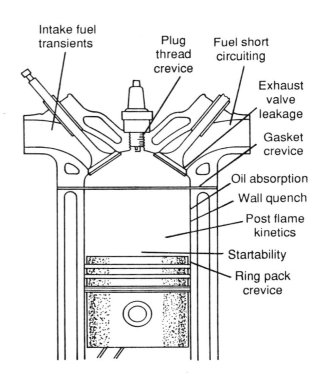

Fig 1 Potential sources of uHC considered

Intake fuel transients
Plug thread crevice
Fuel short circuiting
Exhaust valve leakage
Gasket crevice
Oil absorption
Wall quench
Post flame kinetics
Startability
Ring pack crevice

light-off performance of the exhaust catalyst has focussed attention on the first 2 minutes of operation when catalyst conversion efficiency is low. During this period the engine undergoes considerable thermal changes and these affect the balance between the uHC sources and may also introduce new sources. Project P4 of the UKEEC was therefore a structured 3-year programme of work which sought to confirm the sources of uHC and reassess their relative importance in the warm-up phase. A third objective was to evaluate some practical measures to reduce uHC emissions during the emissions drive cycles. The project management comprised Ford Motor Company (project leader), Jaguar Cars, Lotus Engineering and Rover Group. The main research activity was carried out by the National Engineering Laboratory with support from Queen Mary and Westfield College.

It was recognized from the outset that to achieve such an ambitious set of goals it would be necessary to conduct tests on vehicles and on both single and multicylinder engines. It was also recognized that if a disparate set of experiments on a range of facilities was to produce the coherent results needed, it would be necessary to employ a common combustion chamber and head design throughout. The core of the work was therefore carried out on the Rover M16 combustion chamber with the main test facilities comprising a Rover 820i car, a 4-cylinder M16 test-bed engine and a Ricardo Hydra modified to have the same bore and stroke as the M16 and fitted with a modified M16 production cylinder head. To reinforce the findings from these facilities, work was also carried out on a Ford 1.4 litre CVH Escort, a 5.7 litre Lotus engined Corvette and a single cylinder engine based on the Lotus engine.

To achieve the project objectives of identifying the problem areas of engine operation, quantifying the contributions of the uHC mechanisms during warm-up, assessing practical measures to reduce uHC production and providing a demonstration of the progress achieved, the experimental work was divided between the facilities as shown in Fig 2. In total the project produced some 46 research reports. This paper will describe the work to quantify the sources of uHC, which generated 25 of those reports.

2 TEST EQUIPMENT AND PROCEDURES

All the engines used were comprehensively instrumented for the measurement of warm-up rates and emissions. One instrument - the Cambustion Fast Response Flame Ionization Detector (FRFID) (1) - was of particular value in studying changes in uHC production during an engine cycle that came about as various modifications were made to the engine. Considerable work was necessary to validate this instrument and to interpret its signals and this work has been published elsewhere (2). Other important measurements included intake manifold pressure by Druck pressure transducer, fuel flow by an electronic balance, air/fuel ratio by NTK meter, uHC by Ratfisch FID. The instrumentation was backed up by a high-speed data acquisition system based on the Thorn SE2550 transient recorder which allowed data to be collected from 4 channels at 2 deg crankangle intervals from 90 successive engine cycles.

As the main operating regime of interest was that immediately following a cold (20°C) start it was necessary to develop a test procedure that would ensure repeatable results and which would be representative of the engine's use in service. The test condition was chosen to give an engine speed (2000 r/min) typical of emission cycle driving and a manifold absolute pressure (360 mmHg) which gave a rate of warm-up on the 4-cylinder engine comparable with that recorded during the first 2 minutes of the ECE 15 drive cycle. The single cylinder 'M16' had much higher thermal inertia and electrical heat was provided to match the coolant and oil warm-up rates to the 4-cylinder engines while maintaining combustion conditions by duplicting intake manifold pressure.

Having selected the speed and load, a series of tests were conducted to assess the factors which were likely to influence the repeatability of the results. The factors considered included:

    manifold pressure setting,
    mixture strength,
    starting temperature,
    spark plug condition,
    fuel build-up in the oil on a
    day-to-day basis, and
    long term degration of the oil.

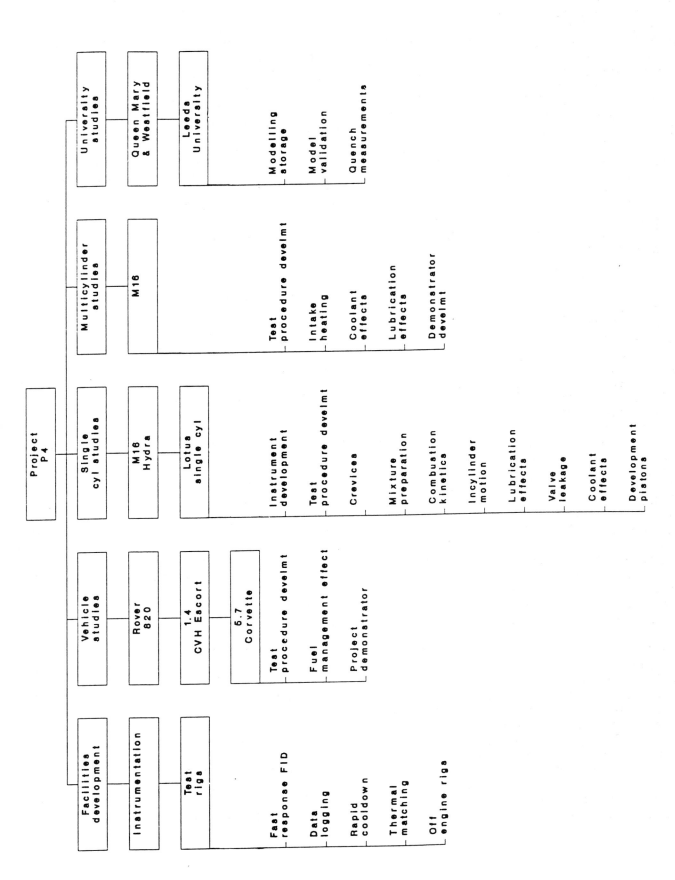

Fig 2 Overall structure of the project

The tests showed that manifold pressure variations of ±5 mmHg could impair repeatability and steps were taken to control the pressure to ±2 mmHg. This was achieved by locking the throttle and making minor adjustments using a needle valve to bleed air into the manifold. Mixture strength was also a major contributor to uHC variance. The fuel control system was able to hold mixture strength to within ±0.25 air/fuel ratios of stoichiometry but even this contributed ±50 ppmC$_3$ to the variation of uHC. This was overcome by conducting an experiment to obtain the uHC v. mixture strength correlation for each engine and then correcting the result of each run to a common air/fuel ratio of 14.5:1 - the stoichiometry point for the unleaded fuel being used. Any tests with a mixture strength more than 0.5 AFR from the desired setting were rejected. Spark plug and oil ageing were shown to have no significant effect. The potential problem of progressive contamination of the oil with fuel was overcome by ensuring that the oil was taken to its maximum operating temperature at the end of each day's testing. From these tests the following procedure was developed.

(a) The engine was started and warmed up for 100 seconds.

(b) The intake manifold absolute pressure was set to the reference value, this eliminated the effect of day-to-day variations in barometric pressure on the combustion conditions.

(c) The throttle was locked, the engine was stopped and force cooled back to the starting condition.

(d) A series of runs was carried out with the same manifold absolute pressure and the engine was forced cooled between runs.

(e) During each run the engine was started and run for 2 minutes with data being collected for the first 100 seconds. In the remaining 20 seconds minor adjustments to the throttle were made to allow for changes in barometric pressure between runs and so ensure that the subsequent run was within the specified ±2 mmHg of the desired manifold absolute pressure setting.

(f) Fuel was drawn from a sealed drum at the beginning of each day and the fuel system was sealed to prevent loss of light fractions by evaporation between runs.

(g) Between runs the battery was recharged to ensure repeatable cranking performance and the intake manifold was vented with compressed air to ensure that variations in the amount of residual fuel vapour did not affect pre-firing emissions of uHC.

(h) At the end of each day the engine was turned to maximum valve overlap and the inlet manifold was pressurized with compressed air. The oil and coolant were circulated using their external pumps and were heated electrically to 90°C. The engine was then held in this condition for 20 minutes. This daily purging of the oil had initially been done by running the engine at 2000 r/min half load but this was not possible with all the experimental builds and so the non-running procedure was adopted as standard.

Even with these precautions, there was a spread of ±5 per cent in the results for any engine configuration. To ensure an accurate assessment of the effect of each engine modification, it was therefore necessary to compare the means of series of repeat runs on a statistical basis.

3 RESEARCH INVESTIGATIONS

3.1 Crevices

The first researcher to attempt to quantify the influence of piston ring crevices was Wentworth (3), who showed that hydrocarbon emissions were dependent on the width and depth of the top ring land. He estimated that under most warmed-up operating conditions, this top ring land contributed about 50 per cent of the engine's total unburnt hydrocarbons. He postulated that the cause was the failure of the flame to penetrate the narrow gap between the cylinder and the piston. This led others (eg 4,5,6) to conduct experiments with zero crevice bombs and artificial crevices to try to quantify the conditions which prevented flame penetration. Meanwhile Namazian and Heywood (7) modelled the flow into and out of the ring pack.

There are three basic ways in which the top ring can act as a crevice for the storage of air/fuel mixture. Firstly the top land region between the piston crown, cylinder bore, piston and top ring can act directly as a crevice. Secondly the space within the top ring groove not occupied by the top ring can act as a subsidiary crevice. Lastly the mixture can be squeezed through the ring gap

60

and into the region between the top and second rings to re-emerge later when the cylinder pressure falls below the ring pack pressure. To be effective in eliminating the crevice action of the top ring and so allow that action to be quantified, an experimental piston had to be developed to address all three aspects. Fig 3 shows the design adopted. A Shamban carbon-impregnated PTFE "Glydring" seal was placed above the normal top ring position. By raising the top of the seal to approximately 1 mm from the piston crown the top land effect was greatly reduced. The sealing ring was backed by three small section 0-rings rather than the one larger section ring recommended for this type of seal by the manufacturer. This together with careful sizing of the groove, ensured that the volume of the groove not occupied by the seal was minimized. Lastly the seal employed was a continuous ring which greatly reduced the amount of mixture being squeezed into the inter-ring space to return later in the cycle.

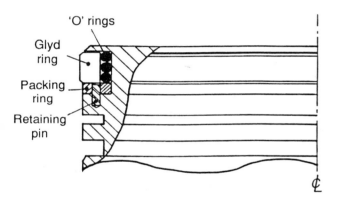

'O' rings
Glyd ring
Packing ring
Retaining pin
¢

Fig 3 PTFE seal design

The design was first tried out on a motored engine without a cylinder head so that ring pressure against the bore could be selected to ensure adequate sealing whilst maintaining friction levels close to those of the standard ring pack. After this the piston was installed in the firing single cylinder M16 Hydra engine and attempts made to run the engine. To guard against overheating the seal while still allowing sufficient data to be collected, all runs with the PTFE sealed piston were limited to 2 minutes. Despite this, initial runs were marred by the failure of the very thin land above the plastic seal and this had to be increased to 1.5 mm to obtain satisfactory piston life. This had

the effect of slightly reducing the amount of ring pack crevice removed but as Fig 4 shows there was still a very substantial reduction. Fig 4 also shows that the removal of such a large amount of crevice had a very significant effect on the uHC levels recorded 100 secs after a 20°C start - 950 ppmC$_3$ against the standard piston value of 1350 ppmC$_3$.

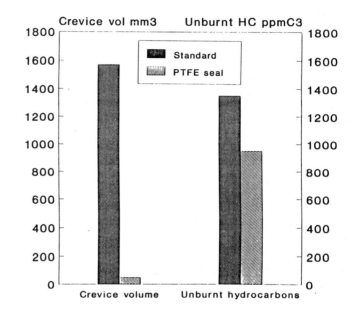

Fig 4 Effect of PTFE seal on ring pack crevice and uHC emissions

The fast FID unit was used to study the uHC v. crankangle signature and Fig 5 shows that the reduction was not uniform throughout the cycle. The main features of the signature in Fig 5 have been discussed elsewhere (2) but it is useful to reiterate those points here.

(a) The high levels during the valve closed period are due to the probe being close to the valve and drawing its sample from the stagnant gas stream which was exhausted from the cylinder immediately before valve closure.

(b) As the valve opens there is a small peak in hydrocarbon concentration which has been shown to be important as it is associated with the high mass flow of blowdown. One possible source of this peak is that small amounts of mixture, which leak past the valve during the high pressure part of the cycle and

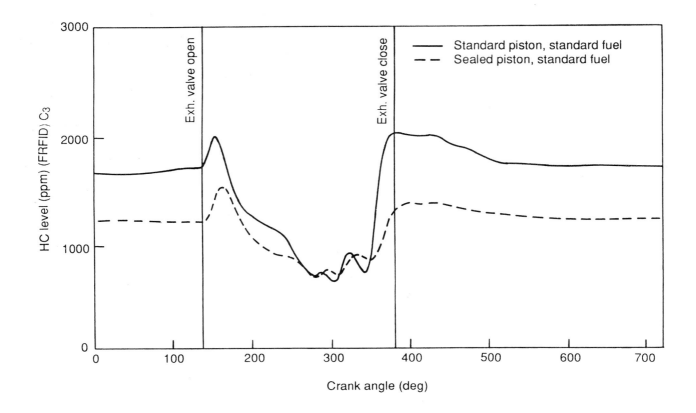

Fig 5 Unburnt hydrocarbons v. crankangle signatures for standard and sealed pistons

accumulate behind the valve, are pushed into the probe as the valve opens. A second possibility is that unburnt mixture is retained in the crevices around the valve seat and this is expelled as the valve opens.

(c) UHC levels then drop as the piston reaches bottom dead centre and remain at a low level almost until the valve closes. Although the levels are relatively low, this is the main exhaust period and flowrates are high making this a very significant period for uHC release from the cylinder.

(d) As the piston approaches TDC uHC levels rise very steeply as the mixture emerges from the various storage mechanisms within the cylinder. The final level leaving the cylinder then sets the level for the closed valve period before the history is repeated in the next cycle.

Fig 5 shows the differences in the signature with the standard and sealed piston. Clearly there is little difference during the main

exhaust stroke and the significant change comes as the piston approaches TDC. Here the elimination of the ring crevice has resulted in a large reduction in uHC level and it is this which is reflected in the reduced mean level after 100 secs. When the flow history during the exhaust period is linked to the uHC signature of Fig 5 (2), the mass reduction due to crevice sealing is only 20 per cent compared with the 30 per cent reduction in mean concentration. To compare results from different engines, and to put the emissions into perspective, it is helpful to express the reductions in terms of a hydrocarbon index. This is defined as the ratio of the mass of uHC leaving the engine to the mass of fuel entering. In the case of the piston crevices the uHC index was reduced from 3.0 per cent (standard engine) to 2.4 per cent (sealed).

A second investigation of the effect of crevices was conducted on the Lotus single cylinder engine. In this engine the head gasket was set well back from the cylinder bore and protected from combustion temperatures by a step in the top of the

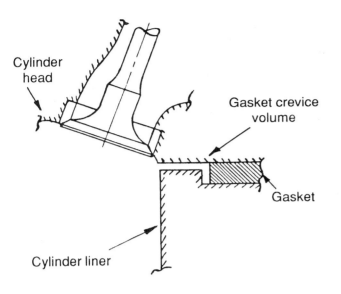

Fig 6 Cylinder head gasket design on
Lotus single cylinder engine

cylinder block. This formed a significant
crevice (Fig 6) which was filled with an
aluminium ring to remove 0.7 cc of the
crevice volume. The effect on uHC levels was
again large with levels dropping from 1700 to
1450 ppmC$_3$ - a reduction in uHC index from
3.9 to 3.4 per cent. The piston ring crevice
and the gasket crevice will be reached by the
flame at about the same point in the engine
cycle. It is therefore not surprising that,
when allowance is made for the differing
amounts of crevice removed, the percentage
reductions in uHC levels in the two
experiments are comparable.

Two other crevices were investigated, the
spark plug thread and the gap between the
centre electrode and the plug body. Neither
had any detectable effect on uHC. In the
first case this was probably because the
volume involved was very small. In the
second case the gap was close in size to that
indicated in the literature as the limit into
which a flame could penetrate. The lack of
any reduction when this was sealed suggested
that the size was on the right side of the
limit and the flame was able to penetrate the
gap and consume the mixture inside.

3.2 Mixture preparation

Poor mixture preparation has several effects
on combustion and uHC production. Robison
and Brehob (8) suggested that liquid fuel
entering the cylinder may be deposited on the

cylinder walls causing maldistribution within
the cylinder. With port injection the
problem is intensified by the lack of time
for mixture preparation and this is further
compounded during engine warm-up by the lack
of hot surfaces within the port to aid
vaporization.

The next stage of the project was therefore
to address the influence of mixture
preparation on the uHC production. To ensure
full vaporization and good mixing of the
resulting vapour with the air, the engine was
fitted with an electrically heated vaporizer
(9). In this the fuel was deliberately
forced onto the walls of a heated brass tube
which was vented into the engine's intake
ports by a flow of compressed air. This air
flow was set below the engine's idle air
requirement.

As one of the mechanisms by which mixture
preparation may influence uHC production is
the storage of liquid fuel in ring crevices,
the tests were run both with the standard
piston and with the sealed one. The
hydrocarbon indices recorded are shown in Fig
7 and tend to support this theory as fuel
vaporization has less effect when the ring
crevices are sealed. Similar changes in
concentration were recorded when the
experiments were repeated on the Lotus

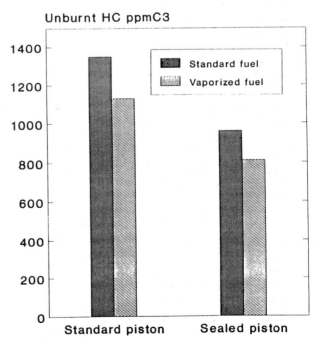

Fig 7 Influence of fuel vaporization on uHC
emissions from sealed and standard
pistons

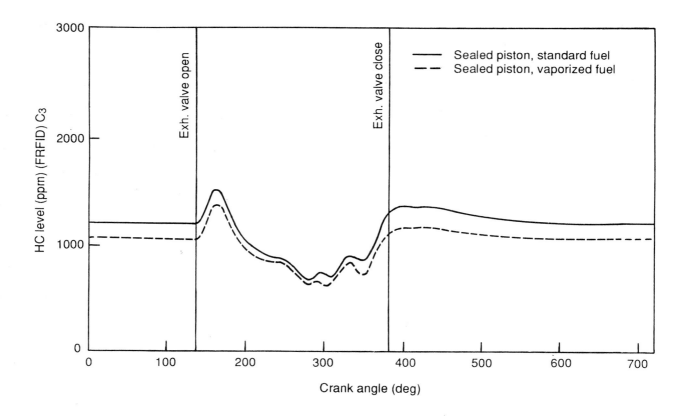

Fig 8 Influence of fuel vaporization on the unburnt hydrocarbon v. crankangle signature with piston crevices sealed

engine.   Fig 8 shows the influence of fuel vaporization on the uHC signatures and again highlights the fact that the dominant change is to the late release storage mechanisms in the cylinder, with little effect on the bulk charge uHC levels.

One of the periods when mixture preparation is absolutely crucial to the production of uHC emissions is the cranking-first fire period.   Combustion will not occur until a combustible mixture is present in the spark plug gap at the appropriate time. In the normally fuelled engine, fuel is supplied partly by flash evaporation of the light fuel components, partly by evaporation of the mid-range components from the wetted port and valve surfaces and partly by the direct flow of liquid fuel into the cylinder either as droplets or as a wall film. Liquid fuel entering the cylinder can only take part in the initiation of combustion if it evaporates during the compression stroke. Meanwhile the mid-range evaporation is only significant when sufficient surface area has been wetted. This wetting process is normally speeded up by providing cranking enrichment but even

this can take several engine cycles and during this time high levels of uHC are emitted as all fuel leaves the engine unburnt.

With vaporized fuel the fuel is entirely airborne and the fuel system designer has the ability to short circuit the wall wetting and surface evaporation delays and so supply the engine with a stoichiometric mixture from the first cycle. This was done on the 4-cylinder M16 test-bed engine. The vaporizer was pre-heated to its working temperature before the engine was cranked. The vaporizer air flow, electrical power input, cranking fuel pulse width and preheat temperature were adjusted, in a short factorial experiment, to give optimum starting. The uHC and air/fuel ratio histories obtained are shown in Fig 9 and demonstrate that, with vaporized fuel, running engine uHC levels can be obtained from the first turn of the engine. The rise in uHC between 1.0 and 1.5 secs after starting is a consequence of the rich excursion of the simple fuel control system employed in these experiments.

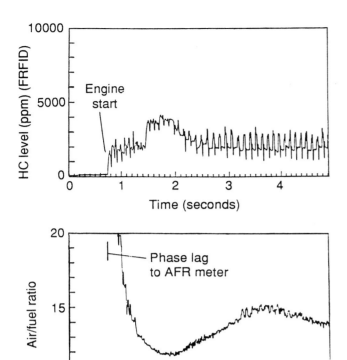

Fig 9 UHC levels achieved during cranking
when using vaporized fuel

3.3 Oil films

The first rigorous investigation of oil film
effects was that by Kaiser et al (10) who
used a low crevice volume bomb and introduced
measured quantities of oil. They were thus
able to show that, for oil layers up to 25
microns thick, uHC emissions increased
linearly with oil film thickness. By using
gas chromatography they were also able to
show that almost all the extra uHC was
composed of fuel molecules, a result
confirmed by Ishizawa and Takagi (11). This
pointed to fuel absorption into the oil as
the most likely mechanism for the increase.
Several workers have produced models of this
mechanism (12-14). More recently Schramm and
Sorensen (15,16) and Trinker et al (17) have
sought to quantify the oil absorption effect
by selecting fuel-oil combinations of very
different solubilities.

The use of specially selected fuel-oil
combinations introduces a number of
additional factors into the study of oil film
effects. The fuel selected will inevitably

be of restricted composition and this will
alter its evaporation characteristics,
stoichiometry and burn rate. These changes
will alter both the absorption of the fuel
into the oil and the post-flame oxidation
after desorption. To avoid these problems,
it was decided to use a full range gasoline
fuel and to select a lubricant which would
have a radically different solubility without
considering its suitability for production
engines. The lubricant selected was a
polyglycol fluid normally used for the
lubrication of natural gas compressors, an
application for which a very low solubility
for low carbon number paraffins is essential.
As the viscosity of this fluid (SAE 20W60)
was rather higher than normal and would have
resulted in thicker than normal oil films on
the cylinder walls, a base mineral oil of the
same viscosity index and normal solubility
characteristics was tested to provide a
direct comparison of the effect of
solubility.

The results of the tests were disappointing
in that there was no discernible difference
in the uHC level or signature shapes obtained
from the two lubricants. The most likely
cause is that described by Trinker et al (17)
whose work was published just as our tests
were completed. In general solubility
increases with carbon number and the
solubility of aromatic compounds is very much
higher than that of paraffins. Trinker also
showed that the solubility of aromatics in a
phosphate ester is almost the same as that in
mineral oil. Discussions with the suppliers
of the polyglycol suggested that a similar
effect may have been present in the current
tests. As the fuel was a full formulation
gasoline with about 45 per cent aromatic
content the absorption of aromatics would
dominate the oil film uHC production. If
almost unaltered by the change of lubricant,
aromatic absorption would thus ensure that
there was no observable difference in overall
uHC.

In a second attempt to demonstrate the effect
of the oil film, the Hydra engine was run
with water as the lubricant. Previous work
by one of the authors (18) had employed a
water-based hydraulic fluid with a 10 per
cent mineral oil content. This proved
inconclusive. Lubricant consumption was high
and it was thought that the water content was
evaporating from the cylinder walls leaving a
film of much higher oil content than
expected. To avoid this problem the engine
was flushed very thoroughly to ensure that

there was negligible oil content in the lubricant. These tests showed a reduction of about 16 per cent with water lubrication. This probably underrepresented the lubricant effect for two reasons. Firstly the piston was more effectively cooled by the water lubricant and this would have increased both the size of the crevices and the density of the mixture stored within them. Secondly reduced NO_x levels indicated that the evaporation of water within the combustion chamber had reduced temperatures and so reduced post-flame oxidation rates. Both mechanisms would tend to increase uHC emissions and mask any reductions due to reduced absorption. When these factors are taken into account the 16 per cent reduction measured is comparable to the 30 per cent reduction reported by Ishizawa and Tagaki (11) in a fully warmed up engine with its combustion chamber running completely dry.

3.4 Flame quenching and combustion kinetics

One of the first theories about the sources of uHC emissions was that the flame was quenched as it approached the cool wall of the chamber and the unburnt boundary layer was subsequently exhausted (19). The advent of detailed models describing the mixing and oxidation processes (eg 20,21) showed that the combustion boundary layer is completely oxidized by the rapid mixing and post-flame reactions in the cylinder. However the same is clearly not true of the mixture that is stored by late or slow release mechanisms such as crevices and oil films. It is also possible that, under cold start conditions, the quench layer is thicker and so less readily consumed. To assess the extent to which improved reaction kinetics could reduce the final engine emission levels, a series of experiments was conducted in which post-flame oxidation, both in cylinder and in port, was increased.

The reaction rate equation (22,23)

$$W = A \exp(-E/RT) \; Y_f Y_O^{0.6} \; (P/RT)^{1.6} \quad (1)$$

shows that, at any point, the rate depends on the current concentration of the two reactants. The concentration of oxygen after the main combustion period could have been increased by running lean but this would have altered the heat release per unit mass of charge and so altered gas temperatures to which the reaction rate is particularly sensitive. Instead the air flow to the engine was reduced and the total gas flow

(and so the mass of charge) made up by supplying pure oxygen. In this way residual oxygen levels were varied from zero to per 6.5 per cent and uHC levels were measured. Fig 10 shows that at 6.5 per cent residual oxygen, uHC levels were reduced by 47 per cent. This demonstrated that the final levels of uHC are strongly dependent on combustion kinetics and that, if practical means can be found to increase the post-flame reaction rates, uHC levels can be greatly reduced.

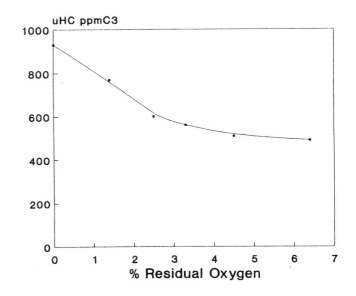

Fig 10   Influence of increased oxygen on residual uHC concentration

In a second set of experiments a more practical method of improving the final reaction rates was sought. Equation 1 shows their dependence on pressure and temperature. By elevating the exhaust backpressure with a simple valve the blowdown expansion of the cylinder contents was restricted and port pressure and temperature were raised. Fig 11 shows that the impact of this on uHC was as dramatic as increasing the oxygen content. When backpressure is raised the pumping work increases and the engine load falls. Since the main point of interest is the reduction of uHC mass flow during the emissions cycle, when brake loads are fixed, it was necessary to open the throttle to restore the engine load. This increased the gas flow through the engine and to allow for this the ordinate of Fig 11 has been plotted as uHC mass flow rather than concentration. These tests were conducted with a fully warmed up engine and

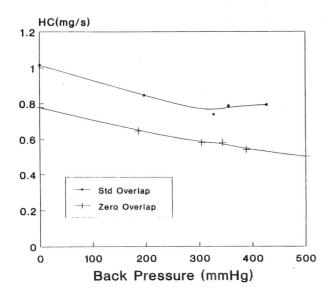

HC(mg/s)

Fig 11 Influence of exhaust backpressure on uHC emissions with various levels of valve overlap

under cold running conditions the elevated exhaust temperature would have the additional benefit of improving heat transfer to the catalyst.

One undesirable consequence of increasing the exhaust backpressure is that it increases the pressure driving force for the flow of exhaust gas residuals into the intake system at inlet valve opening. While the expansion of the cylinder contents from a higher pressure cannot readily be controlled, any tendency for the exhaust pipe to feed the back flow process through the partially open exhaust valve during the valve overlap period can be controlled by adjusting the inlet cam timing to give zero valve overlap. This was done and Fig 11 shows the impact of valve overlap on the uHC v. backpressure characteristic. The reduction in uHC mass flow due to the elimination of valve overlap can be seen to be about 20 per cent and is largely independent of exhaust backpressure. The combination of valve overlap and exhaust backpressure can be seen to be a powerful method for the control of uHC before catalyst light off. Beyond that point however the increased pumping work required represents an efficiency loss which cannot be justified.

3.5 Engine warm-up

Many of the mechanisms for uHC production described so far display a dependence on temperature. If the rate of warm-up of the engine is increased the extra contribution of temperature to these mechanisms can be reduced. At a fixed load warm-up depends on thermal inertia and heat loss. Thermal inertia is largely a function of the mass of water in the cooling system (24) while, until the thermostat opens, heat loss is a function of the way in which coolant flow distributes heat through the engine, transporting it away from the crucial areas of the cylinder head and ports and into the lower cylinder walls, passenger heater etc. To assess the impact of improving the warm-up of the engine two experiments were run. Firstly the engine was run with its coolant pump disconnected to reduce the transport of heat away from the combustion chamber and exhaust ports and to remove the water outside the engine from the thermal inertia. Secondly the engine was run without coolant. This further reduced the thermal inertia and heat flow from the crucial regions.

Fig 12 shows that the reductions in uHC which resulted were significant. The initial experiments were carried out on the Lotus single cylinder engine and the 'no flow' test was repeated on the M16 4 cylinder engine with similar results.

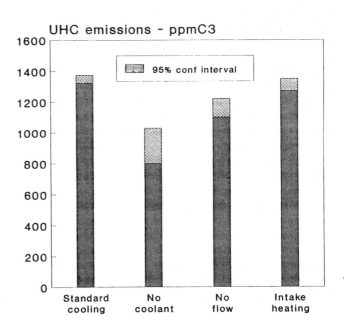

Fig 12 Influence of various engine warmup modifications on uHC emissions

Fig 12 indicates that uHC is strongly dependent on cylinder wall temperature. To assess the extent to which this can be attributed to quench, a series of tests was carried out on a rapid compression machine. In this a mixture of air and butane was compressed by a hydraulic ram and then ignited by a spark. As the machine had virtually zero crevice, ran without lubricant and used a premixed charge of gaseous fuel, it provided an environment from which all the major uHC sources identified so far had been eliminated. The machine had a large volume separated from the combustion chamber by a thin metal diaphragm. The diaphragm could be ruptured at a preset point. This caused a rapid expansion of the residual charge freezing the post combustion reactions. The residual charge was then analysed for uHC content. The results, obtained at various wall temperatures and with two levels of mixture motion, are shown in Fig 13. It is difficult to relate these data directly to the engine measurements. However, while they support the view that, at normal engine temperatures and with strong mixture motion, wall quench layers are largely consumed in the post combustion reactions, they also indicate that at low wall temperatures quench can be a very significant source of uHC emissions.

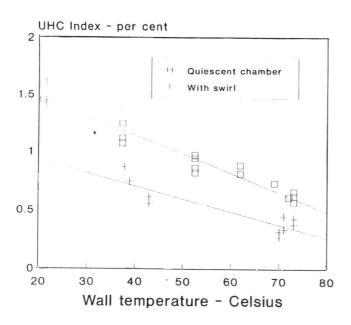

Fig 13   Effect of wall temperature and mixture motion on quench uHC levels

As an engine warms up the increase in port temperatures leads to an increase in intake air temperature and this is normally supplemented by drawing air across the exhaust manifold to reach a stable air temperature as quickly as possible. This has two effects on the in-cylinder processes. Firstly it will aid mixture preparation by increasing fuel evaporation. However the coupling between air temperature and fuel evaporation is known to be weak (25) and in a port injected engine the time available for that coupling to act is very short. Secondly a higher intake air temperature will raise the end of compression temperature which will increase the reactivity of the mixture. To investigate the second effect two series of experiments were run in which the 4-cylinder M16 engine was fed with air, firstly at 20°C and secondly at 60°C. Although expected to be small the influence of mixture preparation was eliminated by running the engine on fully vaporized fuel in both series of tests. In the high temperature tests the air was heated electrically and a reservoir of air was heated before the engine was started to ensure that it was supplied with hot air throughout the first 100 seconds of operation.

As with all the earlier experiments, several runs were carried out at each condition and the difference between the mean levels was tested statistically. The reduction in mean uHC level when the intake air temperature was elevated by 40°C was only 50 ppmC$_3$ (Fig 12) and this was not significant at the 5 per cent level. The tests were run at a constant ignition setting and the increased reactivity of the hot air tests would probably have allowed the ignition to be retarded slightly which would have brought about a further reduction in uHC (26). However it is doubtful whether the uHC reductions achievable by increasing intake air temperature represent an adequate return for the 600 watts of electrical power needed to raise the air temperature.

3.6 Valve leakage

In an earlier section it was suggested that the small rise in the uHC-crankangle signature at valve opening might be due to valve leakage. The measurement of valve leakage under light load running conditions is relatively straightforward in a four-valve engine. One exhaust valve was disabled and its port was plugged (Fig 14). In this way the disabled valve was subjected to a very similar pressure time history to that seen by

68

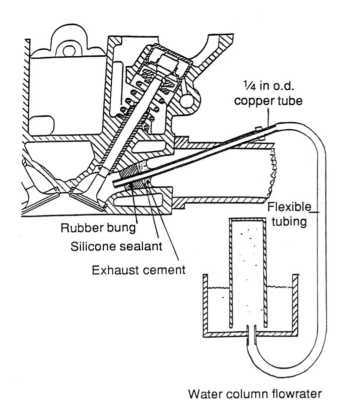

Fig 14 Method of measuring valve
leakage flows

an active valve. During most of the normal
valve open period the pressure difference
across the valve is very small and any flow
past the closed valve in this period is
likely to be trivial compared with the flow
during the normal valve closed period.
During the closed period the pressures on the
disabled valve are identical to those
experienced in normal operation. The sealed
port was vented to a water column and the
volume displaced was then measured as a
function of time. The concentration of uHC
in the leakage flow was also measured. This
was done using the Cambustion FRFID which
draws a very low flow and by setting that
flow well below the measured leakage it was
possible to ensure that no air was drawn into
the instrument. At the test condition of
2000 r/min, 2 bar bmep, a gas flowrate of
65 cc/min was measured. The engine was
running at the stoichiometric air/fuel ratio
and the concentration of uHC in the leakage
flow was measured at 22000 $ppmC_3$. As the
unburnt mixture would have a concentration of
46000 $ppmC_3$, this measured value suggests
that the leakage flow is divided roughly
equally between the period of compression and

early combustion before the flame traverses
the exhaust valve and the period after the
flame has passed when the uHC concentration
will be that of the exhaust products at about
1000 $ppmC_3$. When the measured flowrate was
combined with this concentration measurement
and a correction made for the back flow
during the intake stroke, the leakage flow
was seen to account for perhaps 25-30 $ppmC_3$.

This leakage is clearly very small but as the
valves and their seats become worn the
leakage path may increase. To examine valve
leakage as a mechanism for the increased uHC
production observed in high mileage cars,
tests were run to compare the leakage from
cylinder heads which had been subjected to
endurance running. For these tests the head
face was blanked, air supplied through the
spark plug hole under pressure and the flowrate through the port was recorded.

This flowrate was then corrected for the
different pressure history and combined with
the uHC concentration measured on the running
engine to arrive at an equivalent uHC
emission level for each cylinder. Fig 15
shows the results obtained from three heads
and shows that potential emissions are very
variable and many times higher than those
seen with the newly cut valves and seats of
the M16 Hydra engine. The average uHC
emission level recorded was 250 $ppmC_3$ and
valve deterioration could well therefore make
a significant contribution to the increased
uHC production of high mileage vehicles.

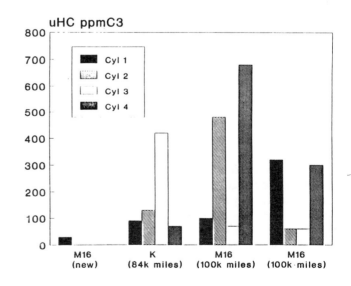

Fig 15 Equivalent uHC emissions due to
valve leakage from high mileage
cylinder heads

Table 1 Contribution of various uHC mechanisms to uHC index after 100 seconds

Unburnt hydrocarbons mechanism	Engine build	UHC index per cent
	Standard engine	2.95-3.06
Crevice action	PTFE seal	2.23-2.48
Fuel preparation	Vaporized fuel	2.57-2.75
Combined crevices and fuel preparation	PTFE seal and fuel vaporization	2.06-2.26
Rate of warm-up	No coolant	2.12-2.56
Oil film effects	Water lubrication	2.48-2.57
Kinetics of post flame oxidation	Oxygen enrichment	≈ 1.56

uHC index = (mass flow of uHC leaving engine)/(mass flow of fuel entering)

3.7 Summary

Table 1 summarizes the results of the various experiments to quantify the uHC sources. The results are expressed in terms of the uHC index defined earlier and the bands indicated are the 95 per cent confidence intervals for the various results. The results are based on uHC levels measured 100 seconds after engine start-up. Mixture preparation has been shown to be of greater importance at start-up and the importance of wall quench has been shown to increase at lower temperatures.

4 CONCLUSIONS

1 A three year research programme has identified a number of important sources of uHC in cold engines though no single source dominates the production of unburnt hydrocarbons from spark ignition engines during warm-up.

2 Careful design attention is needed to minimize crevices associated with the piston ring pack, cylinder head gasket and spark plug.

3 Prior to the intake ports reaching their full operating temperatures, means should be provided to ensure substantial fuel evaporation.

4 Variable valve timing should be employed to minimize exhaust residuals at light load.

5 Prior to catalyst light off, exhaust pressure should be elevated to enhance port oxidation.

6 Coolant flow should be minimized to enhance warmup and reduce wall quench effects.

7 Valve leakage is not a significant contributor to uHC emissions in a new engine but could be a significant cause of the increase in uHC emissions with engine age.

8 Oil companies should be encouraged to develop oil/fuel formulations to minimize the fuel absorption/desorption storage mechanism.

5 ACKNOWLEDGEMENTS

The authors would like to acknowledge the invaluable contributions made to this work by the experimental teams at NEL, by those engineers in the sponsoring companies who have helped with the supply of components and advice and by Shell Research, Lubrizol and AE Pistons who provided materials and components for testing. We would also like to acknowledge the contribution of Leeds University to the work on flame quenching and the efforts of the various members of the DTI's

Environment Unit who have monitored our progress in a sympathetic and helpful manner. Finally we would like to record our thanks to the Directors of Ford, Jaguar, Lotus, Rover and NEL for permission to publish this paper.

REFERENCES

(1) Collings, N. A new technique for measuring HC concentrations in real time in a running engine. SAE Paper 880517, Society of Automotive Engineers, Warrendale, Pa., 1988.

(2) Finlay, I.C. , Boam, D.J., Bingham, J.F. and Clark, T.A. Fast response measurements of unburned hydrocarbons in the exhaust port of a firing gasoline engine. SAE Paper 902165, Society of Automotive Engineers, Warrendale, Pa., 1990.

(3) Wentworth, J.T. Piston ring variables affect exhaust hydrocarbon emissions. SAE Paper 680109, Society of Automotive Engineers, Warrendale, Pa., 1968.

(4) Adamczyk, A.A., Kaiser, E.W., Cavolowsky, J.P. and Lavoie, G.A. An experimental study of hydrocarbon emissions from closed vessel explosions. 18th Symposium on Combustion, The Combustion Institute, 1981.

(5) Adamczyk, A.A., Kaiser, E.W. and Lavoie, G.A. A combustion bomb study of the hydrocarbon emissions from engine crevices. Combustion Science and Technology, Vol 33 pp 261-277, 1983.

(6) Sellnau, M.C., Springer, G.S. and Keck, J.C. Measurements of hydrocarbon concentrations in the exhaust products from a spherical combustion bomb. SAE Paper 810148, Society of Automotive Engineers, Warrendale, Pa., 1981.

(7) Namazian, M. and Heywood, J.B. Flow in the piston-cylinder-ring crevices of a spark ignition engine: effect on hydrocarbon emissions, efficiency and power. SAE Transactions Vol 91, pp 261-288, 1982.

(8) Robison, J.A. and Brehob, W.M. The influence of improved mixture quality on engine exhaust emissions and performance. Journal of Air Pollution Control Association, Vol 17, Pt 7, pp 446-453, 1967.

(9) Finlay, I.C., Boyle, R.J. and Boam, D.J. The NEL fuel vaporizer. To be published at SAE Congress 1993.

(10) Kaiser, E.W., Adamczyk, A.A. and Lavoie, G.A. The effect of oil layers on the hydrocarbon emissions generated during closed vessel combustion. 18th Symposium on Combustion, The Combustion Institute, 1981.

(11) Ishizawa,S. and Takagi, Y. A study of HC emissions from a spark ignition engine JSME International Journal, Vol 30, pp 310-317, 1987.

(12) Carrier, G., Fendell, F. and Feldman, P. Cyclic absorption/ desorption of gas in a liquid wall film. Combustion Science and Technology, Vol 25, pp 9-19, 1981.

(13) Shyy, W. and Adamson, T.C. Analysis of hydrocarbon emissions from conventional spark ignition engines. Combustion Science and Technology, Vol 33, 1983.

(14) Dent, J.C. and Lakshminarayanan, P.A. A model of absorption and desorption of fuel vapour by cylinder lubricating oil films and its contribution to hydrocarbon emissions. SAE Paper 830652, Society of Automotive Engineers, Warrendale, Pa., 1983.

(15) Schramm, J. and Sorenson, S.C. Effect of lubricating oil on hydrocarbon emissions in an SI engine. SAE Paper 890622, Society of Automotive Engineers, Warrendale, Pa., 1989.

(16) Schramm, J. and Sorenson, S.C. A model for hydrocarbon emissions from SI engines SAE Paper 902169, Society of Automotive Engineers, Warrendale, Pa., 1990.

(17) Trinker, F.H., Anderson, R.W., Henig, Y.I., Siegl, W.O. and Kaiser, E. W. The effect of engine oil on exhaust hydrocarbon emissions. IMechE Seminar C435 on Worldwide engine emission standards and how to meet them, London 1991.

(18) Wallace, S. and Warburton, A. The control of CO, HC and NO_x emissions and the application of lean burn engines. IMechE Conference C359 on Vehicle emissions and their impact on European air quality. London 1987.

(19) Daniel, W.A. Flame quenching at the walls of an internal combustion engine. 6th Symposium on Combustion, The Combustion Institute, 1957.

(20) Westbrook, C.K., Adamczyk, A.A. and Lavoie, G. A. A numerical study of laminar flame wall quenching. Combustion and Flame, Vol 40, pp 81-99, 1981.

(21) Carrier, G.F., Fendell, F.E., Bush, W.B. and Feldman, P. S. Non-isenthalpic interaction of a planar premixed flame with a parallel end wall. SAE Paper 790245, Society of Automotive Engineers, Warrendale, Pa., 1979.

(22) Adamczyk, A.A. and Lavoie, G.A. Laminar head-on flame quenching - a theoretical study. SAE Paper 780969, Society of Automotive Engineers, Warrendale, Pa., 1978.

(23) Lavoie, G.A. Correlations of combustion data for SI engine calculations: laminar flame speed, quench distance and global reaction rates. SAE Paper 780229, Society of Automotive Engineers, Warrendale, Pa., 1978.

(24) Boam, D.J. Energy audit on a two litre saloon car driving an ECE 15 cycle from a cold start. Proc Instn Mech Engrs, Vol 200, No D1, p 61-67, 1986.

(25) Boam, D.J. and Finlay, I.C. A computer model of fuel evaporation in the intake manifold of a carburetted petrol engine. I.Mech.E. Conference on Fuel economy and emissions of lean burn engines. London 1979.

(26) Hagen, D.F. and Holiday, G.W. The effect of engine operating and design variables on exhaust emissions. SAE TP-6, 206, Society of Automotive Engineers, Warrendale, Pa., 1964.

C448/003

Emission-optimized fuel injection for large diesel engines – expectations, limitations and compromises

J VOLLENWEIDER, Dipl-Ing, PhD, MSAE
New Sulzer Diesel Limited, Winterthur, Switzerland

SYNOPSIS It is shown in a morphological scheme that a very suitable way of modulating the fuel injection in a large diesel engine is the application of a two-spring injector. A set of such devices was built for this study, tuned on the fuel injection test rig (also by means of high-speed photography), and run on a Sulzer 6S20 diesel engine. The results (compared to the standard injection) showed hardly any change in specific fuel consumption nor in noise, a small reduction of the NOx emission by 0 to 20% (depending on the injector setting, as also on the engine load and speed), and an up to tripled smoke opacity. These results, combined with the fact that a two-spring injector makes the fuel injection system significantly more complicated, led to the conclusion that rate modulated fuel injection is not the first choice in an exhaust emissions control concept for large diesel engines.

1 INTRODUCTION

In all the past years, since the existence of the diesel engine, a countless number of engineers have worked on its development. One topic which has always been in focus, concerned the fuel injection. Rudolf Diesel already mentioned in his patent in February 1892 [1] the possibility of controlling the combustion by the fuel injection – that is exactly the concept, which shall be discussed also in this paper. Hence, it would certainly be unrealistic to think that major discoveries could be made in this study. Nevertheless, a new aspect can be brought into the discussion by examining how far the results of earlier works can be transferred to large diesel engines.

2 EVALUATION OF AN EMISSION-OPTIMIZED FUEL INJECTION SYSTEM

2.1 Formulation of a Research Strategy

The goal of this research is to achieve minimum exhaust and noise emissions at a maximum thermal efficiency. However, the exact function and design of the fuel injection system which is supposed to yield those results are yet to be determined. Hence, it is obvious to base the investigation on the known output parameters, and then to calculate backwards, always taking the given boundary conditions into account. This way, a definition of an emission-optimized fuel injection system appears to be possible, although a number of assumptions and simplifications will have to be made, such that quite some uncertainty will be inherent in that solution. Yet, this problem can easily be solved by experimentally examining the proper functioning of the envisaged fuel injection system.

2.2 Specification of the Expression "Emission-optimized"

Large diesel engines are always optimized for low fuel consumption, since due to the commonly very long yearly running hours, the costs for the fuel make up quite a substantial portion of the operating costs. In view of the importance of a low fuel consumption, researchers have spent a lot of time optimizing that parameter, so that the remaining room for further improvement is probably no longer very significant. On the other hand, low emissions were not that important in the past, thus focussing on that parameter definitely appears to be worthwile. Still, it has to be emphasized, that optimizing for low emissions should always aim at an overall improvement of the engine, i.e. there should be no trade-off with the thermal efficiency.

2.3 Optimizing the Engine Process

The effects of the various parameters of the engine process (scavenging pressure, compression ratio, rate of heat release, etc.) onto the exhaust emissions are generally known, and must therefore not be rediscussed here.

The starting point of this investigation will be an efficiency-optimized, i.e. isobaric combustion. Obviously, this will not lead to an optimum result in terms of emissions. Yet, since any deviation from the isobaric process in order to lower the emissions (particularly the NOx) will entail serious economical drawbacks, it is advisable to incorporate the emission requirements only at a later stage of this study.

2.4 Definition of a Suitable Heat Release

An isobaric combustion can be achieved by the heat release shown in Figure 1:

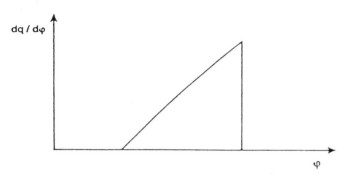

Fig 1   Ideal heat release

2.5 Evaluation of the Fuel Injection Rate Based on the Ideal Heat Release

In practice, an exact reproduction of the heat release shown in Figure 1 is impossible, irrespective of the employed injection system. In particular, this is true for the end of the combustion, since that is primarily dependent upon the local oxygen concentration in the combustion zone (micro-turbulences!), which can hardly be controlled in an easy way. Also the smooth rise of the heat release rate at the beginning of the combustion is quite difficult to control. It is known, however, that an uncontrolled fuel injection during the ignition delay leads to a pronounced premixed combustion with its typically high NOx- and noise-emission. That premixed combustion can be reduced though, by the pre-injection depicted in Fig. 2.

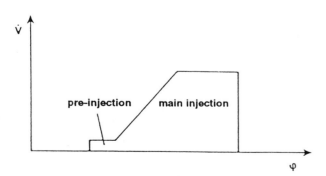

Fig 2   Emission-optimized fuel injection rate

The pre-injection aims at the conditioning of the combustion chamber for the main combustion. Yet, only the start of the combustion can be influenced, i.e. the rise of the heat release rate can be made smoother. Once the actual combustion is on its way, however, the amount of heat released is so big that the conditions in the cylinder are no longer influenced by the fact whether or not a pre-injection has ever taken place. Hence, a pre-injection cannot shorten the combustion duration, and thus also the thermal efficiency cannot be improved. Moreover, there should be no mechanical force generated by the pre-injection; (clearly, it would not make sense to do that before the piston has reached its top dead centre). Hence, it becomes obvious that the amount of pre-injected fuel should be very small - but also not too small, in order not to harm the dispersion quality. In this study, the best results have been achieved with a pre-injected fuel quantity which was approximately equal to 50% of the injected fuel when the engine was idling. As far as the start of the pre-injection is concerned, it must be stated again that the purpose of the pre-injection is to condition the combustion chamber for the main combustion. To achieve that, the start of the pre-injection must be set such that the pre-injected fuel is about to burn when the main injection starts. That is, the time between the start of the pre- and main-injection should be about as long as the ignition delay. Therefore, an ideally modulated fuel injection exhibits a pre-injection which is immediately followed by the main injection, resulting in the step-injection shown in Figure 2. In the same figure we notice that the injection rate during the main combustion is not continuously increasing as might perhaps be concluded from Figure 1. Yet, this particular injection rate is needed, because not only a well-defined amount of fuel should be injected in any given time interval, but also, a permanently high injection pressure is needed to ensure a thorough fuel dispersion. This requirement can presumably only be satisfied in a nozzle having a variable injection hole geometry. For example, this is the case in a pintle nozzle. However, that particular atomizer has also serious drawbacks:

- It is hardly possible to get a well defined spray.
- The whole nozzle design is rather fragile and thus not really suited for heavy duty applications.

Since the results should eventually be useful also in practice, it is certainly wise to drop that variable geometry concept. In order still to obtain the high injection pressures, the injection rate curve must therefore be shaped according to Figure 2.

2.6 Design Aspects of an Emission-optimized Fuel Injection System

Of course, a variety of different designs exists with which the stated goals can be reached. Nevertheless, to make sure that at the end the optimum design is obtained, the various systems should first be discussed in a morphological scheme; (see Table 1).

In the last column of Table 1, a cross indicates the optimum solution. Of course, this particular valuation only applies for large diesel engines with their specific requirements. That is, the situation for another engine type could well be different. In this investigation, however, it appears to be best to carry out the experimental part of our work with a so-called two-spring

Feature	Execution	Advantages	Disadvantages	Optimum Solution
Driving force	mechanical	simple, stiff system	limited variability	X
	electro-hydraulical	extensive variability	complicated, elastical system	
Position of the injection pumps	nozzle separated from pump	simple design, easy to service	hydraulic system prone to oscillations	X
	nozzle flanged onto the pump	high hydraulic stability	complicated design, laborious to service	
Number of injection pumps	one	simple, inexpensive	limited variability	X
	two	extensive variability	complicated, expensive	
Position of the inj. nozzle	pre- and main-injection through the same nozzle	symmetrical spray pattern	compressed design, sub-optimum nozzle geometry for the pre-injection	X
	pre- and main-injection through adjacent nozzles	virtually symmetrical spray, optimum nozzle geometries for both injections	complicated and very compressed design	
	pre- and main-injection fully separated	optimum nozzle geometries for both injections	non-symmetrical spray, requires a new cylinder head design	
Location of the injection modulation	in the nozzle	compact design	only for step-injection, variation of the flow pattern at spray hole inlets	X
	in the injector body (split injection device)	true pre-injection possible	complicated, poor fuel dispersion during pre-injection	
	in the pump	good variability	compressed design, prone to cavitation, poor dispersion during pre-injection	
Nozzle geometry	constant geometry	simple design	poor fuel dispersion at the start of the injection	X
	variable geometry	always good fuel dispersion	uncontrolled spray geometry, fragile design	
Control of the injector-needle closing	conventional (via pressure diff. in the inj. chamber)	simple	deviation from the ideal end of fuel injection	X
	with additional force	close to the ideal end of fuel inj.	complicated design	

Table 1 Morphological scheme to the modulated fuel injection

injector, which is schematically shown in Figure 3.

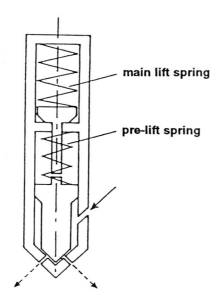

Fig 3 Schematic of a two-spring injector

3 EXPERIMENTAL PART

3.1 Test Engine

The tests were run on a Sulzer 6S20 engine. This medium-speed four-stroke diesel engine is particularly suited for ship propulsion, auxillary propulsion on board ship, electricity generation and industrial applications. The main data are:

Bore	:	200	mm
Stroke	:	300	mm
Power per Cyl.	:	145	kW
Nominal Speed	:	1000	r/min
Mean Eff. Press.	:	18.5	bar

Today's large diesel engines are mostly operated on heavy fuel oil. However, even before starting the actual experiments, it was quite clear that major difficulties would be encountered when trying to run the two-spring injector on heavy fuel oil. That is why it was decided to start the tests on diesel fuel, and then depending on the result, to stop the experiments or to continue on heavy fuel oil.

3.2 Design of the Two-spring Injector

For economical reasons it would have been unwise to start with the design of a two-spring injector from scratch. Therefore, the standard S20 fuel injector was simply modified, as shown in Figure 4. Due to all the modifications, the injector of course became more complicated, more fragile and thus no longer really suited for a commercial application. This shortcoming was not cor-

rected though, since this two-spring injector was anyway designed for experimental purposes for the time being. But it is clear that there is still plenty of room for improvement. For example, the inlet of the spray holes could be rounded off or the guide of the rod between injector needle and pre-lift spring could still be improved.

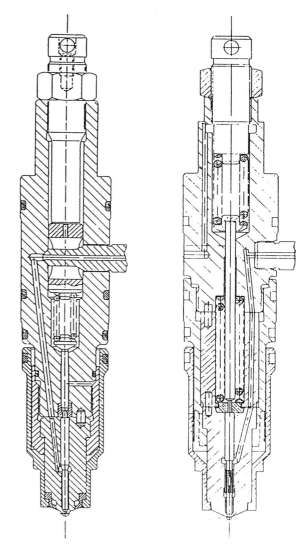

Fig 4a Standard fuel injector of the Sulzer S20 engine (left)
Fig 4b Two-spring injector for the Sulzer S20 engine (right)

3.3 Preliminary Experiments on the Injection Test Rig

The injection test rig allows to check the principal functioning of the injectors and to make a provisional adjustment of the injection parameters. In this device the fuel is injected into the atmosphere rather than into an engine, which is of course advantageous in terms of costs. Table 2 shows a summary of the works done on the injection test bed.

Parameter	Main Influence	Provisional Setting
Pre-lift of injector needle	Amount of pre-injected fuel	0.01 and 0.03 mm
Main lift of injector needle	Injection pressure	0.75 mm
Pretension of pre-lift spring	Amount of pre-injected fuel	670 N $\rightarrow$ p_{op1} = 400 bar
Pretension of main lift spring	Injection rate	650 and 1100 N $\rightarrow$ p_{op2} = 470 and 630 bar
Spray hole diameter	Injection rate and spray pattern	0.35 mm
Number of spray holes	Spray pattern	12
Diameter of injector needle	Dynamics of needle opening and closing	6/3.8 mm
Plunger diameter	Injection duration	19 mm
Fuel back-flow	Dynamics of needle closing	Stagnation pressure valve removed

Table 2 Injection parameter settings

3.4 Examination of the Spray Pattern Using High-speed Photography

Whether or not the pre-injection yields the desired results is strongly influenced by the spray pattern. It was therefore decided to check this aspect via highspeed photography on the injection test bed prior to the rather expensive engine tests. The following conclusions could be drawn from those tests:

- The importance of a symmetrical spray pattern could be visualized. Such symmetrical spray patterns could only be achieved when the pre-lift of the injector needle was larger than 0.02 mm.

- Also the problems attached to the post-lift of the injector needle could be demonstrated: It is obvious that fuel molecules injected after the main combustion will encounter adverse conditions in the combustion chamber. That fuel is therefore wasted for the most part.

3.5 Engine Results

After having run reference tests with the standard injector, a systematical variation with the two-spring injector was carried out (with the parameters 0.01 and 0.03 mm for the pre-lift and 650 and 1100 N for the pretension of the main lift spring). Measurements at the following loads were taken: 100, 75, 50 and 25% (each time at constant speed as well as according to the propeller law, i.e. seven measurements per injector setting). In order to make sure that the results can be compared among each other, the start of injection was always adjusted at each load during the tests with the two-spring injector such that the same maximum cylinder pressure was obtained as in the corresponding case with the standard injector. Once these test conditions were established, the maximum cylinder pressure was kept constant. The thermodynamical effect of reducing the maximum cylinder pressure by retarding the start of fuel injection was investigated in earlier experiments though. The results for the Sulzer S20 engine showed a decrease in NOx of about 15 ppm and an increase in BSFC of 0.25 g/kWh when reducing the max. cylinder pressure by 1 bar (which corresponds to an injection retardation of 0.25 °CA). Since lowering the max. cylinder pressure and using a rate modulated fuel injection are two completely different low-NOx strategies, which are based on two different physical mechanisms, the two combustion control techniques can be investigated separately and the results can then be superimposed. Hence, this study does not include results of the two-spring injector in conjunction with a modified timing plan. The best overall results with the two-spring injector were achieved with a pre-lift of 0.03 mm and a pretension of the main lift spring of 1100 N. In the following, only these results are shown in comparison with the results of the standard injector.

The heat release rates are shown in Figure 5. It has to be stressed though, that these curves can only be used for qualitative considerations. (Heat release rates which also give quantitative information require quite some work in terms of smoothing and averaging when processing the data; it was decided however, not to go that far in this study). Still, Figure 5 is quite revealing:

- As expected, the premixed combustion is particularly pronounced at part load. This combustion phase can effectively be reduced, although not eliminated, with a staged injection. A complete elimination of the premixed combustion is certainly possible, however only by an extremely careful and thus

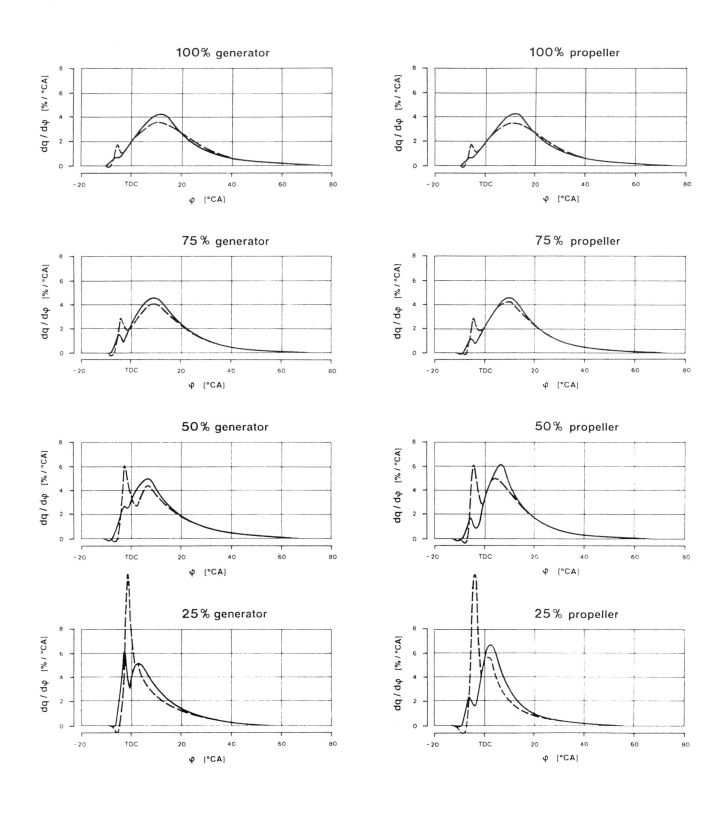

Fig 5 Rates of heat release

time-consuming adjustment of the fuel injectors.

- Particularly interesting is the comparison of the heat release (Fig. 5) with the NOx emission (Fig. 6b). There is a clear correlation between the NOx reduction obtained by the pre-injection and the diminution of the premixed combustion. (By the way, a comparison of the two 25% load points shows that the strength of the premixed combustion is of course not the only factor influencing the NOx emission).

In Figures 6b - 6h the following line code applies:

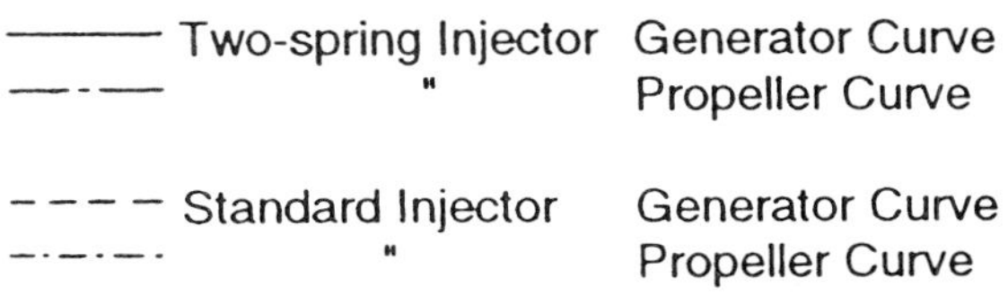

Two-spring Injector Generator Curve
" Propeller Curve
Standard Injector Generator Curve
" Propeller Curve

Fig 6a Line code to Figures 6b - 6h

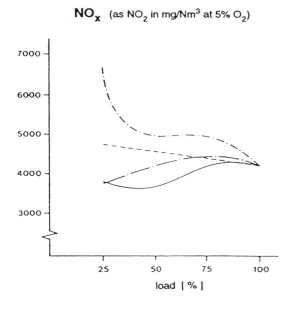

NOx (as NO2 in mg/Nm3 at 5% O2)
7000
6000
5000
4000
3000
25
50
75
100
load [%]

Fig 6b NOx values

The NOx values in Figure 6b show that the nitric oxides can only be lowered at part load, but not at full load (which is an important load point in large diesel engines).

The specific fuel consumption (Fig. 6c) could be lowered at full load by approx. 1% when modulating the injection rate. At part load, however, the figures with the two-spring injector are up to 7% higher. Possible explanations for this result are:

- The standard nozzle has in its tip a cavity volume, which is not sealed towards the combustion chamber by the injector needle. Therefore, it is possible that this volume is partly swept out and thus lost during the scavenging. This effect cannot occur in the two-spring injector due to its different nozzle tip design.

- For simplicity reasons, the two-spring injector was essentially only tuned to full load on the injection test bed. Yet, already the high-speed photography tests showed, that the spray patterns at part load were not optimum, some of them even definitely looked bad, which could very well be the reason for the 7% increase in fuel consumption.

Hence, the different fuel consumption figures in Fig. 6c should by no means be overinterpreted, since the differences probably stem from rather trivial effects. In addition, there was also no indication of a decrease in fuel consumption caused by a potentially shortened combustion (in case of the modulated fuel injection).

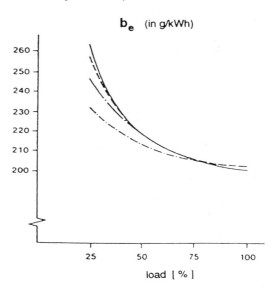

be (in g/kWh)
260
250
240
230
220
210
200
25
50
75
100
load [%]

Fig 6c Specific fuel consumption

Also the low HC emission with the two-spring injector (Fig. 6d) is not directly caused by the modulated fuel injection. Rather, it can again be explained by the particular nozzle tip design featuring a zero sac volume.

The smoke opacity (Fig. 6e), particulate- (Fig. 6f) and CO-emission (Fig. 6g) is considerably worse - above all at part load - with the modulated injection (compared to the standard injection). However, this result should not be attributed to a basic shortcoming of the modulated fuel injection, since the reason for these poor results of the two-spring injector probably lies in the fact that this injector was not optimized as thoroughly as the standard injector. Nevertheless, it must be emphasized that the modulated injection is considerably harder to tune, above all at part load, simply because more parameters have to be adjusted than in a standard injection system.

HC (as CH$_4$ in mg/Nm^3 at 5% O$_2$)

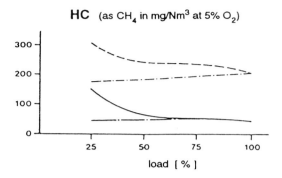

Fig 6d HC emissions

Smoke opacity (in % after Bosch 1l)

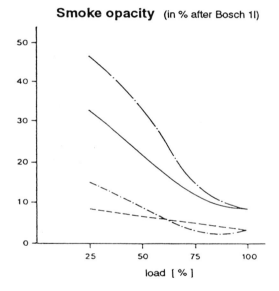

Fig 6e Smoke opacity

Particulates (in mg/Nm^3 at 5% O$_2$)

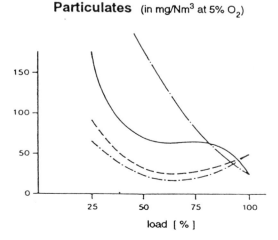

Fig 6f Particulates

CO (in mg/Nm^3 at 5% O$_2$)

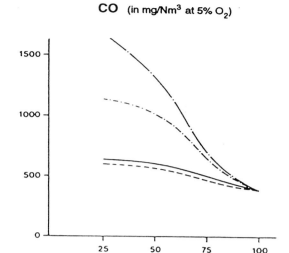

Fig 6g CO emission

Eventually, Figure 6h shows the noise levels. A detectable (3 dB(A)) difference between the two injection systems exists only at 25% load on the propeller curve. The reason being the relatively little contribution of the combustion noise to the overall noise in large diesel engines.

Noise (in dB(A))

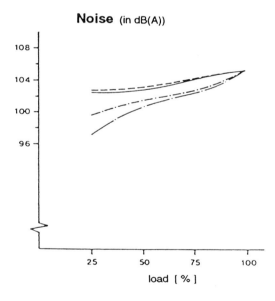

Fig 6h Noise level

3.6 Summary of the Results

The results are summarized in a graphic form in Table 3. The results are split into full and part load, since it became obvious in the experiments that the results are strongly dependent on the load. Moreover, in a third column, it was tried to assess the potential for improvement.

	Full Load	Part Load	Improvable ?
NO$_x$	→	↘	no
b$_e$	↘	↗	yes
Opacity	↗	↗	yes
PM	↘	↗	yes
HC	↘	↘	no
CO	→	↗	yes
Noise	→	→	no
Total		→	yes

Table 3 Summary of the results of the modulated injection (compared to the standard injection)

4 CONCLUSIONS

(a) The modulated fuel injection primarily aims at a reduction of the thermodynamically adverse premixed combustion. Since the duration of that combustion also depends on the amount of fuel which is injected during the ignition delay, the premixed combustion can be effectively reduced with a low injection rate at the start of the injection.

(b) Theoretical considerations have shown, that a modulated injection is most suitably achieved by employing a so-called two-spring injector.

(c) The efficiency of such a device is strongly dependent on the engine size, simply because a number of physical parameters and processes in large engines are completely different from small engines. In particular, the engine speed is an essential parameter, due to the fact that the premixed combustion increases proportionally with the engine speed. Therefore, it turned out that the room for improvement with regard to exhaust- and noise-emission was rather limited on the employed medium-speed test engine. Furthermore, there is a basic difference in the market for small and large diesel engines: For example noise is an important criterion in small diesels, whereas this aspect is not in the foreground in larger engines.

(d) The emission measurements for the modulated injection showed a small reduction of nitric oxides in the range between 0 (at full load) and 20% (at part load) and a marginal improvement of the noise level of 3 dB(A) at best (at 25% load on the propeller curve). At the same time however, a massive (of up to 200%) increase in smoke opacity could be observed.

(e) There was no theoretical nor experimental indication of an increase in thermal efficiency due to a combustion which was potentially shortened by the modulated fuel injection.

(f) Theoretically, an advantage of the pre-injection when burning fuels with an extremely low ignition quality could be expected. Yet, earlier tests showed that Sulzer engines can run with no problem at all even on fuels of super-low qualities. Thus, there is certainly no point in modifying the injection system for this reason.

(g) An essential drawback of the modulated fuel injection lies in the fact that it will always require a design which is more complicated and thus less reliable than the one of a standard injection. Particularly in large diesel engines this is a strong disadvantage, since reliability constitutes an important criterion in such engines.

(h) Hence, the potential of a modulated fuel injection is considerably smaller in large diesel engines than in small ones. Therefore, judging from the results obtained with a two-spring injector, this type of fuel injection will not be the first choice when it comes to primary emissions abatement measures in large diesel engines.

5 ACKNOWLEDGEMENT

This presentation is mainly based on the 2nd report [2] on a project dealing with diesel emissions abatement, which was sponsored by the National Energy Research Foundation (NEFF). This funding, which led to the successful accomplishment of this work is greatfully acknowledged.

6 REFERENCES

[1] DIESEL, R. Arbeitsverfahren und Ausführungsart für Verbrennungskraftmaschinen. German Patent No. 67207, February 1892.

[2] VOLLENWEIDER, J. Untersuchungen zu einer emissionsoptimierten Brennstoffeinspritzung. NEFF project No. 431, 2nd report, New Sulzer Diesel Ltd, March 1991.

C448/010

EGR feedback control on a turbocharged DI diesel engine

A AMSTUTZ, PhD
Institute of Energy Technology, Federal Institute of Technology, Zurich
L DEL RE, PhD
Automatic Control Laboratory, Federal Institute of Technology, Zurich

SYNOPSIS. The increasing importance of environmental protection requires a drastic reduction of NO_x, particulates and CO_2-emissions. This has motivated the development of an exhaust gas recirculation system with closed loop control on a direct injected (DI) Diesel engine. As input for the control system the air/fuel ratio of the exhaust gas is used, measured using an oxygen sensor. It is shown in this paper that by proper choice of the oxygen sensor, control oriented modelling and robust control design based on the linear approximation of the model very good improvements in NO_x can be obtained, while keeping particulates emissions low.

Keywords: Automotive control applications, polluant applications, Economy Improvement, Control Strategies.

1   Introduction

To improve the environmental acceptability of car Diesel engines, the polluants contained in exhaust gas should be reduced, among them especially the NO_x and the particulates emissions. They arise as combustion by-products, and are influenced by different elements of Diesel injection, like injection pressure, injection beginning and preinjection, and are also very sensitive to the oxygen concentration available during combustion. In particular, the PM increases strongly if the oxygen availability falls under an engine specific limit and the production of NO_x increases with the peak burned gas temperature and the oxygen availability.

As it has been shown in an earlier work [1], an engine specific air/fuel ratio can be found that allows reduction of the NO_x and PM emissions without loss in CO- and HC-pollution. A way to achieve this effect is by exhaust gas recirculation , that influences directly both oxygen content and peak burned gas temperature. Exhaust gas recirculation consists in mixing a certain amount of exhaust gazes with the fresh air going to the combustion chamber in order to produce a new gas mixture with lower oxygen content.

This paper presents a new exhaust gas recirculation control philosophy, characterised by

- use of an oxygen sensor as monitoring device

- explicit consideration of the system uncertainties by means of a robust control design method, the LQG/LTR approach [2].

This paper is divided in three parts. First, the plant is introduced and the importance and the appropriate choice of the oxygen sensor are discussed. It turns out that temperature regulated sensors have the necessary static and dynamic characteristics required for our purpose. In a second part, the issues of modelling, linearisation, model reduction and control system design are shortly addressed. A third part presents then the experimental results, among them the impressive reduction of NO_x.

2   The Plant and the Basic Concept

The engine used for the tests is a 2.5 l, five cylinder four stroke Diesel engine with turbocharger (TC) and DI. The engine has excellent fuel consumption and good overall emissions. Although the injection beginning is electronically controlled, the direct injection combustion method leads to high NO_x.

Exhaust gas recirculation represents an efficient approach to reduce them. Figure 1 shows a principle scheme of the recirculation system to explain the working principle. Recirculated exhaust gazes are mixed on the high pressure side with the inlet fresh air going to the engine in order to reduce the oxygen available during combustion. The control unit

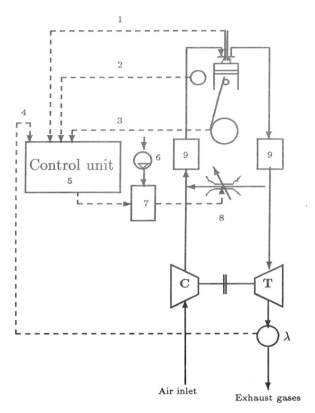

Figure 1   Basic scheme of the plant   *1: Injector opening 2: Engine temperature 3: Engine speed 4: Oxygen sensor 5: Control unit 6: Underpressure pump of the motor 7: Electropneumatical actuator 8: Exhaust recirculation valve 9: Inlet and outlet receiver*

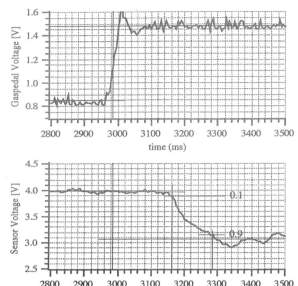

Figure 2   Dynamic characteristics of the chosen oxygen sensor

consists of a basic commercially available injection control and of our EGR control system acting on the recirculation valve. The only measurement required for the basic control of the recirculation rate is that of λ (air/fuel ratio), whereas for feedforward extensions also the injected fuel quantity should be available. However, as the injected fuel is a quantity set by the basic control unit, no new sensor would be required for large scale implementation.

However useful for NO_x reduction, exhaust gas recirculation may lead to an increase of PM especially during engine acceleration, if used without a closed loop control. As direct measurements of NO_x or PM on-line are not yet possible, an indirect measurement must be used. Measurement of residual oxygen content in the exhaust gazes proves a suitable approach.

The choice of the oxygen sensor proves critical due

to its strong temperature dependence. Oxygen sensors are mainly used in automotive industry for spark ignition engines, whose hot exhaust gazes provide a regular heating of the sensor. For lean burn engines internally heated oxygen sensors have been developed, that can be used for exhaust gas temperatures as low as 300°C. However, the exhaust gas temperature of DI Diesel engines may be below 200°C. Therefore special care must be given to the choice of the sensor and to its installation in order to obtain reproducible results and rise times.

Also the dynamic characteristics of oxygen sensors are important. They have an own temperature dependent rise time which is not negligible respect to the other time constants of the system; moreover, in our scheme the sensor is located after the turbine of the turbocharger, so that the transport time increases the delay with which the information on the actual λ-value is available for control processing. Figure 2 presents the dynamic behaviour of the chosen sensor.

Figure 3 compares the outputs of two different sensors as functions of the real λ-value and the engine speed. While for the first one, a sensor with regulated heating, the output of the oxygen sensor is little dependent on the load and on the speed of the engine, *i.e.* indirectly on the exhaust gas cooling, the second one, with a basic, non regulated heating offers a very poor performance and must be discarded.

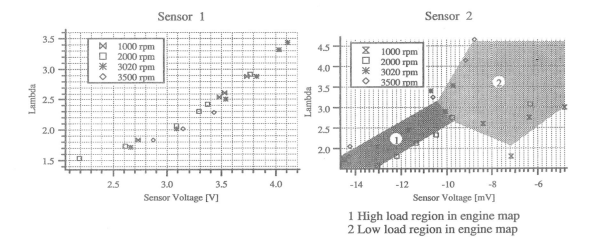

Figure 3 Static characteristics of heated oxygen sensors: sensor 1 with regulated heating, sensor 2 without regulated heating.

Another critical element is represented by the actuator, in our case a underpressure driven valve. The valve is a nonlinear subsystem, and its effect varies strongly with the operating point. Figure 4 shows a photograph of the oxygen sensor and of the valve installation in the test engine.

3 Modelling and Control System Design

Most models of engines are design oriented models, although in recent times more control oriented models have been derived [3]. While exact modelling is quite complex, implementability limitations to the maximum complexity of controls lead to a strong reduction of significance of model details for control purposes [4, 5]. Therefore we prefer to see the engine as a basically simple system, with a few nonlinearities and time variances that should be accounted for by robust control design rather than by more precise models.

The plant can be divided into five parts, the engine, the turbocharger, the two receivers, the recirculation system and last but not least the sensor dynamics. A physical analysis of the plant led to an eight-order nonlinear model described in [6]. Some very strong simplifications, like modelling the power transmission on the turbocharger as a first order low pass, proved quite acceptable for our purposes. Even stronger simplifications were suggested by model/measurement comparisons, so for instance the description of the dynamic of the

Figure 4 Sensor and valve installation in the engine – the upper white dot shows the valve location, the lower one the oxygen sensor

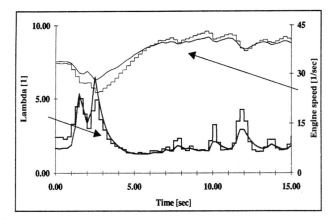

Figure 5   Measurement and simulation of λ and
the engine speed

rotational speed of the engine was modeled by a nonlinear set of equations, but could be approximated amazingly well by a simple linear low-pass model of first order. Figure 5 shows a comparison between simulated and measured behaviour for the rotational speed and the λ .

A linearised model can be obtained by perturbation analysis. It is easy to check that the linear form is not controllable from the recirculation input, as one state, the engine speed, can be controlled only from the fuel input. Actually, this means that, at least in first approximation, the valve opening does not affect the engine speed and acceleration – what proves empirically true only if the recirculation is not too strong.

Furthermore, by considering as outputs both the physical and the measured λ , it can be seen that the fuel injection affects *immediately* the value of the real λ , whereas the measured signal reacts with a certain time lag, due to the sensor dynamics.

The linearised model can be therefore decomposed into a control plant driven by the exhaust gas recirculation and an exosystem driven by the fuel injected. The time constant of the exosystem is one of the largest of the total system, followed by the one of the oxygen sensor. Figure 6 shows the structure of the model.

An important feature of the plant reflected by the model is the low static amplification gain from the control input, especially if compared with the very high control gain from the fuel input, regarded as the disturbance. The amplification is furthermore very depending on the operating point. This, together with the small bandwidth of input signals, requires an integral dynamic extension for low fre-

quency correction [7].

The controllable part of the model obtained by linearisation is of seventh order. By considering the relative pole location, and using the approach proposed by [8], it proves possible to eliminate up to four states, the ones corresponding to the two receivers and the temperatures on the output path from the motor.

So a new model, consisting of an integral block plus a linear model of third order can be used for the LQG/LTR procedure. It turns out that the basic requirement, *i.e.* the minimum phase condition, is fulfilled, although care must be taken during model reduction to avoid shifting zeros into the complex right half plane. As the input quantity is assumed to be constant or slow-varying, first a Luenberger observer is designed acting as a filter to track slow varying errors and reject high frequency noise induced changes. Then a control feedback is obtained using the recovery procedure [2] in order to achieve asymptotically robust behaviour of the chain observer-controller. The introduction of the integral extension proves critical to keep the necessary amplification coefficients acceptable and avoid an undesired bang-bang behaviour.

The control has been realised with a Transputer system programmed in Occam with a sample time of 3.2 ms, while data acquisition has been made on a MacIntosh personal computer.

4   The Results

In order to assess the effects of the controller used, its performance should be compared against the original plant. The figure 7 shows a comparison based on a cycle of the first part of the FTP test. The λ set point for this measure has been set at 2.0. It can be easily seen that the NO_x content is strongly reduced.

The version without exhaust gas recirculation runs with a quite high λ value in the partial load range of the engine outside of the metering range of the oxygen sensor. With closed loop controlled exhaust gas recirculation , the measured λ value is near to the reference value. During a gear change, the amount of injected fuel decreases so that the available exhaust gas recirculation cannot keep λ at the set point. This effect can be observed especially during the acceleration period of the cycle. After the gear change, it may happen that the fuel amount rises quickier than the recirculation rate is decreased, so that λ may fall into a low region where PM produc-

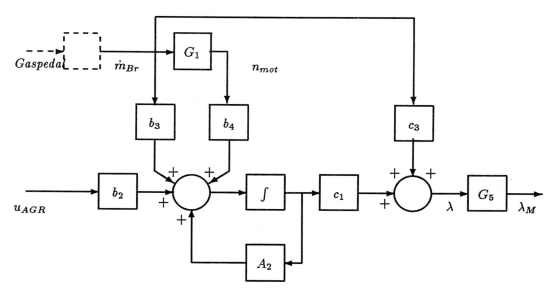

Figure 6   Control scheme of the EGR plant

tion is likely to happen. In certain cases, a great amount of exhaust gas recirculation can also lead to an unsatisfying dynamic behaviour of the turbocharger.

The set point for λ influences both the particulates and the NO_x, and must be chosen taking into account both.

5   Conclusions and Outlook

Some conclusions may be drawn from the experimental results.

1. A closed loop controlled exhaust gas recirculation allows a significant reduction of NO_x emissions of a DI turbocharged Diesel engine. Because of the closed loop control, the drivability of the engine is very good also for high recirculation rates.

2. An oxygen sensor with a regulated heating allows the correct measurements of λ in the whole engine map. To prevent the cooling of the oxygen sensor, it should be protected from the direct exhaust flow. The oxygen sensor dynamics is a dominant time constant in the control system.

3. The plant model can be divided into a path controlled by the exhaust gas recirculation valve and an exosystem fed by the fuel injection, *i.e. in first approximation* the behaviour of the engine is not dependent on the

recirculation. This does not hold if the recirculation rate becomes too high.

4. The LQG/LTR approach gives a very good controller performance in spite of the nonlinearities and time-variances of the plant.

5. Further work is needed to improve the disturbance rejection of the control system. A direct feedforward as well as a dynamic prefilter for the reference input may bring better conditions for a quick acceleration of the turbocharger.

6. The λ set point must be chosen according to the specific characteristics of the engine. The assumption that λ must be kept at a constant value does not hold always, so that engine mapped reference λ-values could lead to further improvements.

REFERENCES

(1)   Alois Amstutz, *Geregelte Abgasrückführung zur Senkung der Stickoxid- und Partikelemissionen beim Dieselmotor mit Comprex-Aufladung*, PhD thesis, Eidgenössische Technische Hochschule Zürich, 1991.

(2)   G. Stein and M. Athans, "The LQG/LTR procedure of multivariable feedback control design", *IEEE Transactions on Automatic Control*, pp. 104–115, February 1987.

(3)   K. P. Dudek and M. K. Sain, "A control-oriented model for cylinder pressure in internal combustion engines", *IEEE Transactions*

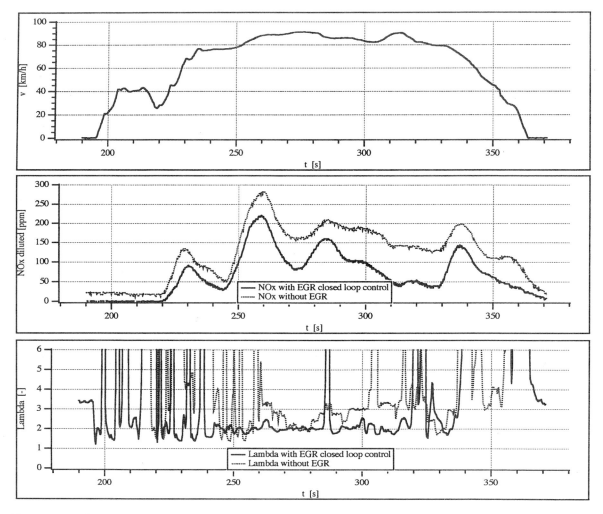

Figure 7   Effect of the controlled exhaust gas recirculation on NO_x and λ in the second cycle of the Federal Test Procedure (FTP)

on *Automatic Control*, vol. 34, pp. 386–397, Apr. 1989.

(4) L. del Re, "Nonlinear modelling and black-box identification of a hydrostatic transmission for control system design", *Mathematical and Computer Modelling*, vol. 14, pp. 219–224, 1990.

(5) L. del Re, "Linear bilinear approximation and feedback linearization of a hydrostatic drive", *in Proceedings of the American Control Conference*, pp. 221–225, 1992.

(6) S. Allegro and R. Baumann, "Regelung der Abgasrückführung bei einem aufgeladenen Dieselmotor zur Senkung der Stickoxidemission, Semesterarbeit", Technical report, Institut für Automatik, 1991.

(7) J. M. Maciejowski, *Multivariable feedback design*, Addison-Wesely Publishing Company, 1989.

(8) B. Moore, "Principal Component Analysis in Linear Systems: Controllability, Observability, and Model Reduction", *IEEE Transactions on Automatic Control*, vol. 26, pp. 17–32, 1981.

The development of the high performance IDI and low emission diesel engine

Y KONISHI
Engine Experiment Department, Nissan Motor Company Limited, Japan

SYNOPSIS   Analytical studies were carried out to investigate how the factors of NOx, smoke formation and ignition can be improved in Indirect-Injection (IDI) diesel engines in order to increase power and reduce exhaust emissions and idle noise.

As a result, it was found that controlling the heat release ratio between the swirl-and main-chambers was effective for improving engine power without increasing NOx and that making the air jet flow counter to the fuel spray injected into swirl-chamber was effective for ignition delay.

In addition, it was experimentally clarified that the improvement of smoke level was achieved by modifing the shape of the injection nozzle throat, and that fuel injection rate had great effects on NOx emissions, noise and engine output power.

1. Introduction

Diesel engines must meet societal requirements for improved air quality and at the same time provide greater product competitiveness by delivering higher power and other attractive performance features.

A great deal of research has already been done in an effort to upgrade the marketability of diesel engines and improve their exhaust emission performance.

However, it has proved to be very difficult to enhance both marketability and exhaust emission performance at the same time, since improvement in one performance factor invariably requires a trade-off-in the other. Thus a major focal point of recent research efforts has been to establish the technology that would allow improvements to be made in both areas concurrently.

2. Development Aims and Major Engine Specifications

The new CD20 diesel engine has been developed to meet the aims outlined below. The major specifications of this IDI engine are given in Table 1.

(1) To reduce exhaust emission levels, especially NOx and particulates, which is strongly required of diesel engines, and thereby contribute to air quality improvements.
(2) To provide high power output and good fuel economy.
(3) To lower noise levels at the sound source and also improve sound quality.

3. Combustion Chamber Improvements

The new engine has been designed with a dual-throat jet swirl chamber in order to facilitate control over the heat release ratio of the swirl chamber and the main chamber. A second objective of this design is to shorten ignition delay through the formation of a high-temperature air jet flow counter to the fuel spray. In conjunction with the shortening of the ignition delay, the geometry of the nozzle throat was improved. These measures were adopted in order to achieve the dual goals of reducing exhaust emissions and improving power output in the new diesel engine with its swirl chamber combustion system.

3.1 Analysis of combustion characteristics effective for reducing exhaust emissions, raising power output and lowering noise levels

3.1.1   NOx suppression methods

The mass model used in the analysis to represent NOx with formation was created reference to Nightingale's mass model [1], in which the fuel spray was divided into elements on the basis of time units. The fuel injection rate was given taking into consideration the factors shown in Table 2. Two assumptions were made in determining the heat release rate. It was assumed that the air-fuel ratio of each fuel spray element would change exponentially. Secondly, it was assumed that the mixture would ignite when the stoichiometric air-fuel ratio was reached. After finding the heat release rate, calculations were made of the NOx concentration and the indicated specific fuel consumption.

Conventional methods for suppressing NOx include (1) retarding the fuel injection timing (i.e., delaying the heat release timing) and (2) lengthening the heat release duration. In addition to these methods, (3) control of the heat release ratios of the swirl chamber and main chamber was investigated in this work as a new concept of combustion control. This approach involves increasing the percentage of excess air, λ s [2] in the swirl chamber for an equivalent fuel injection quantity, a technique that is effective in reducing smoke. A second aspect of this approach is to suppress NOx formation by lowering the gas temperature in the swirl chamber.

Using the mass model, calculations were made of the NOx reduction obtained with each of these methods. An example of the calculated results is given in Fig. 1. While the first and second methods lowered the NOx level, they had the undesirable effect of increasing the indicated fuel consumption rate. In contrast, the third method, involving control of the heat release ratios of the swirl chamber and main chamber, suppressed NOx formation without adversely affecting the indicated specific fuel consumption.

These results suggested that control of the heat release ratios would make it possible to increase the volume ratio of the swirl chamber, since suppression of NOx formation would preclude any rise in the NOx level. Consequently, along with the increase in excess air as a result of burning less fuel in the swirl chamber, it could be expected that the smoke level would be reduced even further.

3.1.2   Reduction of idling noise

It has been reported that shortening the ignition delay is an effective way of reducing idling noise in a diesel engine with a swirl chamber combustion system [3]. One approach to improving mixture ignitability is to diffuse and atomize the fuel spray better by narrowing the throat between the swirl chamber and main chamber so as to strengthen gas flow in the swirl chamber. With this approach, however, there is the possibility for a decline in power output owing to increased throttling losses and heat transfer. In this work, holographic images were obtained with the aid of a ruby laser system to investigate what effect the formation of an air jet flow counter to the fuel spray would have on improving the diffusion and atomization of the fuel spray.

A comparison of the results with and without the air jet flow is given in Fig. 2. When no air jet flow was provided, it is seen that few droplets broke away from the edges of the fuel spray and virtually no turbulence was observed in the mainstream of the spray. On the other hand, when an air jet flow with a velocity of 4 m/s was provided counter to the fuel spray, diffusion of the entire spray was promoted even at this small velocity. (A velocity of 4 m/s is about 1/5 to 1/8 of the swirl velocity in the swirl chamber during idling operation [4]. It was also confirmed that there was stronger turbulence at the boundary between the fuel spray and the ambient air.

These results indicated that fuel spray diffusion and atomization can be enhanced by providing an air jet flow counter to the fuel spray. As a result, it should be possible to shorten the ignition delay and thereby reduce idling noise.

3.2 Construction and aims of dual-throat jet swirl chamber combustion system

The foregoing analyses showed that two methods are effective for increasing the output of a diesel engine with a swirl chamber combustion system while simultaneously reducing its idling noise. One is to control the heat release ratios of the swirl chamber and main chamber. The other is to provide an air jet flow counter to the fuel spray. A dual-throat jet swirl chamber combustion system was devised, as illustrated in Fig. 3, which incorporates these two techniques. The main features of this system are as follows.

(1) An auxiliary throat is provided above the centreline axis of the injection nozzle so that a portion of the fuel injected into the swirl chamber can be discharged directly into the main chamber. This dual-throat design serves to increase the excess air for an equivalent fuel injection quantity, thereby improving power output while reducing smoke.

(2) During the compression stroke, a high-temperature swirling flow is forced through the auxiliary throat to form an air jet flow counter to the fuel spray. This works to promote better diffusion and atomization of the fuel spray and it also raises the ambient temperature of the fuel spray. As a result, the ignition delay is shortened, resulting in reduced idling noise.

3.2.1 Analysis of characteristics of dual-throat jet swirl chamber combustion system

High-speed photography was the main method used to analyze the characteristics of the dual-throat jet swirl chamber combustion system. The characteristics analyzed included the condition of the fuel spray discharged into the main chamber through the auxiliary throat, flame propagation behavior in the main chamber, gas flow, and fuel spay and flame behavior in the swirl chamber. The shape of the swirl chamber and the total throat area (main throat + auxiliary throat) of the dual-throat jet swirl chamber combustion system were the same as those of the swirl chamber combustion system of the previous engine model, which was analyzed for comparative purposes.

(1) Control of heat release ratios of main and swirl chambers

Using a dummy cylinder head, the condition of the fuel spray discharged into the main chamber through the auxiliary throat was recorded on high-speed video tape. Examination of the video images confirmed that a good quality spray was discharged through the auxiliary throat into the main chamber (Fig. 4).

A single-cylinder test engine was then used to photograph flame behavior in the main chamber and the images obtained were analyzed. The experimental apparatus used to photograph flame behavior is shown in Fig. 5. Using a quartz glass piston, photographs were taken from below of the combustion condition in the main chamber of the two combustion systems. An example of the photographed results is given in Fig. 6.

Flame photographs like those shown in Fig.6 were subjected to image analysis to find the duration of the flame eruption from the throat, a time history of the flame area, and the integral of the flame area. Typical results obtained are shown in Fig. 7.

The results indicated that flame erupted twice from the throat into the main chamber in both the dual-throat jet swirl chamber combustion system and the conventional swirl chamber combustion system. The first eruption was caused by the pressure rise in the swirl chamber due to combustion. The second flame eruption was attributed to the drop in pressure in the main chamber as a result of piston descent.

Examination of the change in the flame area showed that combustion in the main chamber of the conventional swirl chamber combustion system was characterized by a two-stage process. While flame diffusion was observed in the main chamber following the first flame eruption, the flame area soon contracted and the flame was extinguished. Subsequently, the second flame eruption began. With the dual-throat jet swirl chamber combustion system, the flame spread vigorously following the first flame eruption and continued for a long duration. Flame was present until the second flame eruption

occurred. Flame diffusion as a result of the second flame eruption was smaller than that observed in the conventional swirl chamber combustion system and the flame was extinguished faster.

Based on these analytical results, it was presumed that unburned fuel vapor was discharged through the auxiliary throat into the main chamber along the flame at the time flame eruption occurred. As a result, the proportion of heat released in the main chamber increased.

(2) Improvement of mixture ignitability

A shadowgraph technique was employed to investigate the effect of the dual-throat jet swirl chamber combustion system on improving mixture ignitability. Figure 8 shows a schematic of the experimental apparatus used to photograph the behavior of the fuel spray and flame in the swirl chamber, which was shaped in the form of a cylinder. Examples of the shadowgraphs obtained are shown in Fig. 9.

With the conventional swirl chamber combustion system, it was seen that the mixture ignited from the bottom of the swirl chamber after the fuel spray had reached the bottom of the chamber. In contrast, with the dual-throat jet swirl chamber combustion system, the high-temperature air jet flow from the auxiliary throat caused the front of the fuel spray to turn toward the centre of the swirl chamber before the spray reached the bottom. Immediately thereafter, the mixture ignited. As a result, it was confirmed that ignition delay was shortened by approximately three crank angle degrees.

The foregoing analysis of the combustion characteristics of the dual-throat jet swirl chamber combustion system thus verified that the system is an effective way to improve engine power output and reduce idling noise. The specific techniques employed to obtain these improvements are control of the heat release ratios of the swirl chamber and main chamber and shortening of ignition delay by providing a high-temperature air jet flow counter to the fuel spray.

3.3 Effect of nozzle 'throat' geometry

The length and width of the injection nozzle 'throat' were varied and the resulting effect on the smoke emission characteristic was investigated. The configuration of the swirl chamber used in this investigation is shown in Fig. 10. It was found that the 'throat' diameter had little effect on the smoke level. On the other hand, the smoke level was reduced as the 'throat' length was shorted, as seen in Fig. 11. However, after the length was reduced to a certain extent, no further change was seen in the smoke level.

An investigation was then made into the reason why the nozzle throat length had such a large effect on the smoke emission characteristic. That was done by using the colur schlieren photographic apparatus in Fig. 8 to analyze combustion behaviour in the swirl chamber. A comparison of the combustion photographs obtained with a conventional nozzle 'throat' and an improved nozzle 'throat' with a shorter length is shown in Fig. 12. An examination of the photographs reveals virtually the same combustion pattern, i.e., the flame spreads upstream through the fuel spray from the onset of ignition. However, with the conventional nozzle 'throat', what appears to be smoke forms in the vicinity of the throat from around a crank angle of 33 ° ATDC following the completion of combustion. This phenomenon is not observed in the photographs for the improved nozzle throat.

Based on these results, the following assumption can be made about the effect of the nozzle throat geometry on the smoke emission characteristic. Following ignition, the flame spreads upstream through the fuel spray. However, because of the influence of the spray velocity, the flame cannot go beyond a certain distance in the direction of the nozzle [5]. As a result, the 'throat' length determines whether the flame penetrates into the nozzle 'throat', a location unsuited to combustion because of the weak swirl there. Above a certain length, the flame enters the throat region, thereby causing the smoke level to increase. Below that length, the flame does not penetrate the throat area and so the smoke emission characteristic is not affected.

4. Fuel Injection System Improvements

Improvements made to the fuel injection system include the adoption of a variable injection pattern system (VIPS), using a two-stage cam, and expansion of the range of control over the injection rate. The latter improvement was achieved by using a flat-cut injection nozzle, in which a flat area is provided in the nozzle tip.

VIPS is characterized by two major features. First, it employs a two-stage cam that is designed to provide both low and high injection rates. Second, it uses a delivery valve in which a flat area is provided for controlling the residual pressure inside the pipe. The aim of this system is to reduce NOx and the distinctive diesel knocking noise in the low load region, such as during engine idling. This is accomplished by lowering the injection rate under low load operation, thereby suppressing any sharp increase in combustion pressure. In the high load region, VIPS provides the same injection rate as conventional fuel injection pumps. The operating principle of VIPS is shown in Fig. 13 in comparison with a conventional system.

As shown in Fig. 14, the flat-cut injection nozzle has a flat area in the nozzle tip, a feature that is not found in the conventional nozzle configuration. This feature works to suppress deterioration in nozzle performance over time by preventing carbon buildup at the nozzle tip. It also makes it possible to reduce the quantity of fuel flow in relation to nozzle lift during throttling. As a result, it enables the low injection rate interval during initial injection to be prolonged. This expands the range of control over the low injection rate during idling and low load operation, when fuel injection tends to be completed within the throttling interval. The flat-cut design has no effect on the injection rate in the high load region after throttling, and the nozzle provides the same injection rate as conventional devices.

Figure 15 compares the cylinder pressure and rate of heat release obtained with a conventional fuel injection system and the improved system featuring VIPS and the flat-cut nozzle. The results indicate that the improved system is more effective in suppressing a sharp rise in cylinder pressure than the conventional one. The NOx and HC emission characteristics of the two fuel injection systems are compared in Fig. 16. The improved fuel injection system reduces NOx emissions by approximately 20% at an equivalent HC level.

Idling noise characteristics are compared in Fig. 17. The improved fuel injection system achieves a large noise reduction in the vicinity of 2 kHz, which is known to have a strong correlation with the diesel knocking sound.

5. Volumetric Efficiency Improvement

The intake port geometry was the focus of efforts to improve volumetric efficiency in connection with the goals of increasing power output and reducing smoke.

5.1 Adoption of aerodynamic (AD) intake ports

The new CD20 engine features the first application in a diesel engine of the AD intake ports that have been used previously in Nissan gasoline powerplants. The AD intake port is designed with a narrower port diameter relative to the throat diameter of the valve seat. This works to increase the flow velocity of the inducted air and it also enhances the effect of induction inertia created by induction pressure pulsations. On the negative side, there is also the possibility that this port design might result in greater induction resistance.

The relationship between the reduction ratio of the port diameter and the rate of increase in volumetric efficiency is shown in Fig. 18 for various engine speeds. In the low to intermediate speed ranges, where the flow velocity of the inducted air is relatively small, narrowing the port diameter further has a large effect on improving volumetric efficiency. However, excessive narrowing of the port diameter in the high speed region where the flow velocity is large tends to have the opposite effect of reducing volumetric efficiency.

5.2 Adoption of slanted intake ports

One reason for the decline in volumetric efficiency in the high speed region when the intake port diameter is narrowed excessively is due to increased induction resistance caused by separation that occurs at port curves. Figure 19 shows the effect of a slanted port angle (θ) on the flow coefficient (Cv). Increasing the slanted angle makes it possible to obtain a favorable Cv value in the region of high valve lift where the working fluid has a large velocity. In view of this result, it was decided to adopt slanted intake ports together with the aerodynamic design.

The use of these techniques made it possible to obtain a high lift characteristic for the intake valves. In addition, as a result of optimizing the cam profile, volumetric efficiency was improved over the entire engine speed range, as indicated in Fig. 20. An improvement of approximately 3% is obtained at the engine speed (4800 rpm) where maximum power is produced.

6. Conclusion

As a result of the improvements made to the combustion chamber, fuel injection system and air induction system, the new IDI diesel engine with a swirl chamber combustion system provides an optimum balance of lower exhaust emissions and higher power output. The main results of this work are summarized below.

(1) Controlling the heat release ratio of the swirl chamber and main chamber has been shown to be an effective technique for boosting power output without increasing the NOx level.

(2) Providing a high-temperature air jet flow counter to the fuel spray has proved to be effective in shortening ignition delay, thereby reducing idling noise.

(3) A dual-throat jet swirl chamber combustion system has been designed which incorporates these two techniques. Tests of this system verified that it improves power output while at the same time reducing idling noise.

(4) The nozzle throat geometry was improved in conjunction with reduced ignition delay achieved with the auxiliary throat. The improved nozzle throat was confirmed to be effective in reducing the smoke level.

(5) A variable injection pattern system (VIPS) has been shown to be effective in reducing NOx and idling noise without sacrificing power performance.

(6) A slanted port angle was incorporated into the aerodynamic intake port design and was shown to be an effective technique for improving power output and reducing smoke.

References

[1] D. R. Nightingale, "A Fundamental Investigation into the Problem of NOx Formation in Diesel Engines," SAE Paper 750848.

[2] M. Tamura et al., "An Analysis of Combustion Characteristics During Supercharging in a Diesel Engine with a Swirl Chamber Combustion System," Journal of JSAE (in Japanese), No. 26, 1983.

[3] N. Miyamoto et al., "Research on Low Compression Diesel Engines," Trans. of JSME (in Japanese), No. 344.

[4] S. Kajiya, "Turbulent Air Flow in the Swirl Chamber of a Diesel Engine," Toyota Technical Review (in Japanese), Vol. 18, No. 3.

[5] N. Yokota, T. Kamimoto, H. Kobayashi and K. Tsujimura, "A Study of Fuel Injection and Flame Behavior in a Diesel Engine Using an Image Processing Technique" (2nd Report), Pre-print of the 8th Joint Symposium on Internal Combusti on Engines (in Japanese).

Table 1   Engine Specification

Model	CD20
Type	4 stroke undersquare
No of cylinders	4 in line, full siamese
Valve train	OHC, Timing belt drive
Cooling	Water cooling
Combustion system	Swirl chamber
Bore × Stroke	84.5 × 88 mm
Displacement	1973 cc
Compression ratio	22.2
Fuel injection pump	VE type, Variable Injection Pattern System
Fuel injection nozzle	Flat-cut type
Maximum output power	56 kW / 4800 rpm
Maximum output torque	132 Nm / 2800 rpm

Table 2 Outline of Mass Model

Heat loss

$$\alpha_M = 0.0033(1+0.15)de^{-0.3}(C_m \cdot P_M)^{0.7} T_M^{-0.2}$$

$$\alpha_S = \frac{(S_M + S_S)\alpha - S_M \cdot \alpha_M}{S_S}$$

$$\alpha = 0.0033(1+0.30)de^{-0.3}(C_m \cdot P)^{0.7} T^{-0.2}$$

Throttling loss

$$P_S \geqq P_M$$
$$dG = \mu \cdot F \cdot \varphi_{SM} \cdot \sqrt{2gP_S/Vs} \cdot dt$$
$$P_S < P_M$$
$$dG = \mu \cdot F \cdot \varphi_{MS} \cdot \sqrt{2gP_M/V_M} \cdot dt$$

Specific heat, ratio of specific heats: related to gas temperature and composition

Change in air-fuel ratio
  of combustion gas : $A(t) = A_0 \cdot \{1 - \exp(-ht)\}$
NOx formation : expanded Zeldovich mechanism

where
α : heat transfer coefficient
P : pressure
V : volume
S : surface area of combustion chamber
μ : flow coefficient
A : air-fuel ratio
t : time
ϕ : relation between ratio of specific heats and pressure that determines gas flow
Cm : mean piston velocity
T : temperature
G : gas mass
F : cross section area of throat
de : cylinder bore diameter
g : acceleration of gravity
h : constant that determines change in air-fuel ratio
subscripts : S = swirl chamber ; M = main chamber

ENGINE SPEED $N_E = 1600$rpm
EQUIVALENCE RATIO $\phi = 0.32$

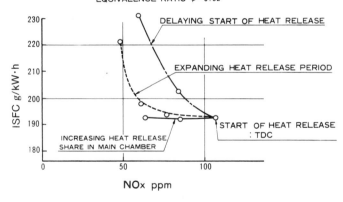

Fig. 1 Calculated Results Obtained With Mass Model

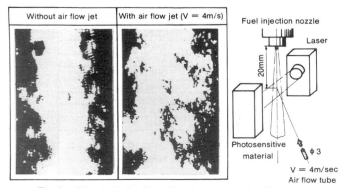

Fig. 2 Effect of Air Flow Jet Counter to the Fuel Spray on Fuel Diffusion and Atomization

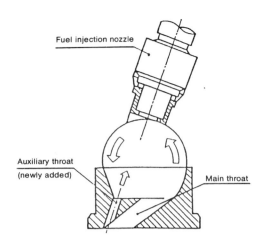

Fig. 3 Configuration of Dual-Throat Jet Swirl Chamber Combustion System

$N_E = 2000$rpm
$Q = 10$mm³/st.cyl.
Ignition timing= 5° BTDC

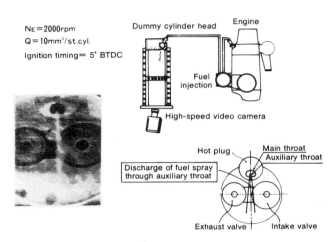

Fig. 4 Condition of Fuel Spray Discharged into Main Chamber Through Auxiliary Throat

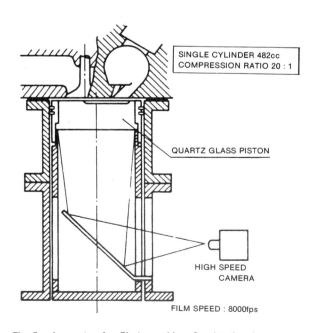

SINGLE CYLINDER 482cc
COMPRESSION RATIO 20 : 1

FILM SPEED : 8000fps

Fig. 5 Apparatus for Photographing Combustion in Main Chamber

92

Single-cylinder test engine
Displacement : 482cc
Compression ratio : 20 : 1
Equipped with dual-throat jet swirl combustion chamber
Nε = 1000rpm   Q = 15mm³/st · cyl
Ignition timing : 3° BTDC
Crank angle : 20° ATDC

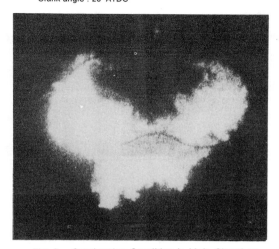

Fig. 6   Combustion Condition in Main Chamber

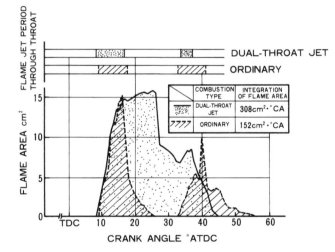

Fig. 7   Image Analysis Results for Flame Behavior

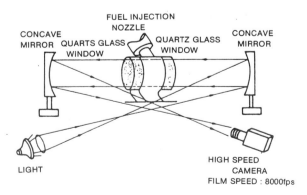

Fig. 8   Schematic of Experimental Apparatus

Single-cylinder test engine
Displacement : 482cc
Compression ratio : 20 : 1

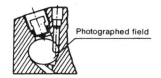

Photographed field

NE = 1200 rpm , Q = 10 mm³/st

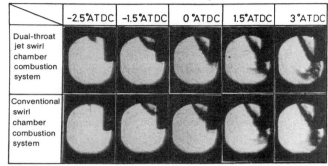

	-2.5°ATDC	-1.5°ATDC	0°ATDC	1.5°ATDC	3°ATDC
Dual-throat jet swirl chamber combustion system					
Conventional swirl chamber combustion system					

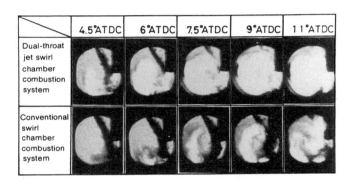

	4.5°ATDC	6°ATDC	7.5°ATDC	9°ATDC	11°ATDC
Dual-throat jet swirl chamber combustion system					
Conventional swirl chamber combustion system					

Fig. 9   Shadowgraphs Showing Fuel Spray and Flame Behavior

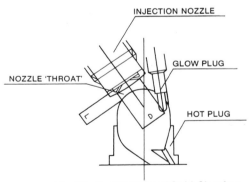

Fig. 10   Configuration of Swirl Chamber

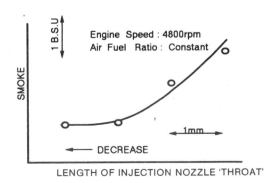

Fig. 11   Effect of Injection Nozzle Throat Length on Smoke Characterisitics

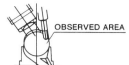

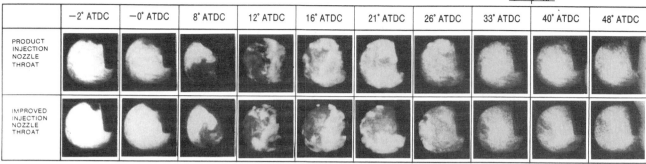

Test Conditions
Engine Speed : 1200rpm
Injection Quantity : 15mm³/st.
Injection Timing : 2° BTDC.

OBSERVED AREA

	−2° ATDC	−0° ATDC	8° ATDC	12° ATDC	16° ATDC	21° ATDC	26° ATDC	33° ATDC	40° ATDC	48° ATDC
PRODUCT INJECTION NOZZLE THROAT										
IMPROVED INJECTION NOZZLE THROAT										

Fig. 12   Effect of Injection Nozzle Throat on Combustion Phenomena

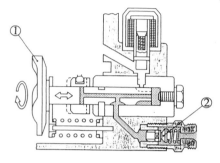

① Adoption of two-stage cam
② Delivery valve with flat-cut design

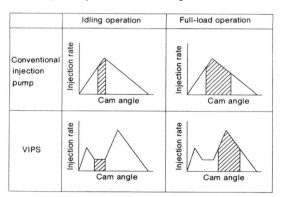

	Idling operation	Full-load operation
Conventional injection pump	Injection rate / Cam angle	Injection rate / Cam angle
VIPS	Injection rate / Cam angle	Injection rate / Cam angle

Fig. 13   Operating Principle of VIPS

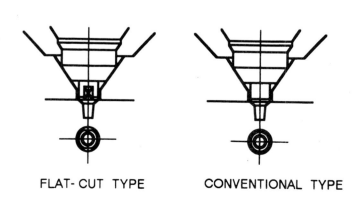

FLAT-CUT TYPE   CONVENTIONAL TYPE

Fig. 14   Configurations of Injection Nozzle

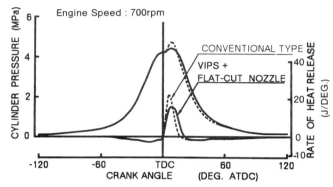

Engine Speed : 700rpm

CONVENTIONAL TYPE

VIPS + FLAT-CUT NOZZLE

Fig. 15   Combustion Characteristics

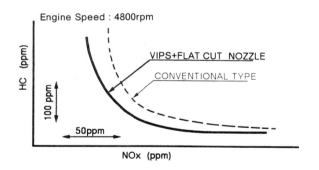

Engine Speed : 4800rpm

VIPS+FLAT CUT NOZZLE

CONVENTIONAL TYPE

100 ppm

50ppm

Fig. 16   Effect of Improved Injection System on Emission Characteristics

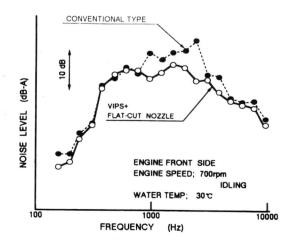

CONVENTIONAL TYPE

10 dB

VIPS+ FLAT-CUT NOZZLE

ENGINE FRONT SIDE
ENGINE SPEED; 700rpm
IDLING
WATER TEMP; 30℃

Fig. 17   Idling Noise Characteristics

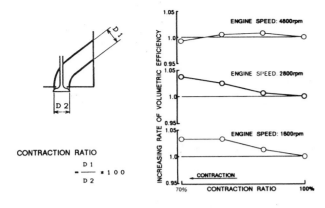

Fig. 18 Effects of Contraction Ratio on Volumetric Efficiency

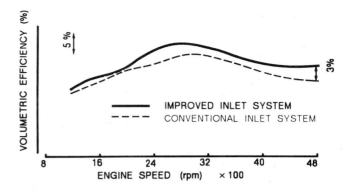

Fig. 20 Effect of Improved Inlet System

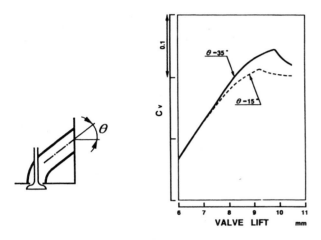

Fig. 19 Effect of Intake Port Angle on Cv

Measurement and prediction of transient NO$_x$ emissions in DI diesel engines

C ARCOUMANIS, PhD, DIC, CEng, MIMechE, MASME, MSAE and C-S JOU, BSc, MSc
Department of Mechanical Engineering, Imperial College of Science, Technology and Medicine, London

SYNOPSIS An experimental and theoretical investigation was undertaken in a turbocharged vehicle direct-injection diesel engine to evaluate oxides of nitrogen (NO$_x$) emissions during transient operation. The experimental approach makes use of a reconstruction technique based on a phenomenological model which allows recovering of the distorted NO signal from the output of a conventional slow-response chemiluminescent analyzer during transient conditions. The theoretical model predicts NO concentration based on the kinetic NO mechanism associated with the fuel burning rate pattern derived from measured cylinder pressure diagrams. Both methods were evaluated against systematically obtained experimental data; the reconstruction model was validated by an engine-simulation experiment while the NO prediction model was validated against steady-speed engine data obtained at various speeds, loads and injection timings.

1 INTRODUCTION

Transient operation is the dominant mode of vehicle diesel engines in urban transportation systems. Since the major increase in diesel engine ratings has come about by turbocharging which is associated with poor transient response, attempts have been made to improve the transient performance of turbocharged diesel engines in terms of their response, fuel consumption and smoke levels.

Particulate and NO$_x$ emissions have been over the last few years the major concern of diesel engineers in their efforts to meet the increasingly more stringent emission standards [1]. In view of the well known trade-off between NO$_x$ and smoke, it is becoming important to monitor NO$_x$ emissions not only under steady-speed but also during transient operation. Unfortunately, although fast-response analyzers have recently been employed to measure transient hydrocarbon and CO/CO$_2$ emissions in the exhaust manifold [2-4], similar instrumentation is not available to evaluate transient NO$_x$ emissions due to the fact that the system that transports the sample gases to the analyzer for measuring their transient concentration is inherently complex and, thus, slow. Fast sampling valves and spectroscopic techniques have been used to measure transient NO$_x$ emissions [5,6] but they can provide only local concentrations and are far from straightforward to quantify the volume-averaged NO$_x$ emissions in the engine exhaust.

Conventional chemiluminescent analyzers are suitable to measure steady-state NO$_x$ concentrations, but are of limited use in transient engine testing since the dynamics of the analyzer cause distortion to the exhaust signal. Recently signal reconstruction methods have offered some promise to provide information of transient emissions [7,8] and this approach will be explored further in this study. A phenomenological model was developed based on the mode of operation of conventional chemiluminescent analyzers and used to reconstruct the distorted signals at the output of the analyzer during transient experiments. The proposed model was validated first against a series of bench experiments and then compared with transient NO concentrations acquired under real engine operating conditions. In the latter case, NO emissions were calculated based on the Zeldovich kinetic mechanism and the fuel burning rate pattern, derived from cylinder pressure diagrams measured during transient engine operation. This indirect method of predicting in-cylinder NO concentration and comparing it with the measured signal after its reconstruction by the phenomenological model was validated under steady-speed conditions at various engine speeds, loads and injection timings.

2 NO$_x$ EMISSIONS IN DIESEL ENGINES

Oxides of nitrogen (NO$_x$) are formed during the combustion process mainly as a result of chemical reactions between atmospheric oxygen and nitrogen, and consist of nitric oxide (NO) and nitrogen dioxide (NO$_2$); NO$_2$ concentration in the exhaust is generally small compared to NO concentration. Although the NO$_2$/NO$_x$ ratio may reach 30% under very light load conditions, this ratio reduces to less than 5% at medium loads [9]. Relatively large concentrations of NO$_2$ can be formed in the combustion zone, but they are subsequently converted back into NO in the post-flame region. NO$_2$ is thus generally considered to be a transient intermediate species existing only at flame conditions [10] unless the NO$_2$ formed is suddenly quenched by mixing with cooler fluid. This may be the reason why higher NO$_2$/NO ratios occur at light load operating conditions. Therefore, NO$_2$ emissions may be neglected while the equivalence ratio is higher than 0.6. During engine acceleration, which is the period of most concern to engine developers for transient engine operation, the equivalence ratio easily exceeds 0.6, especially in turbocharged engines, since the fuel pump delivers maximum fuelling while the governor setting is far away from the position corresponding to the actual engine speed and NO_2 levels are expected to be very small. Therefore, only transient NO emissions will be considered in this study.

Nitric oxide can be formed in three ways : (i) at the high temperatures present in the flame front and burned gas region, atmospheric N_2 reacts with oxygen to form thermal NO; (ii) when the fuel has nitrogen-containing compounds, this nitrogen is released at comparatively low temperatures to form fuel NO; (iii) additional NO is formed in the thin flame front region other than that formed from atmospheric N_2 and O_2 and is referred to as prompt NO. Although diesel fuel contains more nitrogen than gasoline, current levels are not significant. Prompt NO is considered to be formed via CN-group containing intermediates in turbulent diffusion flames where maximum temperature levels may be as low as $1300°C$ [10]. From the above three mechanisms, thermal NO is widely accepted to be the dominant source of oxides of nitrogen under diesel engine conditions.

The principal reactions governing the formation of thermal NO from molecular nitrogen and oxygen during combustion of lean or near stoichiometric fuel-air mixtures are given by the extended Zeldovich mechanism:

$$O + N_2 \Leftrightarrow NO + N \qquad (1)$$
$$N + O_2 \Leftrightarrow NO + O \qquad (2)$$
$$N + OH \Leftrightarrow NO + H \qquad (3)$$

The third reaction which was added by Lavoie et al. [11] becomes important only in near stoichiometric and rich flames that remain at high temperature long enough to produce significant amounts of nitric oxide.

Vioculescu and Borman [12] have shown that the NO decomposition during the expansion stroke is very slow due to the reducing temperatures which confirms that the thermal NO formation rate is much slower than the combustion rate and, thus, most of the NO is formed after completion of combustion. Since the interest lies generally in quantifying the emission levels in the engine exhaust, predictions can be made by approximating the kinetics of nitric oxide formation and assuming that the time required to achieve equilibrium concentrations of nitric oxide is greater than the time required to attain equilibrium for the other compounds in the combustion products. Details of the formulation will be described in section 4.

3 RECONSTRUCTION MODEL

The reconstruction technique is a method which acquires distorted emission signals from slow-response analyzers during transient conditions and then recovers the true signal by analytical methods. The reconstruction model developed by McClure [7] is based on about 100 pulse responses which were analysed to characterise the measurement system empirically. Another reconstruction model, which was developed by Beaumont et al [8], is based on generalized predictive control theory to develop a reconstruction filter. These transfer function models, however, seem to have limited physical meaning as methods for enhancing the accuracy of the reconstructed signals through rearrangement of the configuration of the measurement system.

A phenomenological model has been developed which not only can be applied to correct transient NO_x signals but can also be used as a guide to rearrange the instrumentation in order to reduce the distortion of NO_x signals measured by conventional chemiluminescent analyzers under transient conditions. Details of the proposed model will be presented in the next sub-section followed by a description of the validation tests.

3.1 Mathematical formulation

The principle of chemiluminescence is based upon the gas phase reaction between O_3 and NO to give NO_2 and O_2. About 10% of the NO_2 produced is in an electronically-excited state and its transition to the ground state gives rise to light emission whose intensity is measured by a photomultiplier tube and is proportional to the mass flow rate of NO_2 and, thus, NO in the reaction chamber [13].

Since both the chemical reaction and photomultiplier response are very fast, the transport time seems to be the major cause of the instrument's delay in transient testing. The main aspects of the distortion in the mixture concentration pattern within the sampling system were examined by considering sample gases passing through a series of surge volumes and pipes (Fig.1a). The assumptions that were made to set up this model are : (i) The mixture passes through pipes at constant speed and the concentration gradient is in the axial direction only. The distortion of the concentration pattern during this pipe flow is modelled as an unsteady, one-dimensional mass diffusion process with the moving coordinate being at the same speed as the flow velocity. (ii) The sample gases enter into the surge volumes being well mixed and the changes of the concentration pattern during its passage through the surge volumes are modelled by the mass conservation equation. The various governing equations are as follows:

(a) Diffusion model (pipe flow)

$$\frac{\partial C(x,t)}{\partial t} = D \frac{\partial^2 C(x,t)}{\partial x^2} \qquad (4)$$

where D is the diffusivity.
x is axial direction.

(b) Mixing model (surge volume)

$$A u C_i = A u C_o + V \frac{d C_o}{dt} \qquad (5)$$

where A is the area of inlet and exit port.
u is the flow velocity.
V is the surge volume.
subscripts i and o represent inlet and outlet, respectively.

The solutions of equations (4) and (5) are given by

$$C_o(t) = \frac{1}{\sqrt{\pi}} \int_{-\infty}^{\infty} C_i(t + b\ AC)\ \exp(-b^2)\ db \qquad (6)$$

$$C_o(t) = \frac{1}{B}\exp(-t/B) \int_{0}^{t} C_i(t)\ \exp(t/B)\ dt \qquad (7)$$

where $B = \dfrac{V}{A u}$, $AC = \sqrt{4 D t_o}$
t_o = delay time.

The whole flow system including pipes and surge volumes can be represented by a system of equations having the same form to that of equations (6) and (7).

The reconstruction model, shown in Fig.1a, consists of a flow network of pipes and surge volumes which is responsible for the distortion of the concentration signal within the NO_x measurement system. B_1-B_3 represent characteristic coefficients of each surge volume and AC_1-AC_3 are the characteristic coefficients of the pipes.

Before using this model to reconstruct measured signals, the above coefficients have first to be determined by correlating measured data at given input conditions. For further validation, three testing procedures were examined to acquire data for correlating the above coefficients:

(i) The <u>inner gas case</u> corresponds to several given concentration patterns added as inputs into the inlet of pipe 3, with the NO_x analyzer output data being used to correlate coefficients AC_3 and B_3 through the governing equations.

(ii) The <u>no prefilter case</u> corresponds to several given concentration patterns added as inputs into the inlet of the heating pipe. The measured data are used to correlate coefficients AC_1, AC_2 and B_2 through the governing equations while B_3 and AC_3 are already known from the previous case.

(iii) The <u>prefilter case</u> corresponds to several given concentration patterns added as inputs into the inlet of the prefilter, surge volume 1; the measured data are then used to correlate the remaining unknown coefficient B_1.

3.2 Experimental system

Figure 1b shows the schematic diagram of the SIGNAL 4000 NO_x analyzer. The three test cases described previously were used to characterise this system with step input concentration patterns generated by a three-way solenoid valve. For the inner gas case described above, the internal control solenoid valve was employed while an additional solenoid valve was used for the other two cases. The data logging system was an eight-channel, 10bit fast analog-to-digital converter system with the sampling rate fixed at 100Hz. The measured output and control signals were recorded and used to determine the characteristic coefficients.

To further verify the mathematical model, a series of prefilter simulators made from acrylic tubes of the same diameter to that of the actual filter but of different lengths, were used to simulate different prefilter volumes. The use of a prefilter upstream of the chemiluminescent analyzer was considered necessary to prevent blocking of the tubes by the particulates present in the exhaust of diesel engines.

3.3 Model validation

From the two physical mechanisms responsible for signal distortion under transient conditions, the diffusion effect has proven to be small by both theoretical calculations and experimental validation and, in addition, its governing equation is an initial-value problem which makes the reverse solution time-consuming and the signal noisy. Therefore, only the mixing effect has been taken into account in the reconstruction procedure and the characteristic coefficients were determined by correlating experimental data using equation (5). Typical examples of distorted and measured signals from the inner gas case, without and with a prefilter simulator in the engine-simulation experiment are shown in Figs. 2(a)-(c); the results in all cases are considered satisfactory especially those which correspond to different sizes of simulated surge volumes. Figure 2(d) shows an example of a reconstructed signal based on the measured one for the 20 cm prefilter simulator case. A reasonably good agreement was obtained between the reconstructed and the input signals which seems encouraging for extending this approach to measurements in the diesel engine. Overall, the reconstruction model offers potential in both directions. For calibration purposes where the NO signal entering the analyzer is known, the model can predict its distortion through the NO_x measurement system. Alternatively, if the true NO concentration is not known as is the case under transient engine conditions, the distorted signal measured at the output of the analyzer can be used as input to the model which can then reconstruct quite faithfully the true NO exhaust signal.

An estimate of the reconstruction error can be obtained by the standard deviation of the ratio of the measured and input total concentrations (obtained by integration). Based on fifteen single-step input tests for different configurations, the uncertainty in the mean and r.m.s. NO concentration was found to be about 0.5 and 2 percent, respectively. It has to be noted, however, that the total concentration obtained by integrating the measured signal is valid only when both measured and input signals decay to zero, since otherwise part of the signal may not be accounted for.

4 NO EMISSIONS PREDICTION MODEL

The Zeldovich mechanism of NO formation from atmospheric nitrogen is generally considered to be applicable to combustion systems with near stoichiometric fuel-air mixtures. Most researchers have incorporated this mechanism into their combustion model to predict NO emissions. e.g. [14-24]. Some adopted the two-equation Zeldovich mechanism, others its extended three-equation version to calculate NO formation rate.

For predicting NO emission levels during transient operating conditions, the approach proposed here is to use the fuel burning rate pattern deduced from the measured cylinder pressure diagram in conjunction with the extended Zeldovich mechanism. The equilibrium concentrations of species required as inputs to the kinetic model were calculated using correlation functions derived from curve fitting of the calculated data [25]. This approach seems appropriate for predicting transient NO emissions in diesel engines since it is based on the instantaneous cylinder pressure which can be recorded accurately during transient engine operation. Equally, it can be easily incorporated into transient engine simulation programs, e.g. [26], which predict quite accurately fuel burning rate patterns.

4.1 Fuel burning rate sub-model

The model employed in the present analysis has been derived by Krieger and Borman [27]. The cylinder contents are considered as a single open system and the only mass flow across the system boundary while the intake and exhaust valves are closed is the injected fuel. The first law for such a system gives

$$\frac{dQ}{dt} - P\frac{dV}{dt} + \frac{dm_f}{dt}h_f = \frac{dmu}{dt} \quad (8)$$

where dQ/dt is the heat-transfer rate across the system boundary, $p(dV/dt)$ is the rate of work transfer by the system due to system boundary displacement, and dm_f/dt is the fuel mass flow rate into the system across the system boundary. Assuming that the properties of the gases in the cylinder are uniform and in chemical equilibrium at pressure P and average temperature T during combustion, then in general

$$u = u(T,P,\phi) \quad \text{and} \quad R = R(T,P,\phi)$$

The fuel burning rate can be derived from equation (8) as follows:

$$\frac{1}{m}\frac{dm}{dt} =$$

$$\frac{-(RT/V)(dV/dt) - (\partial u/\partial P)(dP/dt) + (1/m)(dQ/dt) - CB}{u - h_f + D(\partial u/\partial\phi) - C[\,1 + (D/R)(\partial R/\partial\phi)]}$$

$$(9)$$

where $B = \dfrac{1}{P}\dfrac{dP}{dt} - \dfrac{1}{R}\dfrac{\partial R dp}{\partial P dt} + \dfrac{1}{V}\dfrac{dV}{dt}$

$$C = \frac{T(\partial u/\partial T)}{1 + (T/R)(\partial R/\partial T)}$$

$$D = \frac{(1+FA_0)m}{FA_s m_0}$$

ϕ is equivalence ratio and FA is the fuel/air ratio; the subscript 0 denotes the initial value prior to fuel injection and subscript s denotes the stoichiometric value.

Equation (9) can be solved numerically for $m(t)$ given m_0, ϕ_0, P(t) and appropriate models for the working fluid properties [28] and for the heat transfer term dQ/dt which can be calculated using Woschni's heat transfer coefficient [29]. The cylinder pressure diagrams being used to calculate the fuel burning rate generally employ the curve fitting technique to reduce the error of dP/dt in equation (9). In this study, a least-square cubic-spline approximation was used to fit the cylinder pressure data for calculating dP/dt.

4.2 NO kinetic sub-model

For combustion of near stoichiometric fuel-air mixtures, the principal reactions governing the formation of NO from atmospheric nitrogen are given by the extended Zeldovich mechanism. The forward and reverse rate constants for these reactions are given in Table 1 [30].

The rate of formation of NO is given by

$$\frac{d[NO]}{dt} = k_1^+[O][N_2] + k_2^+[N][O_2]$$
$$+ k_3^+[N][OH] - k_1^-[NO][N]$$
$$- k_2^-[NO][O] - k_3^-[N][OH] \quad (10)$$

where [ ] denotes species concentration in moles per cubic centimetre. Based on the assumption that $\dfrac{d[N]}{dt} = 0$, the kinetic model of NO formation becomes

$$\frac{d[NO]}{dt} = \frac{2R_1\{1 - (\frac{[NO]}{[NO]_e})^2\}}{1 + (\frac{[NO]}{[NO]_e})\frac{R_1}{R_2+R_3}} \quad (11)$$

where $R_1 = k_1^+[O]_e[N_2]_e = k_1^-[NO]_e[N]_e$
$R_2 = k_2^+[N]_e[O_2]_e = k_2^-[NO]_e[O]_e$
$R_3 = k_3^+[N]_e[OH]_e = k_3^-[NO]_e[H]_e$

Reaction	Rate constant$(cm^3/mol\ s)$	Temperature(K)
$O+N_2 \Rightarrow NO+N$	$k_1^+ = 7.6\times10^{13}\exp(-38000/T)$	2000-5000
$N+NO \Rightarrow N_2+O$	$k_1^- = 1.6\times10^{13}$	300-5000
$N+O_2 \Rightarrow NO+O$	$k_2^+ = 6.4\times10^9 T\exp(-3150/T)$	300-3000
$O+NO \Rightarrow O_2+N$	$k_2^- = 1.5\times10^9 T\exp(-19500/T)$	1000-3000
$N+OH \Rightarrow NO+H$	$k_3^+ = 4.1\times10^{13}$	300-2500
$H+NO \Rightarrow OH+N$	$k_3^- = 2.0\times10^{14}\exp(-23650/T)$	2200-4500

Table 1 Rate constants for NO formation mechanism

4.3 Chemical equilibrium calculations

There are two alternative, but equivalent, chemical equilibrium formulations for calculating complex chemical reactions : the first is the equilibrium constant method and the second is the minimisation of Gibbs free energy method [31]. The chemical equilibrium constant method, which is based on atom conservation, mole constraint and equilibrium constant relations to calculate the composition of products at a given temperature and pressure, has been used in this study. The following equation represents the reaction between diesel fuel $C_{14.4}H_{24.9}$ and air at an equivalence ratio ϕ [25].

$$\epsilon\phi C_{14.4}H_{24.9} + 0.21O_2 + 0.79N_2 \Rightarrow$$
$$v_1 CO_2 + v_2 H_2O + v_3 N_2 + v_4 O_2 + v_5 CO$$
$$+ v_6 H_2 + v_7 H + v_8 O + v_9 OH + v_{10}NO \quad (12)$$

where ϵ is the molar fuel-air ratio. Following atom balancing and mole conservation, five equations are obtained which have eleven unknowns, ten species' mole fractions and the total number of moles. Therefore, six more equations are needed which are provided by the criteria of equilibrium among the products and are expressed by the following six hypothetical reactions.

$$\frac{1}{2}H_2 \Leftrightarrow H \quad K_1 = \frac{y_7 P^{1/2}}{y_6^{1/2}} \quad (13)$$

$$\frac{1}{2}O_2 \Leftrightarrow O \quad K_2 = \frac{y_8 P^{1/2}}{y_4^{1/2}} \quad (14)$$

$$\frac{1}{2}H_2 + \frac{1}{2}O_2 \Leftrightarrow OH \quad K_3 = \frac{y_9}{y_4^{1/2}y_6^{1/2}} \quad (15)$$

$$\frac{1}{2}H_2 + \frac{1}{2}N_2 \Leftrightarrow NO \quad K_4 = \frac{y_{10}}{y_4^{1/2}y_3^{1/2}} \quad (16)$$

$$H_2 + \frac{1}{2}O_2 \Leftrightarrow H_2O \quad K_5 = \frac{y_2}{y_4^{1/2}y_6 P^{1/2}} \quad (17)$$

$$CO + \frac{1}{2}O_2 \Leftrightarrow CO_2 \quad K_6 = \frac{y_1}{y_5 y_4^{1/2}P^{1/2}} \quad (18)$$

where y represents the mole fractions of each species. The equilibrium constants K_1 to K_6, as defined above, are considered to be functions of temperature only. Olikara and Borman [32] have curve fitted these constants to the data of the JANAF Table in the following form:

$$\log K = A \ln (T/1000) + B/T + C + DT + ET^2$$

(19)

where T is in degrees K. For the range $600 < T < 4000$ K their results are summarised in Table 2 [25].

	A	B	C	D	E
K_1	0.432168E 0	.112464E+5	0.267629E 1	-0.745744E-4	0.242484E-8
K_2	0.310805E 0	.129540E+5	0.232179E 1	-0.738336E-4	0.344645E-8
K_3	-0.141784E 0	.213308E+4	0.853461E 0	-0.355015E-4	-0.310227E-8
K_4	0.150879E-1	.470959E+4	0.646096E 0	-0.272805E 5	-0.154444E-8
K_5	-0.752364E 0	.124210E+5	-0.260286E 1	-0.259556E-3	-0.162287E-7
K_6	-0.415302E-2	.148627E+5	-0.475746E 1	0.124699E-3	-0.900227E-8

Table 2 Correlation coefficients of equilibrium constants

Because these eleven equations are nonlinear, Olikara and Borman [32] have suggested a method to solve them which allows after several algebraic arrangements to reduce these equations into four unknowns y_3, y_4, y_5 and y_6. The resulting four equations, however, are still nonlinear and can best be solved by Newton-Raphson iteration. A first approximate solution can then be obtained by using a precursory calculation, which uses the same concept but deals with only six species by excluding H, O, OH and NO at low temperatures in order to get the exact solutions. This offers good initial values for solving the nonlinear equations by the Newton-Raphson iteration. Although equilibrium concentrations of species can be calculated following this method, it is very time-consuming when applied to in-cylinder combustion products. The species concentrations can be fit by fifth order polynomial functions of temperature and pressure at a fixed equivalence ratio which can make in-cylinder computations much faster.

4.4 In-cylinder NO prediction model

The local temperature is the dominant factor for NO formation in diesel combustion. The in-cylinder NO prediction model follows the approach of [17] and uses the fuel burning rate sub-model to generate burnt elements which, assuming stoichiometric combustion and adiabatic expansion, provide an estimate of the local temperature; the NO kinetic sub-model based on this local temperature is then used to calculate the NO concentration. The model assumes that (i) NO is formed only in the burnt products region after the flame element has burned during the heat release period of the engine cycle, (ii) mixing or heat transfer between individual burnt product elements and the remaining fresh air does not take place during the computation period and (iii) the burnt products and air are fully mixed at the end of the calculation which allows determination of the mean exhaust NO concentration. The charge air temperature is assumed to follow the measured cylinder pressure adiabatically based on the pressure and temperature set at the start of combustion and derived from their values at inlet valve closure. This air temperature is used as the basic energy level from which the flame temperature is subsequently calculated by carrying out an energy balance involving the heat of formation of the fuel and ten combustion products whose equilibrium concentrations in the flame and the burnt

elements are calculated by either the chemical equilibrium constant method or their correlation functions. Each burnt product element has an independent temperature/time history from the time of its formation from a flame element, and this temperature is updated adiabatically based on the measured cylinder pressure data.

Comparison of measured and predicted exhaust NO emission levels under steady-speed conditions is shown in Fig. 3. Increasing the engine speed at constant load, produces a reduction in the exhaust NO concentration (Fig. 3a) while increasing the engine load at constant speed produces an increase (Fig. 3b). As expected, retarding injection timing reduces NO emission levels significantly (Fig. 3c).

5 TRANSIENT TESTS

Typical transient operating conditions in diesel engines include engine acceleration at constant load and load acceptance at constant engine speed. These tests were carried out in the transient testbed described in the Appendix. The period of demand acceleration and load acceptance are set to be 0.5 s for 1200 rev/min - 2000 rev/min and 0 - 400 NM, respectively. Other typical transient conditions, such as engine deceleration at constant load and load release at constant speed, may not increase emissions, since the conventional diesel engine governor which is a speed sensitive device that automatically controls or limits the speed of the engine by adjusting the amount of fuel supplied, may cut off fuelling while the set speed is far less than the actual engine speed. This means that when the engine is subjected to deceleration, emissions are not a problem since no fuel is injected into the chamber and, thus, there is no combustion. The condition of load release is similar to deceleration except that in the early period engine speed may increase while engine load is far less than the engine generated torque; consequently, the engine is subjected to deceleration.

The reconstructed signals, based on the reconstruction technique described in section 3, and the predicted signals based on the measured cylinder pressure diagrams are compared in Figs. 4 to 6 for different test cases. Figures 4 and 5 represent the NO emissions during acceleration without load and with 200 and 400 NM loads, respectively. These results show that the predicted and reconstructed signals are similar and have the same order of magnitude peak values during acceleration at different loads. Although the duration of the NO emission signal in the zero load case is shorter than that at 200 or 400 NM loads, the NO emissions seem to be a more serious problem at zero load during acceleration since the difference between the peak and steady-state value is larger. Figure 6 shows the predicted and reconstructed NO emissions during load acceptance conditions which confirms that no significant peak occurs during load change.

Under steady-speed conditions, the measured and predicted NO emission levels decrease while engine speed increases at fixed engine load due to the lower overall equivalence ratio caused by higher turbocharger performance and the relatively smaller fraction of diffusion-controlled combustion at higher engine speeds.

On the contrary, during acceleration the NO emission levels increase, since the fuel pump delivers more fuel into the engine to make it accelerate which, in turn, gives rise to higher overall equivalence ratio.

6. CONCLUSIONS

The poor transient performance of turbocharged DI Diesel engines can be improved by different control strategies which aim at improving transient response and reducing, at the same time, smoke levels. The NO_x emission levels, however, should also be considered during engine optimisation in view of the well known trade-off between NO_x and particulates under steady-speed conditions and the need to satisfy the corresponding emission standards. During engine transient operation, NO emissions can be estimated either by reconstructing measured signals which are distorted by slow response analyzers or calculating them based on recorded cylinder pressure diagrams. The accuracy of the reconstructed signals can be further improved by rearranging the sample gases transportation system in order to reduce its length and the prefilter volume.

Based on the results obtained with the reconstruction and prediction methods, an accelerating engine produces not only high smoke levels but also high NO emissions especially under light load conditions; the latter can be reduced by retarding injection timing, albeit with a deterioration in combustion performance.

ACKNOWLEDGEMENTS

The authors would like to thank Lloyd's Register of shipping, the Ford Motor Company Ltd. and the Department of Transport for partial support to this project. Mr. C-S Jou was supported by a Fellowship from Chung Cheng Institute of Technology, Taiwan, R.O.C.

REFERENCES

1. Wachter, W.F., Analysis of Transient Emission Data of a Modelyear 1991 Heavy Duty Diesel Engine, SAE Paper 900443, 1990.
2. Collings, N. and Eade, D., An Improved Technique for Measuring Cyclic Variations in the Hydrocarbon Concentration in an Engine Exhaust, SAE Paper 880316, 1988.
3. Miyatake, K., Ishida, K., Kohsaka, H. and Harvey, N., Fast Response NDIR for Real-Time Exhaust Measurement, SAE Paper 900501, 1990.
4. Horiba, Fast Response Motor Exhaust Gas Analyzer, MEXA-1300FRI, Bulletin HER-2223A, 1991.
5. Aoyagi, Y., Kamimoto, T., Matsui, Y. and Matsuoka, S., A Gas Sampling Study on the Formation Processes of Soot and NO in a DI Diesel Engine. SAE Paper 800254, 1980.
6. Lavoie, G.A., Spectroscopic Measurements of Nitric Oxide in Spark Ignition Engines, Combustion and Flame, vol. 15, 1970.
7. B.T. McClure, Characterisation of the Transient Response of a Diesel Exhaust-Gas Measurement System, SAE Paper 881320, 1988.
8. A.J. Beaumont, A.D. Noble and A.D. Pilley, Signal Reconstruction Techniques for Improved Measurement of Transient Emissions, SAE Paper 900233, 1990.
9. Hilliard, J.C. and Wheeler, R.W., Nitrogen Dioxide in Diesel Exhaust, SAE Paper 790691, 1979.
10. Chigier, N.A., Pollution Formation and Destruction in Flame - Introduction, Prog. Energy Combustion Science, Vol.1, 1975.
11. Lavoie, G.A., Heywood, J.B. and Keck, J.C., Experimental and Theoretical Investigation of Nitric Oxide Formation in Internal Combustion Engines, Combustion Science Technology, Vol 1, 1970.
12. Vioculescu, I.A. and Borman, G.L., An Experimental Study of Diesel Engine Cylinder Average NO_x Histories, SAE Paper 780228, 1978.
13. NO_x Analyser operation manual, The Signal Instrument Co. Ltd., 1985.
14. Khan, I.M., Greeves, G. and Probert, D.M. Prediction of Soot and Nitric Oxide Concentrations in Diesel Engine Exhaust. IMechE Paper C142/71, 1971.
15. Khan, I.M. Greeves, G. and Wang, C.H.T. Factors Affecting Smoke and Gaseous Emissions from Direct Injection Engines and Method of Calculation. SAE Paper No. 730169, 1973.
16. Shahed, S.M., Chiu, W.S. and Lyn, W.T. A Mathematical Model of Diesel Combustion. IMechE Paper C94/75, 1975.
17. Nightingale, D.R. A Fundamental Investigation into the Problem of NO Formation in Diesel Engines. SAE Paper No 750848, 1975.
18. Hiraki, H. and Rife, J.M. Performance and NO_x Model of a Direct Injection Stratified Charge Engine. SAE Paper No. 800050, 1980.
19. Lipkes, W.H. and Dejoode, A.D. A Model of a Direct Injection Diesel Combustion System for Use in Cycle Simulation and Optimization Studies. SAE Paper No. 870573, 1987.
20. Hiroyasu, H., Kadota, T. and Arai, M., Development and Use of a Spray Combustion Modelling to Predict Diesel Engine Efficiency and Pollutant Emissions. JSME Vol.26, No.214, 1983.
21. Nishida, K. and Hiroyasu, H. Simplified Three-Dimensional Modeling of Mixture Formation and Combustion in a D.I. Diesel Engine. SAE Paper No. 890269, 1989.
22. Kyriakides, S.C., Dent, J.C. and Mehta, P.S. Phenomenological Diesel Combustion Model Including Smoke and NO Emission. SAE Paper No. 860330, 1986.
23. Brown, A.J. and Heywood, J.B. A Fundamentally-Based Stochastic Mixing Model Method for Predicting NO and Soot Emissions From Direct-Injection Diesel Engines. Calculations of Turbulent Reactive Flows 1987.
24. Nagai, T. and Kawakami, M. Reduction of NO_x Emission in Medium-Speed Diesel Engines. SAE Paper No. 891917, 1989.
25. Ferguson, C.R. Internal Combustion Engines , John Wiley and Sons, 1986.
26. Arcoumanis, C. and Chan, S.H., Optimisation of a Vehicle/Engine Transient Model through Injection Control. 3rd Int. EAEC Conference on Vehicle Dynamics and Powertrain Engineering, 1991.
27. Krieger, R.B. and Borman, G.L., The Computation of Apparent Heat Release for Internal Combustion Engines, ASME Paper 66-WA/DGP-4, Proceedings of Diesel Gas Power, ASME, 1966.

28. Chan, S.H., Transient Performance of Turbocharged Vehicle Diesel Engines, Ph.D. Thesis, Imperial College, 1991.

29. Woschni, G., A Universally Applicable Equation for the Instantaneous Heat Transfer Coefficient in the Internal Combustion Engine, SAE Paper No. 670931, 1967.

30. Heywood, J.B., Internal Combustion Engine Fundamentals, McGraw-Hill Book Company, 1988.

31. Gordon, S. and Mcbride, B.J., A Computer Program for Calculation of Complex Chemical Equilibrium Compositions, Rocket Performance, Incident and Reflected Shocks and Chapman-Jouguet Detonations, NASA SP-273, 1971.

32. Olikara, C. and Borman, G.L., A Computer Program for Calculating Properties of Equilibrium Combustion Products with some Application to I.C. Engines, SAE Paper No. 750468, 1975.

33. Koustas, J. and Watson, N., A Transient Digital Test Bed with Direct Digital Control, SAE Paper 840347, 1984.

APPENDIX: Test bed specifications

The test-bed used for this analysis has been developed over a number of years for research into transient operation [33], and incorporates the components and features summarised in the following Table.

Engine	Leyland 500 series, 8.2 litre, 6-cylinder DI diesel engine.
Turbocharger	Garret AiResearch, T04 with twin entry turbine.
Fuel pump	Sigma, in-line, with mechanical governor.
Dynamometer	Eddy current, Vibrometer 4WB25.
Charge Cooler	Serk, cross flow, air-to-water heat exchanger.
Speed and load control	Microprocessor-based control as described by Koustas and Watson [33].
Data logging system	An 8-channel fast analog-to-digit converter with a conversion speed of 800 kHz and a medium speed 32-channel ADC with a conversion speed of 50 kHz.
Microprocessors	A network of three microprocessors, one DEC LSI 11 for overall network and two TI TMS 9900 for control speed and load.
Instruments	Shaft encoder for engine speed, a strain-gauge load cell for engine load, a viscous flowmeter with a fast response strain-gauge differential pressure transducer for air flowrate, an LVDT displacement transducer for fuel rack position, etc.

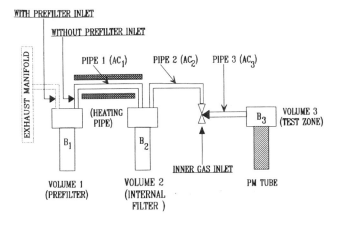

Figure 1a Schematic diagram of the reconstruction model

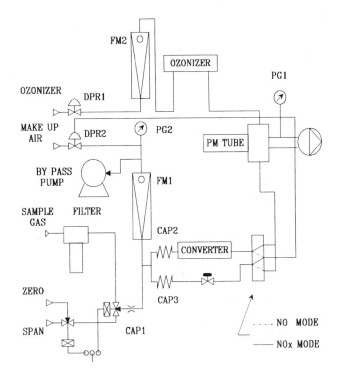

Figure 1b Schematic diagram of NO$_X$ analyser sample gas flow system

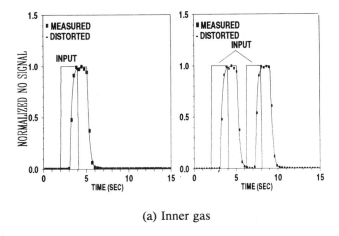

(a) Inner gas

Figure 2 Comparison of measured and predicted NO concentration patterns for various input signals

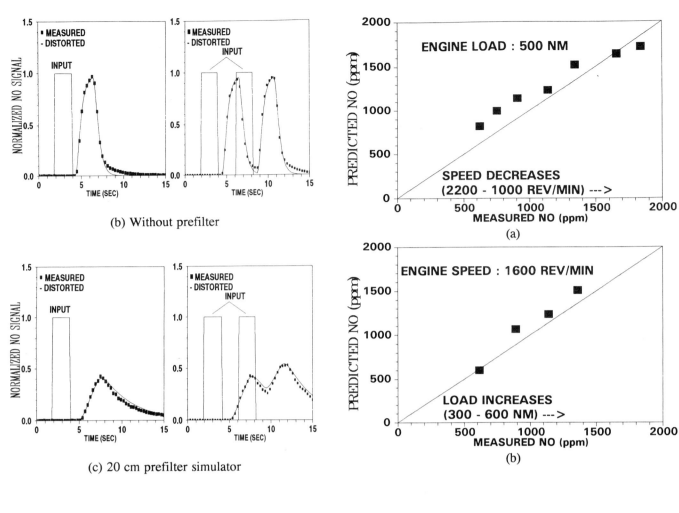

(b) Without prefilter

(c) 20 cm prefilter simulator

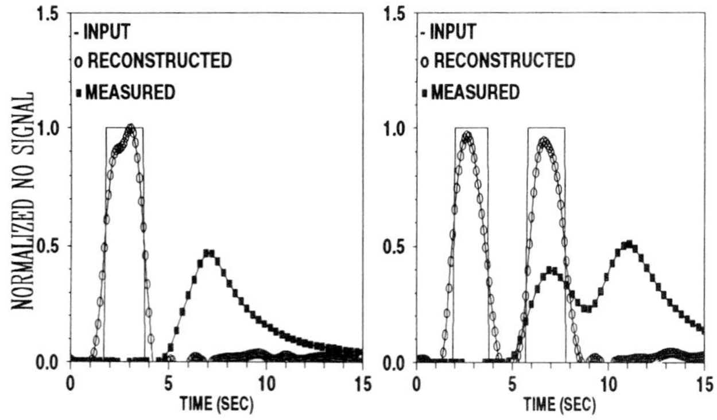

(d) 20 cm prefilter simulator

Figure 2 Comparison of measured and predicted NO concentration patterns for various input signals

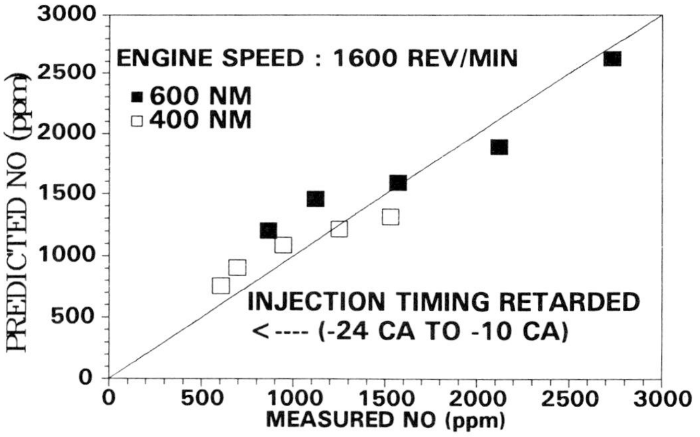

Figure 3 Comparison between measured and calculated NO emissions under steady-speed conditions

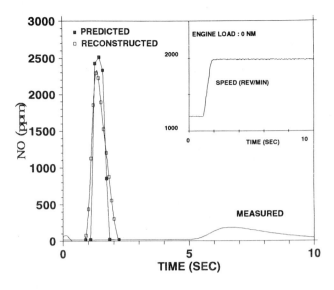

Figure 4 NO emission levels during engine acceleration at zero load

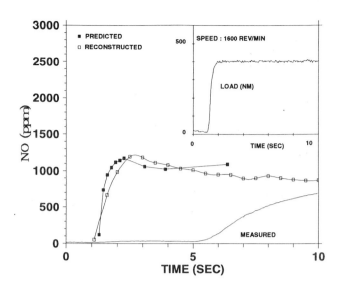

Figure 6 NO emission levels during engine load acceptance at 1600 rev/min

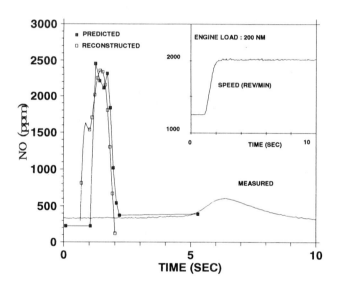

Figure 5a NO emission levels during engine acceleration at 200 NM load

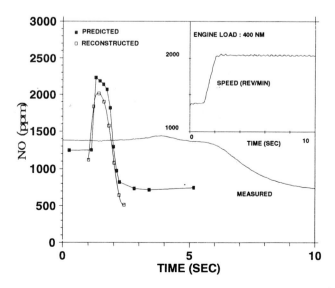

Figure 5b NO emission levels during engine acceleration at 400 NM load

C448/068

In-cylinder sampling technique used in the development of the Ruston RK270 lean burn gas engine

S K SINHA, BSc and A S BRETTON, BSc
Ruston Diesels Limited, Newton-le-Willows, Merseyside
E CLOUGH, MSc, CEng, MIMechE
Department of Mechanical Engineering, University of Manchester
Institute of Science and Technology

SYNOPSIS The paper describes the design and application of a timed sampling valve used to investigate different combustion systems during the development of the RK270 lean burn gas engine.

Gas species concentration data was obtained throughout the cycle from both engine cylinder and exhaust duct. In the cylinder, the valve was either traversed from the centre line to the wall sampling at fixed times or fixed in space. The latter technique was used in the exhaust duct to study the scavenging characteristics.

Data is presented which shows the progression of combustion across the cylinder and the subsequent formation of emissions. Also detailed are comparisons of the effects of combustion chamber shape changes on pre-ignition mixing of fuel and air.

Engine variables considered are piston crown profile, compression ratio and valve timings.

1 INTRODUCTION

In recent years, the Ruston RK270 spark ignited gas engine has been the subject of a program of continued development aimed at reducing both fuel consumption and exhaust emissions while offering greater load carrying capabilities at higher engine speed [1] [2] [3] [4].This process of improvement and evolution had concentrated on obtaining a robust and efficient engine design. The basic configuration of the engine features a four valve cylinder head with centrally located pre-combustion chamber and an individual, mechanically actuated, gas admission valve for each cylinder.

In order to further improve the engine it became necessary to achieve a greater understanding of the in-cylinder processes. Ruston Diesel's participation within an EC Brite Project on the Computer Simulation of In Cylinder Flows presented an ideal basis from which to investigate the conditions within the combustion chamber of the RK270 spark ignition engine. Although the requirements of the Brite project were purely pre-ignition conditions the opportunity was taken to obtain some data during combustion. From this data it is hoped to develop a greater understanding and knowledge of the in-cylinder events which will then lead to further improvements in engine performance.

In order to obtain the in-cylinder data necessary for the EC Brite Project, Ruston Diesels engaged UMIST to produce the necessary equipment and provide expert assistance with the testing. Previous work at UMIST has shown that data obtained using time sampling valves could give valuable insight into the combustion process provided the samples are taken from various positions across the cylinder [5]. A sampling valve and control system was therefore designed and manufactured, by UMIST, to be capable of being traversed across the combustion space from cylinder wall to centre line. A similar valve was also used to traverse and sample in the exhaust tract of the cylinder head. Some of the results obtained during these tests are presented in this paper.

2 TEST PROGRAM

The basis for the work presented was Ruston Diesel's involvement in an EC Brite Project aimed at the investigation of in-cylinder flows in medium speed engines. For its part in this collaborative European research project, Ruston Diesels set out to obtain in-cylinder sampling data from its RK270 spark ignition engine. The data collected was intended to determine the fuel air mixing process within the combustion chamber of a firing cylinder. This information could then be used by other members of the project to validate Computational Fluid Dynamics (CFD) based computer simulations of the in-cylinder flows. The project would however only consider the non-combustion part of the engine cycle, dictating that the in-cylinder sampling would concentrate on the gas exchange and compression phases.

The test program investigated a wide number of changes to the engine in terms of combustion chamber shape, air fuel ratio, ignition timing, engine load and engine speed. The opportunity was also taken to investigate the combustion part of the engine cycle which was outside the scope of the Brite project. These tests provided details of the formation of the emission substances NOx, CO and CO_2, across the combustion chamber. Additionally, a sampling valve was placed in the exhaust port of the cylinder head and information obtained to provide a better understanding of the gas exchange process whilst examining unburnt hydrocarbons in the exhaust.

3 ENGINE SPECIFICATION

The Ruston RK270 spark ignition gas engine is a derivative of the RK270 diesel. Developing the spark ignition engine from the diesel version has a number of distinct advantages. All of the major components, such as the bedplate, crankcase, crankshaft, connecting rod, and cylinder liner been designed to withstand operating conditions far in excess of those ever likely to be encountered with a spark ignition engine. Thus the basic structure of the engine is inherently strong and well proven. As a consequence of sharing many of the major components with the diesel version, the development of the spark ignition engine could concentrate on the important areas of fuelling, ignition and control.

The RK270SI engine has a bore of 270mm, stroke of 305mm and the engine is produced in 6 and 8 cylinder inline and 12 and 16 cylinder vee configurations. The speed range of the engine is between 600 and 1000rpm with production power outputs from 600kWb to 3200kWb. The spark ignition engine has a four valve cylinder head with two inlet and two exhaust valves and is turbocharged using ABB or EGT high efficiency turbochargers. Gas is supplied to each cylinder from a common pressurised manifold, a branch pipe feeds the gas admission valve in the inlet port of each cylinder head. The gas valve is mechanically operated from the engine camshaft. Featured in the centre of the cylinder head is a pre-combustion chamber (PCC) and this receives a small supply of gas via a non-return valve from a separate pressurised manifold. By utilising a PCC it is possible to operate the engine on a "lean" fuel/air mixture in the main chamber which contributes greatly to the engines inherently low fuel consumption and exhaust emissions. The PCC has a volume of just 2% of the main combustion chamber, with the piston at TDC. Ignition is by a single spark plug within the PCC which contains a near stoichiometric fuel/air mixture, and burns readily to generate a flame front

strong enough to ignite the lean charge in the main combustion chamber. A single hole nozzle directs the flame front from the PCC into the main combustion chamber.

The engine is controlled by means of a microprocessor based electronic management system supplied by Regulators Europa which is linked to an Altronic III ignition system. The engine is capable of operating on a wide variety of fuels from natural gas, with a near 100% methane content to landfill or mines gas with a methane content of less than 30%.

For the work described in this paper an inline 6 cylinder engine operating on natural gas, was modified to accept the time sampling valves in both the combustion chamber and the exhaust tract of number 6 cylinder, figure 1.

4 UMIST SAMPLING VALVE

4.1 Sampling valve specification

Although two types of valve, i.e. poppet and needle , were considered, the former was used because not only did it give larger sample flow rates for shorter opening periods but experience had shown that it was more suited to the frequent regrinding necessary when operating for long periods at high loads and off-design conditions. The water cooled valve body which was 9mm in diameter and 150mm long was made of stainless steel and the seat and spindle were manufactured from Inconel 600.

Operation of the valve was based on the pulse hammer system in which a solenoid energised striker knocks open a normally closed spring loaded poppet valve.(A more detailed description can be found in [6]).

Access to the combustion chamber was via a hole drilled through the top of the cylinder liner with a stuffing box to provide a gas seal. The liner was orientated such that the axis of the sampling valve was positioned between the two exhaust valves. The sampling valve was traversed radially across the cylinder from wall to centre line, at an angle of 5° down into the piston bowl, by means of a lead screw and nut driven by a variable speed stepper motor.

The control system for the valve and traversing mechanism used input from a crankshaft mounted, 1/2° encoder and cycle sensor and allowed control over valve opening angle, open period, number of samples and traverse speed. The output of the solenoid pulse, valve lift and valve position was in analogue form. The same valve design was used for the exhaust tract sampling and the relative positions of the two valves are shown in figure 1, at points A and B.

4.2 Sample analysis

The sample was analysed using an emissions analysis system which incorporated a heated sample handling system, FID, and NOx analysers and cold IRGA and paramagnetic analysers. An analogue output from each analyser was fed into a multi-channel pen recorder that was an integral part of the system. Also recorded was the valve traverse position.

The front end gas handling system which incorporated the heated filters, sample pump and valves was designed so that the individual analysis units could be switched to a serial configuration for multi gas analysis at low flow rates. Typical flow rates were between 500 ml/min and 1000 ml/min when sampling over a period of 1 millisecond depending on the sample pressure and temperature. Two different sampling techniques have been used in this work:

Traverse sampling (TS). Continuous sampling at a fixed valve opening angle whilst slowly traversing the valve across the combustion chamber.

Sample and hold (SH). Valve fixed at a position in the cylinder whilst the sampling angle is varied in discrete time steps (e.g. 5° crank angle step every 15 seconds).

5 RESULTS

5.1 Data Format

The in-cylinder sampling data presented is in the form of concentrations of individual gas species on a volumetric basis. Thus the results detail the volume of the respective substance as a proportion of the total volume sampled, which is taken to be representative of the volume around the end of the sampling valve in the combustion chamber.

5.2 Combustion chamber shape effects

The effect on the in-cylinder fuel/air mixing of the shape of the combustion chamber was investigated by testing the engine with two different designs of piston crown profile. The two designs investigated were the simple open bowl shape and the Hesselman bowl, with its raised centre. An example of the results obtained from within the combustion chamber by the TS method during the 180° crank angle period up to TDC ignition are shown in figure 2. In both cases the engine was operated under the same conditions, with an ignition timing of 15° BTDC. The data for the Hesselman design stops short of the centre of the combustion chamber, this was necessary to avoid contact between

the sampling valve and piston at TDC.

Clearly there is a difference in the distribution of fuel across the cylinder at the -180° (BDC) position. Where the open bowl generates a relatively even mixture across the cylinder the Hesselman shows a markedly higher concentration at the cylinder wall than towards the centre of the combustion chamber. However as the piston rises through the compression stroke it can be seen that the hydrocarbon distribution at the -100° is approximately the same for both piston shapes. At the -30° angle the distribution is again identical for both piston shapes. It can also be deduced from the fact that the -30° traces are almost flat that the mixture through which the sampling valve is passing is virtually homogeneous. Since the sampling axis is angled at 5° to the horizontal then it can be assumed that the sampling results represent a two dimensional picture of the conditions within the combustion chamber. From this it could be argued that the entire chamber contains a homogeneous mixture, in which case the piston crown profile is unimportant to the homogeneity of the mixture immediately prior to ignition. However it is most probable that the motion induced in the two chambers is not the same.

As stated the ignition timing of this engine was -15° at the spark plug in the PCC, therefore the TDC traces show the early stage of combustion in the main chamber. Here it is clear that the piston crown profile is influencing the combustion. It should be remembered that the single hole PCC nozzle directed the flame entering the main chamber vertically downwards towards the centre of the piston. In the case of the open bowl piston crown profile the sampling valve detects very little fuel in the centre under the PCC, indicating that complete combustion has taken place in this region. However a small distance away from the centre the mixture is only partially combusted. Further from the centre a region of more fully combusted gas is detected. Towards the edge of the chamber, combustion has not begun and the pre-ignition fuel concentration is measured. In the case of the Hesselman piston crown the combustion appears to spread smoothly from the centre of the chamber towards the cylinder wall. This could be caused by the raised centre of the piston crown guiding the flame more smoothly throughout the chamber. The schematic diagrams on the right hand side of figure 2 show the possible flame front shapes induced by these two different piston crown profiles.

5.3 Open bowl combustion

It is possible to see from figure 3 how the combustion progresses with time with the open bowl piston crown. In this figure oxygen concentration is shown instead of total hydrocarbons (fuel). This substitution is valid since oxygen is known to show the same characteristics both before and during combustion. Also shown in this diagram is the interaction between flame speed and engine speed.

From the TDC instant common to both figure 2 and 3 within 10° the combustion has progressed across the chamber until nearly all of the fuel at the top of the cylinder has been burnt. Within a further 10° (+20°) the combustion is almost complete, however by this time the chamber volume has increased by approximately 30%, thus the fuel adjacent to the piston surface may not have been reached by the flame front.

By comparison of the 750 and 1000rpm traces at the same crank angles it is possible to observe the difference in flame progression brought about by the difference in time for the crank to rotate through a 10° sector. Measurement of the distance covered by the flame front between TDC and +10° reveals a flame speed of approximately 17m/s in both cases.

5.4 Open bowl emission formation

A more complete picture of the combustion within the cylinder of the open bowl engine can be gained from figure 4. These results from three different operating conditions, show the effects of both load and speed on the main products of combustion. In all cases the duration of the period displayed has been increased to 100° after TDC ignition. It is possible now to see the movement of the gases at the top of the chamber as the volume is increased during the expansion stroke.

The CO formation of the 1000rpm high load case reflects the combustion pattern at TDC observed in figures 2 and 3. As CO is indicative of partial combustion it is not surprising to see the greatest concentrations around TDC. Indeed the peak in the curve moves outwards from the centre of the cylinder following the flame front, leaving behind low concentrations, indicating complete combustion. By the same token the corresponding CO_2 traces show how the high values of CO are further oxidised to CO_2 as the combustion progresses. At +50° and then more so at +100° it is possible to see from the CO and CO_2 plots how the products of partial combustion within the PCC emerge into the main chamber towards the end of expansion, indicated by a sharp increase in concentrations at the centre of the cylinder.

The NOx traces do not mirror the other results, showing that the formation of NOx is dependent on different parameters. Here time plays a very important part, along with the temperature of the local surroundings. At TDC the majority of the NOx detected is concentrated under the PCC indicating high temperatures. As the flame front moves further out into the main combustion chamber, +10°, lower concentrations are formed, which then rapidly increase with residence time in the very hot region of the flame at +20°. However once the flame has passed through the chamber and the temperature falls the NOx formation ceases. Thus the variation in the traces shown after +20° is assumed to represent the movement of the gases within the chamber as the piston falls.

By reducing the load but maintaining the engine speed it can be seen that the combustion differs greatly. A major reason for this could be the fuel air mixture obtained prior to ignition. It is known that the fuel/air ratio at the low load condition is higher than that at the high load case. The combustion does not therefore take place as readily or as smoothly under these circumstances. This is indicated most readily by the NOx, as the levels detected are much lower than at the high load condition. The conditions under which NOx is formed, high temperatures and pressures are not available, and it is observed that a large proportion of the NOx remaining at +100° is concentrated in the centre of the combustion chamber, suggesting that this NOx is originating from the PCC. It has been suggested previously by other researchers that in the case of extremely lean mixtures the majority of the NOx formed originates from the PCC [7].

It is also interesting to note from these results that there appears to be some movement of gases taking place towards the end of combustion. This is indicated by the CO_2 level which falls significantly across the whole radius of the chamber between +50° and +100°. Although not shown, this was seen to be due to an increase in the oxygen levels. This indicates that a volume of unburnt oxygen displaces the combustion products at the top of the combustion chamber during this period. However the levels of total hydrocarbons (fuel) do not show any increase, suggesting that there is no unburnt fuel in the intruding gas mixture.

The 750rpm high load results show a comparatively rich fuel/air mixture. Because the engine was configured for 1000rpm operation, the available boost pressure was lower than would normally be desirable, however these results do show the consequences of operation under these conditions. As noted previously, degrees of crank angle at 750rpm are somewhat longer in time than the degrees at 1000rpm. Thus the flame front within

© IMechE 1992 C448/068

the chamber has travelled further at each respective sampling angle.

The levels of CO and NOx have increased significantly (note that the graph scales are double those for the previous two cases). Because of the richer mixture, much greater levels of all three emissions are detected. Once again there is evidence of a jet of partially combusted gas entering the main chamber from the PCC late on in the cycle at +100°, as seen in the CO results. This combustion has obviously been at a lower temperature as no NOx is evident in the jet, indeed the NOx curve for +50° shows a marked reduction at the centre line.

The NOx trace for +100° has dropped significantly (approximately 50%), this along with the slight drop in CO_2 between +50° and +100° could once again be due to a displacement of the gases at the top of the chamber by unused oxygen.

Comparison of the 1000rpm low load case and the 750rpm high load case shows the effect of fuel/air ratio on NOx emissions, the 1000rpm case is very weak and the 750rpm case rich, almost stoichiometric.

5.5 Open bowl full cycle results

Taking a wider view, it is possible to look at the concentrations of a gas species through the full engine cycle of 720° at a fixed point in the combustion chamber using the SH method of sampling. Two such traces are shown in figure 5, for total hydrocarbons in the open bowl piston crown engine at 1000rpm high load. The points within the combustion chamber at which these results are taken from are shown in figure 1 as (1) under the PCC and (2) between the exhaust valves. The reference point of 0°/720° is TDC firing, also shown on this diagram is the valve timing of the engine to aid in the interpretation of these results.

As seen previously, the flame front emerging from the PCC (0°) spreads rapidly across the sampling region of the valve, the fuel is rapidly burnt and the traces drop to a residual level. At 270° the exhaust valves have been open for approximately 150° and the pressure in the cylinder has dropped to that of the exhaust. It should be noted that the fuel gas supply to the PCC is controlled on a simple plate type check valve which is opened and closed dependent on the pressure conditions on either side of the valve. In the particular case shown in figure 5 the PCC supply pressure was almost 5 bar with the mean exhaust pressure approximately 0.5 bar.

Once the cylinder pressure has fallen below that of the PCC gas supply level, gas will begin to flow into the

PCC. This becomes evident soon after AVO when the valve under the PCC detects a rise in the level of hydrocarbons. It is believed that this is due to gas seeping from the PCC nozzle into the main chamber. Prior to AVO the gas from the PCC is carried towards the inlet valves by a vertical vortex created by the exhaust outflow from the cylinder. Thus a volume of gas will be present under the inlet valves at the time of AVO which is then carried towards the exhaust valves, past the sampling valve, as the scavenge flow is initiated. This would explain why the hydrocarbon level between the exhaust valves does not rise until after the flow field across the cylinder is initiated by the opening of the air valves and after the rise is seen in the centre of the chamber.

As the piston reaches 360° (TDC) the upward motion of the cylinder contents is sufficient to partially flush out the gas which has filled the PCC and a sudden high level spike is seen under the PCC. This is quickly carried away by the increasing flow from the air valves into the cylinder and therefore the level falls equally rapidly. At this time the main gas admission valve opens in the inlet port of the cylinder head followed shortly afterwards by the closing of the exhaust valve, which cuts off the flow towards the exhaust valves. By the time the PCC gas peak reaches the sampling valve between the exhaust valves its detection is eclipsed by the flooding of the top of the cylinder by the main gas charge from the inlet port valve. The inlet gas flow reaches a maximum just prior to 450° as the piston falls on the induction stroke, the rise in levels just as the main gas valve closes being due to recirculation of a rich pocket of gas. The levels throughout this period show a downward trend due to the expanding cylinder volume and once the gas valve closes the dilution of the gas with more fresh air.

Following the air valve closing point just after BDC the mixture is compressed and the levels slowly rise to a common value just prior to ignition, as observed in figure 2. Ignition within the PCC takes place at 15° BTDC and the level drops sequentially at the two sampling points as the flame front is detected.

5.6 Further investigation of PCC gas outflow at TDC exhaust

As discussed previously, the large peak detected under the PCC at approximately 15° ATDC exhaust is attributed to a sudden purging of the PCC by gases from the main chamber. In order to investigate this further and to better understand the influencing factors, a closer study was made of the region under the PCC. The TS curves shown on the left hand side of figure 6 demonstrate how the magnitude of the peak is dependent on the PCC gas supply pressure. It can be seen that the peaks of these curves are apparently not in the centre of the cylinder, this is due to the valve sampling gas from the region ahead of the end of the valve, thus when the valve is at the centre line of the cylinder it is in fact sampling from the other side of the centre line. It can be deduced that the peak, although not indicated at the centre of the cylinder is in fact a maximum on the centre line.

It should be noted that the magnitude of the peak indicated under the PCC by the SH method in figure 5 corresponds to the value detected by the TS method for the same 2.5 bar PCC pressure at the centre line, shown in figure 6. Further localised SH tests were performed for a PCC pressure of 2.5 bar at various distances from the centre line out towards the exhaust valves to confirm the results already shown, these are given on the right hand side of figure 6.

This figure shows that the large peak detected under the PCC decays as it moves towards the exhaust valves. Therefore, the gas causing the peak originates from the PCC and the amount of gas contained in the PCC is directly influenced by the supply pressure. From this it is evident that the PCC pressure should be maintained at the lowest pressure to reduce the possibility of gas wastage during the exhaust period.

5.7 Scavenge period – Effect on gas carry over.

Following the work described in the previous two sections relating to the gas emanating from the PCC and the possibility of this unburnt gas being carried through into the exhaust, a sampling valve was fitted in the exhaust port of the cylinder head. The effect of two different valve timings, one with a wide exhaust and air valve overlap and one with a low overlap period were investigated. The results of this investigation are shown in figure 7.

It is possible to see, with reference to figure 5, which shows results for the wide overlap engine, how the rise in hydrocarbons after the air valve opens is carried towards the exhaust valves. Indeed, figure 7 shows that this gas does reach the exhaust port, as indicated by the rapid rise from the exhausting level (between 180° and 270°) up to a level approaching that seen in the cylinder. However, before the cylinder level is reached the exhaust valve closes and the flow in the port stops. It is believed that a concentration gradient exists along the port from the exhaust valve to the sampling valve, however because there is no flow the sampling valve detects a

constant value. This condition remains until just prior to the exhaust valve opening again, at which point a pressure wave within the exhaust manifold disturbs the gas in the port causing a higher concentration of gas to reach the sampling valve, hence the slight rise prior to EVO at 123°. Once the exhaust valve opens and flow is established in the port the concentration rapidly falls to the exhaust gas ambient level at about 180°.

In contrast to the above, if a low overlap period is specified then there is no carry over of gas and the cylinder exhaust value is similar to the constant value throughout the cycle. A problem was encountered during the collection of the low overlap data shown here since the exhaust stack ambient level of hydrocarbons was noticeably higher than the cylinder exhaust level, indicating that one of the other cylinders was not operating correctly and misfire was resulting in significant quantities of unburnt gas being passed through into the exhaust. Hence the low overlap level shown in figure 7 rises slowly during the exhaust valve closed period of the cycle.

6 CONCLUDING REMARKS

The tests described formed part of an EC collaborative project under the Brite scheme. The results were intended to validate theoretical modelling work conducted by other partners in the scheme. In order to achieve worthwhile results it was necessary to test under a wide range of parameters, rather than only those based on the optimum design to date. Accordingly many of the tests were carried out at off-design conditions with a view to providing a more general set of results. The Brite project concentrated on pre-ignition data, however the opportunity was taken to obtain further data during the combustion phase. The theoretical modelling recognises the concepts of motion in the air/fuel mixing process in the pre-ignition phase and it is clear that this motion continues during the combustion phase. Clearly the transportation of gases has a direct implication on engine design for reduced emissions and a theoretical modelling of this phase would be of considerable help. Similarly, engine development has taken into account the motion of unburned hydrocarbons during the scavenge period to prevent carry over into the exhaust system and modelling of this phase would also be beneficial.

The data collected has given a valuable insight into the combustion process, in particular fuel/air mixing and emission formation. This knowledge will be utilised in the future development of the spark ignition engine to provide lower emissions without adversely affecting fuel economy. During the course of this work it has become apparent that certain areas of interest, if investigated further, may yield improvements in engine performance. In particular the interaction of the flame jet from the PCC nozzle and the piston crown geometry has a significant effect on the combustion in the main chamber. However the piston geometry appears to be less critical during the pre-ignition fuel air mixing phase of the engine cycle. Also the investigation of the gas supply to the PCC and its control may result in benefits in fuel consumption by preventing wastage of gas during scavenge. Thus giving the opportunity to optimise the valve timing for maximum scavenge and thereby facilitating the improved engine cooling from this arrangement. Finally it is hoped that some of the findings from this work can be applied to the future development, in terms of fuel consumption and emissions, of the diesel version of the RK270 and also the new smaller bore RK215.

7 ACKNOWLEDGEMENTS

The authors wish to thank the organisers of the EC Brite scheme for their support, UMIST for their collaboration and the management of GEC Alsthom Diesels for their permission to publish this paper. Additionally, the authors would like to express their gratitude to their colleagues who have contributed to the project.

REFERENCES

1. Whattam M. and Sinha S. K. The Ruston RK270 range of spark ignited engines. Institution of Diesel and Gas Turbine Engineers. Publication No. 435.

2. Whattam M., Sinha S. K. and Moylan G. Developing the second generation Ruston RK270 spark ignition engine for low emissions and improved economy. CIMAC, 1987, D79.

3. Whattam M. and Sinha S. K. RK270 Spark ignited engine. IMechE seminar, Gas engines and Cogeneration, May 1990.

4. Whattam M. and Sinha S. K. Further refinement of the Ruston RK270 spark ignited engine. CIMAC, 1991, D25.

5. Whitehouse N. D., Clough E. and Uhunmwango S. O. The development of some gaseous products during diesel engine combustion. S.A.E. 800028. 1980.

6. Whitehouse N. D., Clough E. and Roberts P. S. Investigating diesel combustion by means of a time sampling valve. S.A.E. 770409. 1977.

7. Charlton S. J., Jager D. J., Wilson M. and Shooshtarian A. An investigation of mixing and combustion in a lean burn natural gas engine. IMechE seminar, Gas engines and Cogeneration, May 1990.

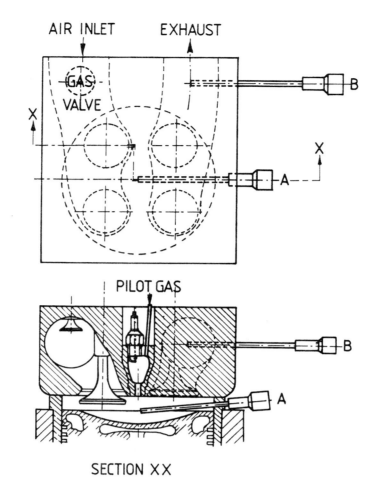

Fig 1. Sampling valve arrangement.

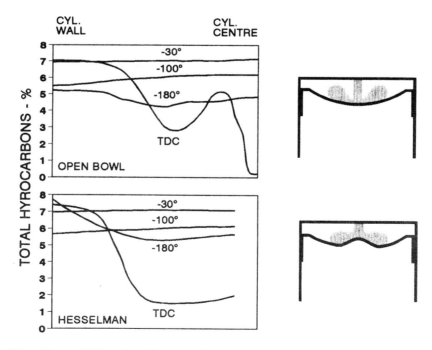

Fig 2. Effect of combustion chamber shape on in-cylinder mixing.

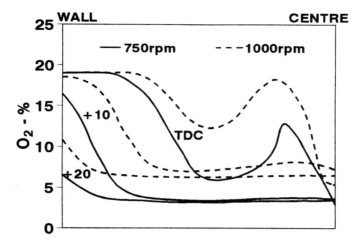

Fig 3.   Open bowl combustion.

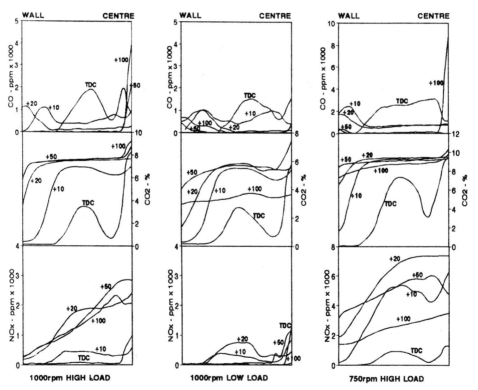

Fig 4. Open bowl emission formation.

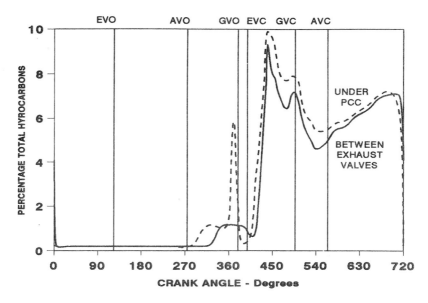

Fig 5. Open bowl full cycle results.

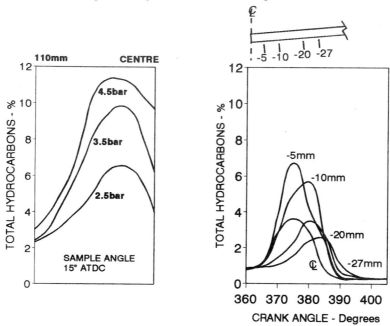

Fig 6. PCC gas flow at TDC exhaust.

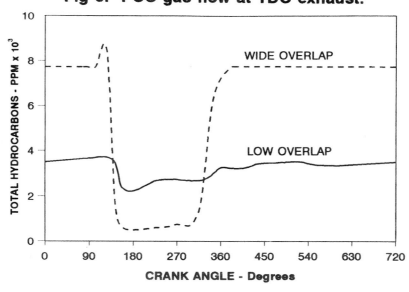

Fig 7. Valve timing effect on scavenge period gas carry over.

C448/070

Combustion and emission studies in high-speed DI diesel engines

K K RAO, BE, MTech, D E WINTERBONE, BSc, PhD, DSc, FEng, FIMechE, MSAE and
E CLOUGH, MSc, CEng, MIMechE
Department of Mechanical Engineering, University of Manchester Institute of Science and
Technology

ABSTRACT

Experimental investigations were carried out on multi cylinder Ford York 2.5L and a single cylinder Hydra engine of the same cylinder geometry but fitted with high pressure electronic unit injector to find the influence of engine operating variables on performance and emissions. The test bed, emission gear, photographic rig and engine instrumentation are briefly described. A synopsis of the experimental results obtained using high pressure electronic unit injector and simulated standard injection equipment are presented. The parameters varied were speed, load, injection timing and swirl ratio. The parameters monitored were cylinder pressure, injection line pressure, needle lift and levels of toxic gases, smoke and particulates in the exhaust. The data is analysed and comparisons are made to identify the influence of varying individual parameters mainly injection pressure on the observed product specie concentration and trends.

High power Pulse laser illuminated spray and combustion photographic studies were conducted to further understanding of the physical processes inside the cylinder and provide a logical explanation for emission behaviour. A sequence of spray and combustion photographs with synchronized data acquisition signals is illustrated for both the injection systems under identical conditions. The suitability of high pressure injection system to small high speed DI diesels to meet future emission standards is discussed.

INTRODUCTION

The introduction of stringent legislative measures for emissions to contain the environmental pollution levels brought about by an ever increasing vehicle population has driven researchers to find energy efficient, environmentally desirable prime movers. Direct injection(DI) diesel engines are higher fuel efficient but inherently noisier and emit higher levels of exhaust emissions, mainly smoke and particulates, than comparable idi diesel engines. Until recently, automotive DI diesels were mainly confined to medium and heavy duty applications. Due to rising fuel costs and stringent emission regulations, efforts are being directed to make them suitable for light duty applications[1,2].

In the past, the emphasis was on specific power and fuel efficiency with little concern about the environment. Recent advances in science and technology have singled out the contribution of automotive exhausts to environmental pollution and pointed to the need for urgent action. The mechanisms which govern the formation of these emissions are not well understood fundamentally, despite extensive research. Diesel combustion is so complex that the emphasis is still on tuning methods rather than fundamental investigations, mainly to reduce development costs. This is more so in the case of high speed DI diesels which require cost effective, sophisticated fuel injection equipment. This led to the evolution of the unit injector[3] to achieve a better emission/fuel consumption trade off[4] and offer a solution to meet emission standards[5]. The advent of microprocessor based flexible manufacturing methods made it possible to produce unit injectors economically, making them viable for small DI diesels in the face of cut throat competition. However, extensive experimental investigations are needed to evaluate them on small DI diesels, and some of these are undertaken here.

Often investigations are conducted to get some reference data and very rarely extend to full scale comprehensive testing over a wide range of operating conditions. The reference data is extrapolated to other operating conditions which at times is misleading. Occasionally parametric studies are supported by more fundamental investigations like combustion photography.

High speed photography[6,7,8,9] is not a new technique to explain the physical processes taking place inside the cylinder and has usually ended in the past with poor quality pictures during the ignition delay period, due to the unavailability of a high energy light source to provide illumination in the restricted environment of a small DI diesel engine cylinder. In the present investigation a copper vapour pulse laser was used to provide illumination of extremely short duration to freeze the frames and give sharply defined spray pictures. Visual analysis of the photographs can reveal the influence of swirl on spray growth, mixing and combustion, all of which are vital to the design of small high speed DI diesel engines.

ENGINE TEST BED AND MEASUREMENT SYSTEM

A schematic diagram of the test equipment is shown in fig.1 and the technical specifications of the engines are given in table 1. The Hydra engine was mounted on a test bed coupled to a DC motoring dynamometer. Speed and load controls were operated from a remote console. Cooling water and lubricating oil circulation were by electrically driven pumps with thermostatic valves for temperature control to maintain constant temperature of oil and water. Engine speed was read from an optical encoder and load was obtained from a load cell. A Kistler 6121 piezo electric transducer was used for cylinder pressure measurement, while a strain gauge transducer was used for injection line pressure and an inductance transducer for needle lift. Air flow was measured using a viscous flow meter which was connected to the variable swirl inlet port. Crank degree signals were acquired from a Ferranti optical shaft encoder. Fuel flow rate was measured using a Pierburg PLU116-H flow meter with bubble separator and P4282 indicating system. Mean exhaust temperature was measured with a K-type thermocouple.

The Ford engine[10] was coupled to a hydraulic brake and operated from the common remote console; it was connected to the same emission gear and data acquisition system. The unburnt hydrocarbons in the exhaust were measured by FID and oxides of nitrogen by chemi-luminescent analyser, both were connected through a heated line to the engine exhaust. Carbon monoxide was measured by NDIR analyser and the smoke density by Bosch pump and digital comparator. The particulates were measured using a Ricardo variable split particulate tunnel and environment control chamber.

The electronic unit injector(EUI) used in the present investigation was of Lucas design and fitted to the single cylinder Hydra research engine. The plunger was driven directly by an overhead camshaft. The start and end of injection were electronically controlled using a fast acting colenoid and spill valve to achieve fast response and flexibility. The injector was instrumented for injection line pressure and needle lift. The required fuel feed line pressure to the injector was 4.5 bars and was maintained by a Bosch electrical fuel pump with closed loop control. Data from the crankshaft sensor was used to calculate speed, timing and fueling. Electronic control gave a high degree of timing control and fueling.

HYDRA PHOTOGRAPHIC RIG

The Ricardo Hydra engine used for combustion photography had the cylinder head left in its standard condition, and the combustion chamber viewed through the underside of a modified transparent piston crown . An extended piston with open slots in the skirt and a quartz piston crown was fitted in place of the standard piston, fig.2. A face aluminised angled mirror was placed through the piston slots and bolted on to the upper cylinder block. A camera mounted on the integral bracket was aligned with the angled mirror to give a clear view of the combustion chamber. A fibre optic light guide of 1

Table 1   Engine specifications

Model:	Ford HSD425	Hydra
Bore/Stroke:	93.67/90.54 mm	←
Cylinders:	4	1
Swept volume:	2.496L	0.624L
Swirl ratio:	2.15 - 4.0	2.65
CR:	18.3 (measured)	←
Piston:	Standard	
	York D.I.	←
Bowl dia.:	46.4 mm	←
Valve timings:	IVO 40 BTDC	←
(Measured)	IVC 50 ABDC	←
	EVO 60 BBDC	←
	EVC 24 ATDC	←
Injection		CAV
system:	Bosch	EUI50
Plunger		
diameter:	8 mm	←
Orifice spec.:	4×0.24	4×0.2
NOP(bars):	250	←
Max injection		
pressure(bars)	750	1200
Rated speed:	4000 rpm	←
Photographic		
build:		3600 rpm

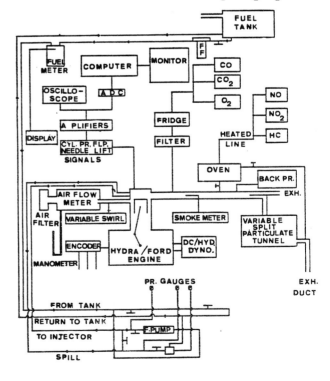

Fig. 1 Schematic layout of engine, instrumentation and emission gear

mm effective diameter was fixed behind the mirror. This was used to direct the light from a copper vapour laser into the combustion chamber: the technical specification of the laser are given in table 2. An illuminated pointer and flywheel degree marker were also within the field of view of the camera via two mirrors and were filmed at the same time as the combustion process. The entire sequence of events was computerised to obtain satisfactory film records of spray and combustion. A schematic diagram of the photographic rig is shown in fig.3.

Table 2   Laser specifications

Model:	CU15-A laser.
Laser Medium:	Copper
Wavelengths(nm):	510.6/578.2
Average Power (W):	15
Green/Yellow Ratio:	2 : 1
Pulse Energy (mj) Max:	2.75
Pulse Width (ns):	10 - 40
Peak Power (kW):	70
Pulse Repetition Frequency (kHz):	3 - 20
Beam Diameter (mm):	25
Power Consumption (kW) Maximum:	3.0

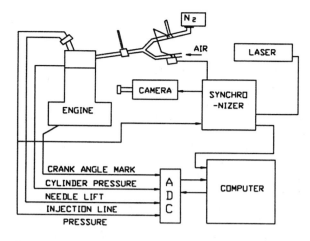

Fig.3 Schematic diagram of synchronizing device for high speed photography

FILMING PROCEDURE

The laser was powered up and the engine was brought to normal operating conditions quickly to avoid build up of deposits on the quartz piston crown before taking the film, by heating the oil and water separately. Then the engine was motored at the test speed with fuel injection turned off and the fuel injection pulse width and injection timing preset according to the test results obtained on the baseline engine. The laser beam shutter was opened and the flywheel marker was illuminated. The computer program controlling the sequence of events was executed. The camera was switched on and the film accelerated to the preset framing rate, at which time fuel injection was switched on and data acquisition was activated. The engine was fired and the events were photographed with the synchronised laser light pulses, controlled by the camera triggering mechanism, over a number of consecutive cycles. The fuel injection was turned off immediately after a preset number of cycles in order to reduce unnecessary deposit build up on the piston quartz window.

COMBUSTION STUDIES

A large number of 400 feet Ektachrome colour films were shot at various speeds, loads, swirl ratios and injection timings for two piston bowl shapes and four injection systems. Some of the films were taken at the very high engine speed of 3000 revs/min; possibly the first time diesel engine photographs have been taken at such a speed. The framing rate of the camera was up to 20kHz. Due to space restrictions in this paper only one example is chosen, and the sequence of photographs for conventional and high pressure injection under identical conditions is shown in fig.4 and fig.4a. The conventional injection system was simulated by fitting the standard Bosch nozzle to the electronic unit injector pump. The difference in hole size (0.2mm) for the EUI and 0.24mm

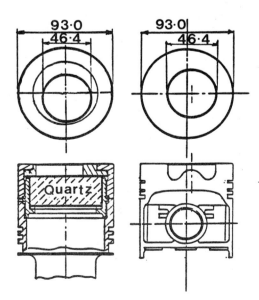

Fig. 2   Hydra photographic piston and standard Ford York piston

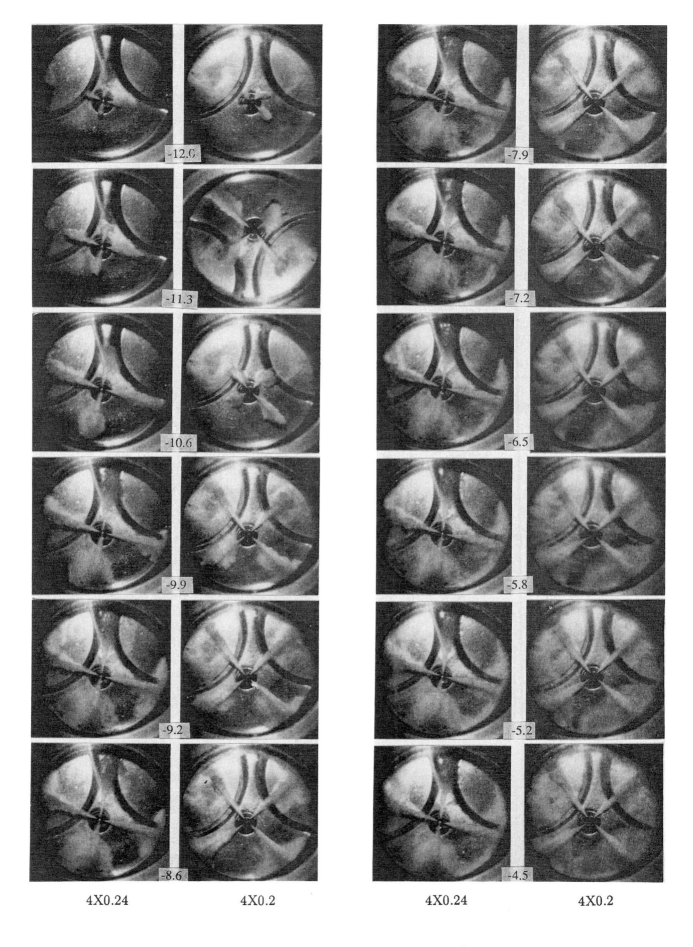

-12.0		-7.9	
-11.3		-7.2	
-10.6		-6.5	
-9.9		-5.8	
-9.2		-5.2	
-8.6		-4.5	
4X0.24	4X0.2	4X0.24	4X0.2

Fig. 4   Sequence of high speed photographs of spray and combustion(originally in colour)

© IMechE 1992 C448/070

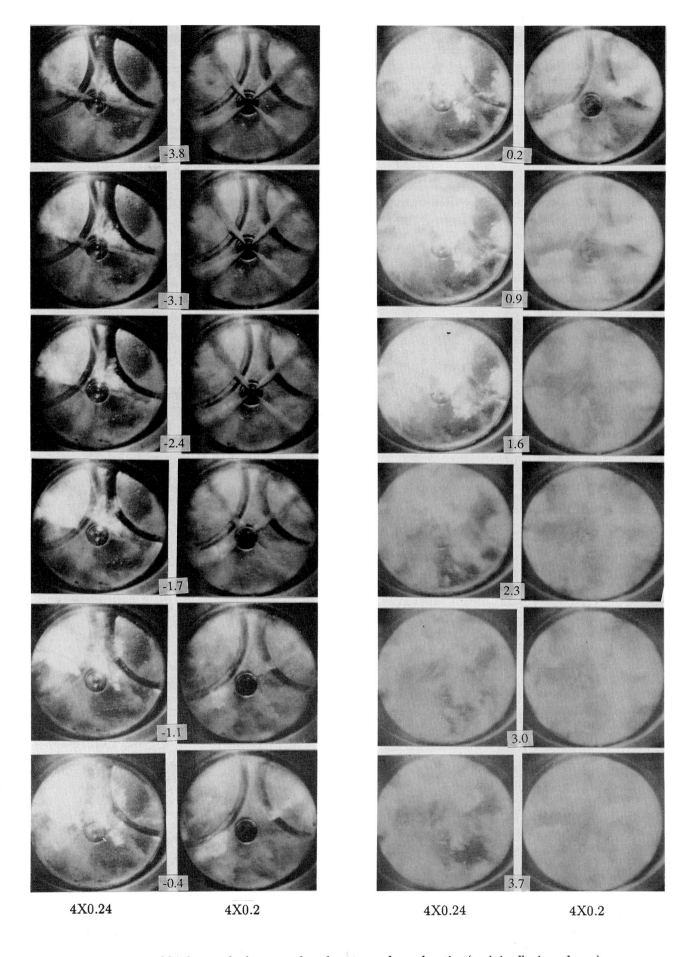

4X0.24   4X0.2   4X0.24   4X0.2

Fig. 4a   Sequence of high speed photographs of spray and combustion(originally in colour)

for the Bosch injector means that the fuel line pressure is significantly lower with the "conventional" injector. It was not possible to run the Hydra engine with the full Bosch fuel injection system because of cylinder head and fuel pump drive arrangement. Synchronised data acquisition signals are shown in fig.5 and the corresponding heat release in fig.6. The figures are self explanatory but some of the salient features of the photographic work, relevant to emission formation, are described below.

The start of injection was very consistent and occurred within 0.2 crank degree for the consecutive cycles. The sprays emerging from a particular nozzle into the high pressure and high temperature environment were all of different shape. Unequal spray penetration was observed and this is inherent with VCO nozzles. The spray impinged on the bowl wall of radius 23 mm

in about 400 μs, irrespective of engine speed and injection timing for the same injector. The sprays appeared to have two distinct zones: a high density core(which is clearly visible on fig.4) and a lower density vapourizing region(which can be seen on the original films). It is apparent from fig.4 that the core is almost independent of the swirl velocity, and travels to the bowl wall in a straight line. Also the cone angle is bigger with the "conventional" injector, with its lower injection pressure. However the situation changes when the vapourizing region of small fuel droplets is considered, as shown in fig.7. Here, the EUI system produces a larger

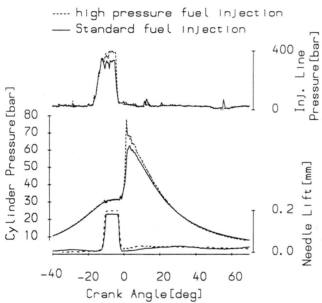

Fig. 5 Cylinder pressure, injection line pressure and needle lift as functions of crank angle for synchronized photography

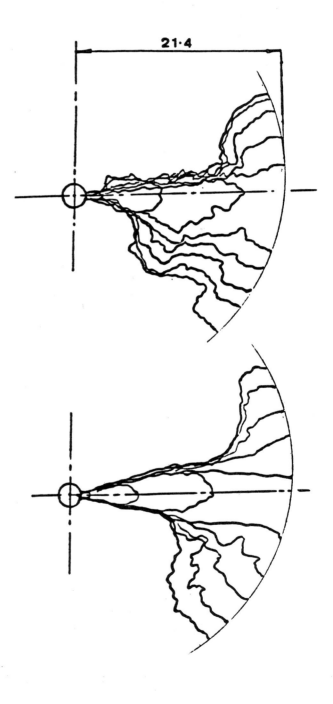

Fig. 6 Rate of heat release calculated from cylinder pressure of Fig. 5.

Fig. 7 Sketches of outer vapour boundary of fuel spray from seven successive frames of fig.4 showing interaction with combustion chamber wall and swirl

volume of vaporizing region than the conventional injector. The low mass of the fuel droplets in this second zone makes them susceptible to the swirl and this part of the jet is deflected, even close to the nozzle. The situation is supported by the initial rate of heat release, indicating a large premixed zone, which is discussed later. The sprays were deflected by swirl as shown in fig.7. The fuel vapour carried away by swirling air even close to the injector tip, fig.7, mixes in to the lean flame out region (LFOR) especially at part loads, and this would contribute to particulates in the form of SOF and higher unburnt hydrocarbons.

It can be seen from fig.4a that with high pressure injection premixed combustion started slowly (there was a long ignition delay) but rapidly engulfed the entire combustion chamber with white or yellow-white flames. This is reflected on the synchronised cylinder pressure diagram fig.5, and this would result in as increase in the formation of oxides of nitrogen figs.10, 11, 12, 13, 14. In contrast the combustion with conventional injector was slow, sparse and brownish mainly due to the larger droplet size which required longer preparation times, and the luminous combustion phase lasted much longer than for the other case.

It is apparent that the swirl motion did not break down to turbulence under part load conditions but simply carried away the fuel rich burning sprays in solid body rotation until late during the expansion process, resulting in higher CO and HC emissions. At high loads 75% of the fuel injected was deposited on the wall of the piston bowl and heavy overlapping of sprays took place mainly due to the small bowl diameter; this resulted in fuel rich mixtures. Subsequent combustion showed soot clouds in the overlapped regions of the spray resulting in increased dry soot, HC, CO and smoke(figs.10, 11, 12, 13, 14). In the case of the open bowl combustion chamber the mixing subsequent to premixed phase of combustion was very poor where as the photographs of a re-entrant bowl showed vigorous mixing and rapid combustion.

Some tests were undertaken with pilot injection then the combustion was smoother but still more turbulence is required to improve air fuel mixing. This basically shows that, for small DI diesels it is important to adopt re-entrant bowl, pilot injection, variable swirl and flexible governing to improve air fuel mixing and combustion. Another aspect is that the fuel entering into burning sprays formed very fuel rich mixtures which appeared in the photographs as bulging black clouds: these remained as patchy clouds until the luminous combustion died down.

At speeds above 2500 revs/min and high pressure injection, the tip of the sprays were fluttered away like a fountain in a storm by swirling air and resulted in explosive premixed combustion. This meandering of the fuel sprays has also been reported by Nishida etal[11] and possibly forms an important role in the atomization process. The rate of cylinder pressure rise during the premixed region was 15 bars/deg crank which results in noisy combustion and a high NO_x production. Due to space restrictions the results of the photographic work will be published elsewhere in detail.

EMISSION STUDIES

A large number of tests were done on both engines and the data collected was analysed for performance and emissions. The variables were speed, load, swirl ratio and injection timing. The parameters recorded were CO, CO_2, O_2, NO, NO_2, HC, smoke, particulates, fuel flow rate, needle lift, cylinder and injection line pressures at each operating condition. The signals were stored on a digital computer through a twelve bit ADC and analysed for heat release using a simple analysis program.

The concentrations of toxic gaseous emissions in the engine exhaust were converted to mass flow rate and normalised by the fuel flow rate to eliminate the effect of number of cylinders because as multi cylinder engines are mechanically more efficient than a single cylinder engine. The calculation procedure is given in appendix **A**.

INJECTION PRESSURES

Since the main emphasis is to study the influence of injection pressures on performance and emissions, injection line pressure data were analysed for maximum and mean pressures. About 80 tests were selected for the "conventional" injection system and 376 tests for the high pressure injection. The fuel line pressure diagrams were analysed on a digital computer and the maximum presure was recorded. The duration of injection in degrees crank angle(θ_{inj}) was found from the needle lift diagram and fuel/stroke was calculated from fuel flow rate. Mean injection pressures (p_{mean}) are calculated assuming a coefficient of discharge of 0.7 and the equation is

$$p_{mean} = (\frac{0.0427 M_f N}{\theta_{inj} N_h D^2})^2 \qquad (1)$$

where

M_f = Quantity of fuel injected/stroke(mm^3)
N = Engine speed (revs/sec)
N_h = number of nozzle holes
D = Nozzle hole diameter(mm)

The calculated mean injection pressures, as a function of maximum fuel line pressures, are shown for the high pressure injection system in fig.8 and for conventional injection system in fig.9. The droplet sauter mean diameter is inversely proportional[12] to mean injection pressures, thus high mean injection pressure gives better atomisation and mixing.

OXIDES OF NITROGEN EMISSION INDEX ($EINO_x$)

Figs. 10, 11, 12 13 compare the variation of $EINO_x$ as a function of relative air fuel ratio(ϵ_R) at different speeds. The injection timing(θ_s) was maintained constant at each speed but differed for different speeds. θ_s was read precisely from needle lift diagrams based on experience gained from synchronised photographic studies.

The $EINO_x$ was directly related to maximum rate of cylinder pressure rise $(dp/d\theta_{max})$, which in turn is a function of fuel available to burn during the premixed phase of combustion. The amount of fuel ready to burn during the premixed phase of combustion rose with increasing fuelling(or load) and fell as the ϵ_R decreased, when the sprays overlapped heavily. The NO_x emission levels were much higher with high injection pressures, or high swirl or both. This has been explained in combustion studies above.

High pressure unit injector

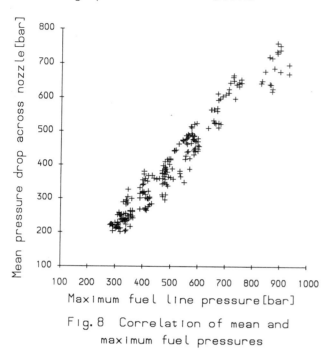

Fig. 8 Correlation of mean and maximum fuel pressures

Conventional fuel injector

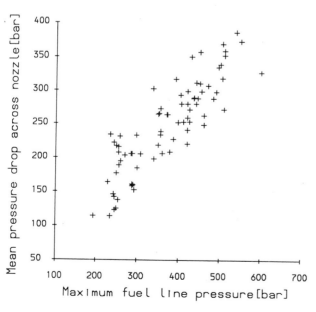

Fig. 9 Correlation of mean and maximum fuel pressures

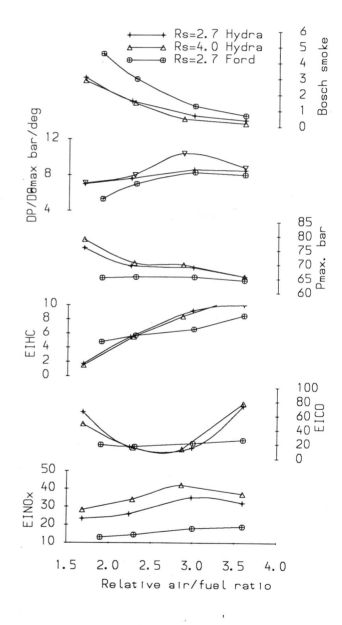

Fig. 10 Comparison of emission data at 1200 rpm as a function of relative air/fuel ratio at SNL = 4 deg BTDC

124

At low loads the rapidly burning sprays were bright white with high injection pressures and yellow-white with the conventional injection system as they burnt slowly. As the fuelling rate increased it was noticed that the sprays impinged on the wall and overlapped creating fuel rich zones, resulting in a reduction of oxygen availability and a low temperature field which was indicated by the colours of the flames. In the case of the conventional injection system the $EINO_x$ levels are much lower under the same conditions. This is supported by visual analysis of the photographs which revealed that the spray and combustion processes were slow and weak, which also resulted in higher mean exhaust temperatures, fig.12. As the swirl ratio increased, under the same conditions, the NO_x levels were significantly higher due to better mixing and rapid combustion resulting in an increased rate of pressure rise $(dp/d\theta_{max})$.

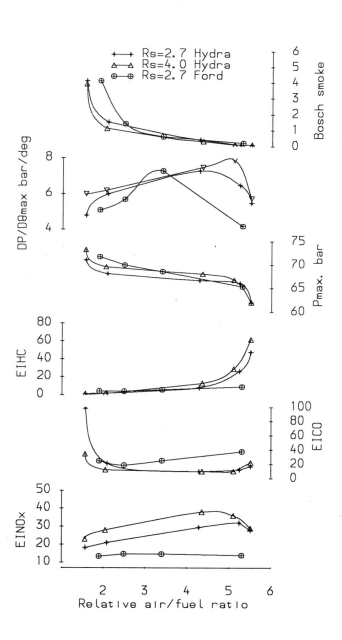

Fig. 11 Comparison of emission data at 2000 rpm as a function of relative air/fuel ratio at SNL = 4 deg BTDC

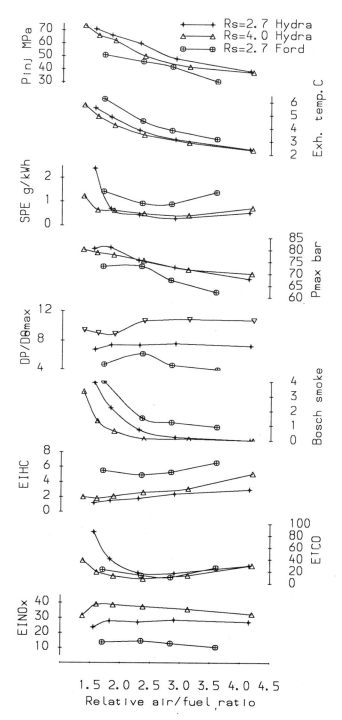

Fig. 12 Comparison of emission data at 3000 rpm as a function of relative air/fuel ratio at SNL = 7 deg BTDC

The variation of emissions as a function of injection timing is shown in fig.14 for different swirl ratios. As expected the oxides of nitrogen increase as the swirl ratio increases and do so more rapidly with advanced injection timing, due to higher cylinder pressures and temperatures. The NO_x levels are consistently lower at all speeds with low injection pressures by two or three times in magnitude.

One of the key observations was that over the entire speed range the NO_2 component of total NO_x measured, which directly contributes to acid rain, was high at low loads and decreased as the load increased.

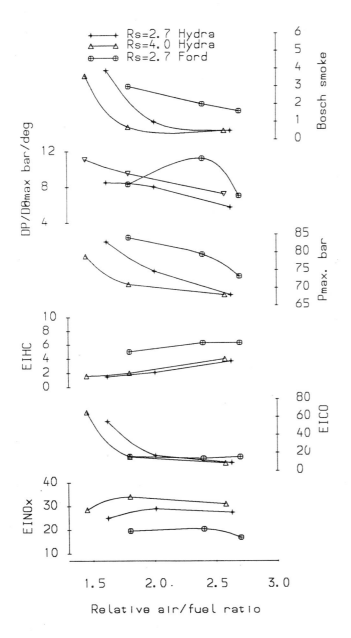

Fig. 13 Comparison of emission data at 4000 rpm as a function of relative air/fuel ratio at SNL = 11 deg BTDC

HYDROCARBON EMISSION INDEX(EIHC)

At low loads and speeds(that is high ϵ_R), the higher injection pressures from the EUI resulted in a higher hydrocarbon index than obtained with the conventional injector. When the engine speed exceeded 2500 revs/min the high pressure injector in reducing the EIHC became apparent, and this is caused by the smaller droplets and improved mixing. A possible explanation of this will now be given. The overall cylinder temperature rises with increased fueling, burning any fuel mixed in LFOR as the combustion progresses. Under part load conditions the fuel is stripped off the sprays during ignition delay and mixed with air beyond lean limit of combustion(fig.7). The mean temperature of the cylinder contents is so low and the flames are so localised that most of this fuel would simply be left unburnt. This contributes to high particulate emissions in the form of soluble organic fractions(SOF). As the ϵ_R decreases, higher temperatures of the cylinder contents burn off any fuel in LFOR resulting in lower HC emissions. At full load the lack of oxygen impairs the combustion, steeply increasing HC, smoke and particulates. Therefore high injection pressures coupled with high swirl are favourable at high loads. With the conventional injection system the HC emissions increase as the speed rises mainly because of poor atomisation and weak mixing.

CARBON MONOXIDE EMISSION INDEX (EICO)

CO emissions from diesels are low and usually unimportant as they always operate well on the lean side of the stochiometric ratio. But one has to be cautious with high pressure injection and small DI diesels. As shown in fig.10, CO levels are higher when operated with high injection pressures and either very lean or very rich mixtures because at low speeds there was not enough turbulence to cause proper mixing and subsequent combustion.

At high loads and all speeds the injection pressures were so high that most of the fuel was deposited on the wall of the piston bowl, forming very rich mixtures and the toroidal bowl was not good enough to further assist the mixing during combustion, resulting in higher CO levels. At constant ϵ_R and θ_s the increase in swirl reduced the CO emissions significantly, fig.14.

SPECIFIC PARTICULATE EMISSIONS(SPE)

A typical case study is presented in fig.12 for a constant injection timing of 7 deg BTDC at varying fuelling rate and two swirl ratios. It is obvious that the NO_x and particulate emissions vary inversely, the larger being two to three times the smaller. The higher $(dp/d\theta_{max})$ and comparatively lower exhaust temperatures with high injection pressures confirm faster burning and high cylinder temperatures during the early stages of combustion, favouring higher NO_x formation and burning of the fuel mixed in LFOR, thus lowering particulates. The combustion photographs showed that luminous combustion ended early with high injection pressures and this is reflected as lower cylinder pressures in fig.5 during the tail end of the expansion process. Under part load conditions(say up to 40% of full load) high swirl under similar operating conditions causes an increase in SPE's, as can be seen from fig 12. However, at full load high swirl is beneficial. This is because, at low load, the high swirl increases mixing of the greater fuel quantity in to LFOR.

From in-cylinder photography it was observed that at full load 70% of the fuel was deposited on the combustion chamber wall most of this fuel emerging into burning sprays with little influence of swirl(Rs=2.7), ignition or combustion on burning spray configuration while in-

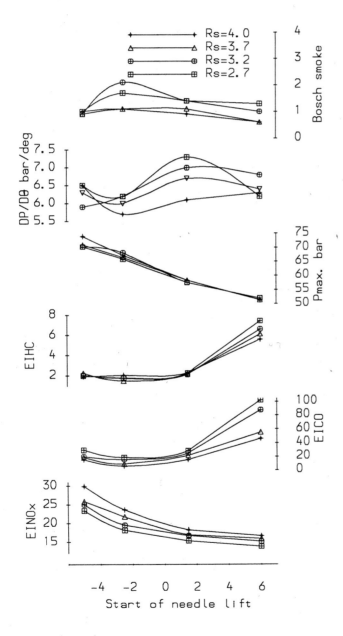

Fig.14 Comparison of emission data at 2000 rpm as a function of injection timing

jection was continuing. As the combustion progressed swirl broke down but not strongly enough to cause rapid mixing of the fuel. There is the point in fig.12 where SPE and smoke emissions rapidly increase even overtaking the conventional injection system. However the mixing is better with increased swirl ratio(Rs=4.0). Therefore, in the case of small DI diesels high injection pressures have negative effects unless used with optimised combustion chamber shape and swirl ratio to achieve better emissions/performance trade off. The trends are similar at all other speeds.

BOSCH SMOKE

The smoke levels are low at high ϵ_R and increase gradually as fueling increases at all speeds. The increase is much more gradual with high pressure injection. The influence of varying different parameters on smoke levels is shown in figs.10,11,12,13,14. With the conventional injection system the smoke levels are relatively high and increase steeply at high loads for the reasons explained above. The smoke limited power could be significantly improved with high injection pressures and better combustion chamber shape.

SUMMARY AND CONCLUSIONS

A comprehensive test programme has been designed and implemented to study the influence of fuel injection pressures on the emission behaviour of small high speed DI diesel engines. The injection, mixing and combustion processes which control the formation of emissions inside the cylinder were visually observed by pulse laser illuminated spray and combustion photography with synchronised data acquisition.

The results of the study can be summarised as follows:

1. Mean injection pressures with the unit injector are 100% higher than conventional injection. Higher injection pressures were achieved with reduced nozzle hole size and pump cam design.
2. At any time during injection the four sprays were observed to be of different shapes and sizes.
3. With high pressure injection swirling air carried fuel vapour away from spray boundaries mixing beyond the lean limit of combustion.
4. At part loads with high injection pressures the premixed combustion was explosive and resulted in a short combustion duration. The burning sprays were carried away by swirling air without further break down, creating high temperature gradients.
5. At high loads and high injection pressures 75% of the fuel was deposited on the wall thus producing fuel rich mixtures.
6. It was found that NO_x and $dp/d\theta_{max}$ showed similar trends. Therefore reducing pressure rise not only reduces NO_x but noise as well.

With high pressure injection,the Hydra engine in its present form was noisier, emitted higher NO_x and HC at low loads and higher CO at maximum load than the existing Ford HSD425, although smoke and particulates were considerably better.

The authors are aware that the combustion chamber of the Ford HSD425 has recently been updated. A re-entrant combustion chamber bowl has been incorporated and the inector(a stanadyne pencil injector) has been relocated close to the bowl centre-line. A more modern electronic governing system has also been fitted.

ACKNOWLEDGEMENTS

The authors would like to thank the Science and Engineering Research Council of the United Kingdom for providing the financial support for the project through Grant GR/F 94071. The authors extend their gratitude to Dr. Hugh Frost and Roger Pruce for their technical assistence during the project. The authors also wish to express their appreciation to Oxford lasers for providing help and advice in operating the laser.

Appendix A

CALCULATION PROCEDURE FOR EMISSIONS

SPECIFIC PARTICULATE EMISSIONS(SPE)

The SPE was calculated using the following equations[13]

$$P_1 = P_4 \frac{NO_{x1}}{NO_{x3}} \frac{V_a}{V_4}(1 + 0.06606\lambda) \qquad (1)$$

Where

P_1 = Mass of the particulates emitted by the engine in time T (mg)

P_4 = Mass of particulates captured by the filter in time T (mg)

NO_{x1} = NO_x concentration in engine exhaust [ppm]

NO_{x3} = NO_x concentration in diluted exhaust [ppm]

V_a = Engine air flow rate corrected to STP conditions $[m^3]$

V_4 = Sample gas flow rate corrected to STP conditions $[m^3]$

λ = Equivalence ratio

$$SPE[\frac{g}{kWhr}] = \frac{3.6P_1}{Power(kW)} \qquad (2)$$

HC EMISSION INDEX(EIHC)

The HC dry concentration from the FID analyser (ppm) was converted to wet-concentration according to SAE J177[14]

$$WBC_{HC} = DBC_{HC}[1 - x(F/A)] \qquad (3)$$

$$\dot{M}_{HC} = \dot{M}_{ex} \times 0.0287 \times WBC_{HC} \qquad (4)$$

$$EIHC = \frac{\dot{M}_{HC} * 1000/60}{\dot{M}_f} \qquad (5)$$

NO_x EMISSION INDEX $(EINO_x)$

NO_x emission index was calculated according to SAE J1003[15]

$$WBC_{NO_x} = DBC_{NO_x}[1 - x(F/A)] \qquad (6)$$

$$\dot{M}_{NO_x} = \dot{M}_{ex} \times 0.0952 \times WBC_{NO_x} \qquad (7)$$

$$EINO_x = \frac{\dot{M}_{NO_x} * 1000/60}{\dot{M}_f} \qquad (8)$$

CARBON MONOXIDE EMISSION INDEX (EICO)

CO emission index was calculated according to SAE J1003

$$\dot{M}_{co} = 0.0580 \times CO(ppm) \times \dot{M}_{ex} \qquad (9)$$

$$EICO = \dot{M}_{co} * 1000/60/\dot{M}_f \qquad (10)$$

where

WBC = Wet basis concentration [ppm]

DBC = Dry basis concentration [ppm]

HC = Unburnt hydrocarbons[ppm carbon]

NO_x = Oxides of nitrogen[ppm]

x = H/C ratio (1.836)

F/A = Fuel-air ratio

$\dot{M}_{NO_x}$ = NO_x emissions [kg/min]

$\dot{M}_{HC}$ = HC emissions in [kg/min]

$\dot{M}_{ex}$ = exhaust mass flow rate [kg/min]

$\dot{M}_f$ = Fuel flow rate [kg/sec]

References

[1] Monaghan, M. L. The High Speed Direct Injection Diesel for Passenger Cars. SAE 810477, 1981.

[2] Wade, W. R., Idzikowski, T., Kukkonen, C. A. and Reams, L. A. Direct Injection Diesel Capabilities for Passenger Cars. SAE 850552, 1985.

[3] Frankl, G., Barker, B. G. and Timms, C. T. Electronic Unit Injector Revised. SAE 891001, 1989.

[4] Kakegawa, T., Suzuki, T., Tsujimura, K. and Shimoda, M. A Study on Combustion of High Pressure Fuel Injection for Direct Injection Diesel Engine. SAE 880422, 1988.

[5] Gill, Alan. P. Design Choices For 1990's Low Emission Diesel Engines. SAE 880350, 1988.

[6] Alcock, J. F. and Scott, W. M. Some More Light on Diesel Combustion. IMechE March 1963.

[7] Hiller, W., Lent, H. M., Meier, G. E. A. and Stasicki, B. A pulsed Laser for High Speed Photography. *Experiments in Fluids 5*, **pp141-144**, 1987

[8] Hiro Hiroyasu and Masataka Arai. Structure of Fuel Sprays in Diesel Engines SAE 900475, 1990.

[9] Kato, T. et al. Spray Characteristics and Combustion Improvement of D.I.Diesel Engine with High Pressure Fuel Inection. SAE 890265, 1989.

[10] Bird, G. L. The Ford 2.5 litre Direct Injection Diesel Engine. IMechE, **Vol 199 No D2**, 1985.

[11] Masahiko Nishida, Toshio Nakahira, Masanori Komori, and Kinji Tsujimura, Ikuo Yamaguchi. Observation of high pressure fuel sprays with laser light sheet method. SAE 920459, 1992.

[12] Elktob, M. M. Fuel atomization for spray modeling. *Prog. Energy Comb. Sci., 8, 1982*

[13] G.Cussons Ltd. Mini Variable Split Particulate Sampler - **P1815**, With Splitter Unit **P 1820**

[14] Measurement of carbon dioxide, carbon monoxide and oxides of nitrogen in diesel exhaust.Technical Report. **SAE J177** *SAE Handbook- Vol. 3*, 1987.

[15] Diesel engine emission measurement procedure. Technical Report **SAE J1003** *SAE Handbook- Vol. 3*, 1987.

C448/007

Further development of a small diesel engine for low cetane number fuels

A W E HENHAM, BSc, PhD, CEng, FIMechE, FInstE, MRAeS and
S J M KAWAMBWA, MSc, AMIMechE
Department of Mechanical Engineering, University of Surrey, Guildford

SYNOPSIS Earlier work in the Energy & Thermodynamics Group at Surrey has shown that a small, direct-injection, four-stroke, stationary diesel engine can be modified to run on both ethanol and methanol. The only changes were the addition of a simple spark-ignition system and a larger capacity fuel pump. This operation was successfully achieved over a range of loads at brake thermal efficiencies equal to those obtained with the original configuration on gasoil. The engine operated unthrottled, so retaining the part-load efficiency advantage of the true diesel. The current work explores the possibility of replacing the spark-assistance by two other methods of providing the increased energy needed to ignite alcohol fuels. These are a higher compression ratio, to give a higher air temperature at fuel injection, and exhaust gas recirculation (EGR), to increase the temperature at the beginning of compression. The paper reports the effects of fitting pistons giving a number of different compression ratios, initially using gasoil as a preliminary to operation with alcohol. The effect of EGR is also reported. Characteristics of brake specific fuel consumption against brake mean effective pressure are shown. Limited results with ethanol as a fuel are shown on the same basis. Developed originally as an engine to utilise indigenous bio-sustainable fuels, this may be considered an indicator of possible wider use as environmental pressures demand a particulate-free diesel engine.

1 INTRODUCTION

From the point of view of the diversification of energy resources available to provide convenient small-scale shaft power, biomass-derived fuels are worthy of investigation. These are generally of two broad categories - vegetable oils used in original or esterified forms and alcohols produced by distillation of a wider range of crops (1). The cetane values of vegetable oils can often be comparable with those for hydrocarbon diesel fuels but those for ethanol and methanol are much too low for the fuels to be used in a conventional diesel engine. Further, the autoignition temperatures of the alcohols are higher than that of diesel fuel. The low cetane value of the fuels, together with the high autoignition temperatures, necessitate changes in the composition of the fuel or the configuration of the engine in order to get the fuels to burn satisfactorily. A summary of the important properties is given in Table 1. Various methods have been used by a number of researchers in an attempt to ignite and maintain combustion of alcohols in diesel engines (2).

Table 1 Properties of alcohol fuels compared with standard hydrocarbon fuel

Fuel	Gasoil	Methanol	Ethanol
Calorific value MJ/kg*	41.8	19.9	27.2
Energy density MJ/litre*	36.4	15.9	21.6
Octane number RON	na	114	111
Cetane number	45	3	8
Spontaneous ign temp °C	245	385	365

* net or lower calorific value and energy density given.

2 SPARK-ASSISTED DIESEL

Experience has been gained at the University of Surrey in operating a small stationary diesel engine on alcohol fuels using spark assistance to initiate combustion. The conversion of the 7.4 kW Petter PH1 engine, involving spark assistance and a larger capacity fuel pump, has been reported with the test results obtained (3). The engine details are given as table 2. A long-electrode plug was selected from a range of special NGK plugs developed for use in lean-burn gasoline engines. This enabled the spark to be placed so that it is in contact with the fuel spray as this is deflected towards it by the swirl. The operating envelope of the engine on alcohol differs somewhat from that for gasoil in that the smoke limit does not apply for oxygenate fuel but the authors' experience is that care must be taken with thermal limits. To monitor this restriction a thermocouple has been imbedded in the head material of the test engine. Work has progressed to optimise injection and ignition timings at full-load for both methanol and ethanol as sole fuels and to analyse combustion performance data.

Table 2 Experimental engine specification

Make and type	Petter PH 1
Cycle	Four-stroke
Bore x Stroke	87.3 x 110 mm
Cylinders	one
Swept volume	658 cm^3
Compression ratio (in original form)	16.5
Speed range	750-2200 rev/min
Continuous rating (gasoil) power torque bsfc	7.4 kW 32.1 Nm 0.262 kg/kWh

Parallel programmes of tests were undertaken using ethanol and methanol as fuels. Ranges of injection and ignition timings were explored and then optimised on ethanol for an output 4 kW at 1500 rev/min. Optimisation plots of brake specific fuel consumption contours on axes of injection and ignition advance angles were made. These show that the optimum values are at about 50° btdc for injection and 30° btdc for ignition. This produced a minimum brake specific fuel consumption, based upon gasoil equivalent lower calorific value, of 255 g/kWh compared with 259 g/kWh for the unconverted engine on gasoil. Optimised at 5.6 kW and 1800 rev/min, the best ignition timing was somewhat later at 20-24° btdc, indicating that the flame propagation speed (which varies with load) affects the optimum timing. The tests at 4kW were repeated using methanol when the optimum timings were 50° and 35° btdc for injection and ignition respectively. The minimum bsfc was slightly improved at 240 g/kWh gasoil equivalent. The bsfc curve deteriorates by comparison with the gasoil standard as the load is reduced. The temperatures appear to become too low at part loads for reliable ignition to be guaranteed.

New problems occurred during this latter test series, however, because of the more corrosive nature of methanol. Despite the addition (as with ethanol) of 2% Castorene to the fuel, there was a lubrication failure of the fuel pump. This was found to result from a seizure which corrected itself after a few minutes. Another manifestation of the effect of methanol on metal-to-metal interfaces was leakage between the injector body and its seat which was not cured by fitting a new body. For this reason tests at other loadings were not completed and new methods of overcoming this problem will be explored.

3 COMBUSTION ANALYSIS

At the same time as the practical performance data are produced the combustion data from each test can be analysed. The standard procedure is to record, using a microcomputer, the in-cylinder pressure for each degree of crank angle on line. These data are then transferred to the mainframe for analysis using a number of analytical programs which have been developed. These include evaluating the geometry of the engine volumes and surface areas at each crank angle (4) and involve assumptions about the rate of heat transfer. Statistical analysis of 50 consecutive cycles is also undertaken to give cyclic dispersion values which indicate the reliability of combustion. For the cycle approximating closest to a typical cycle from the 50 recorded, the burning rate of fuel is calculated by two methods - based on those used by Rassweiller & Withrow and Krieger & Borman. There has been found to be good correlation of these two methods in determining the times (expressed in degrees of crank angle) to burn 10% and 50% of the total fuel per cycle. The time to burn 1% is, of course, much more sensitive to small changes in the assumptions and the correlation is lower for this stage. This part of the combustion process, representing the establishment of a flame kernel, has been found from this analysis to be dependent more upon the timing of the spark than of the fuel injection. Injection timing, however, has a more dominant effect on the later stages of combustion since the homogeneity of the charge at the time of ignition depends upon the difference in advance between injection and ignition.

Two distinct operating regimes are evident from the analysis of combustion. Early injection allows considerable mixing of fuel and air and the engine behaves rather like a port-injected, lean-burn, homogeneous-charge, engine. Late injection creates stratified-charge conditions more akin to those in a true diesel engine. It is possible

that some of the effects observed are characteristic of the inlet swirl, particular geometry of the combustion chamber, spray pattern and plug location.

4 CURRENT DEVELOPMENTS.

Following successful running of the engine on alcohol as a spark-assisted diesel, this project proposed to increase the compression ratio of the engine in order to attain a high enough air temperature at injection to ignite alcohol spontaneously without a spark. The original engine compression ratio of 16.5:1 was, therefore, raised to 25:1. This was achieved by modifying a standard PH1 piston to take an insert that produces the desired combustion chamber volume (Fig 1).

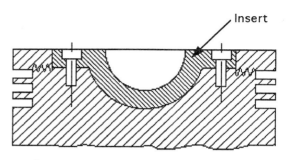

Fig 1 High compression ratio piston.

The resulting high compression ratio engine was initially tested on gasoil to establish baseline performance at the higher compression ratio. Experiments showed that engine power output was highly limited by the increase of compression ratio to 25:1. At an engine speed of 1650 rev/min., while the original engine is specified to achieve 5.1 kW, the power output of the engine at the higher compression ratio is limited to 2.7 kW. Similar limitations have been observed at an engine speed of 1500 rev/min (Fig 2) and at speeds higher than 1650 rev/min. Attempts to load the high compression ratio engine above this limited power output stalled the engine. Further, the brake specific fuel consumption was much higher for the same loads.

Engine performance was markedly improved when the compression ratio was lowered to 20:1 (Fig 2). Baseline tests on the engine at this compression ratio with and without exhaust gas recirculation (EGR) showed the considerable influence of exhaust gas on the combustion of diesel fuel at the higher compression ratio of 20:1 (Fig 3). At low levels of recirculation, EGR has the effect of improving the performance of the engine. This observation has also been reported by Takada et al (5) and Hikino et al (6). It will be observed from the results of tests at 1500 rev/min shown in Fig 3 that, based on brake specific fuel consumption, the performance of the engine at the 20:1 compression ratio with low levels of EGR is virtually identical to that of the original engine.

5 EXHAUST GAS RECIRCULATION SYSTEM.

Hikino et al (6) reported results from engine tests with a spark-assisted alcohol fuelled diesel engine. They showed that with spark-assistance it was difficult to achieve a good combustion especially at light loads, given the fuel distribution and the single ignition point at the spark plug. This observation seems to confirm earlier work by Newnham (7) where relatively good performance on a spark-assisted methanol/ethanol diesel was obtained only at near full-load conditions. Following on Newnham's work, this project proposes autoignition of alcohol fuels in order

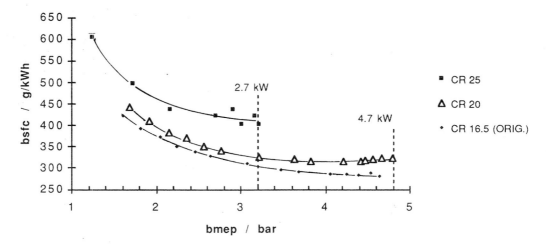

Fig 2   Engine performance on gasoil at compression ratios of 16.5:1 (original), 20:1 and 25:1; at a constant engine speed of 1500 rev/min.

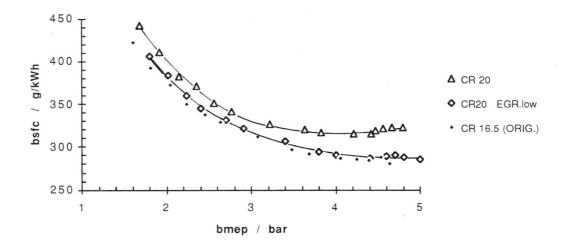

Fig 3   Engine performance on gasoil at 1500 rev/min. without EGR and with a low rate of EGR.

to overcome the poor low-load combustion of the spark-assisted diesel engine and also to achieve a better overall fuel economy. Because of the higher autoignition temperature of alcohol fuels, a higher air temperature in the combustion chamber at the point of injection will be required than with gasoil. This is expected to be achieved by increasing the compression ratio of the engine from the current 16.5:1 to 20:1, and also by fitting an exhaust gas recirculation (EGR) mechanism. Compression air temperature at low loads is known (6) to be lower than that obtained at high loads. Autoignition of alcohol fuels at low loads, therefore, could be expected to be achieved with the help of EGR.

Results from preliminary engine tests on gasoil at compression ratio 20:1 with exhaust gas recirculation are shown on Fig 3, compared with the same test without EGR. These results were obtained with very low levels of recirculation of less than 10 per cent. With such a low rate of EGR, the intake air temperature at the inlet manifold always remained below 50°C. The EGR system was very much restrictive in the sense that it only allowed low rates of recirculation, and as such the intake air temperature could not be raised high enough for possible ignition of alcohol fuels in a pure diesel engine. This system was therefore changed to the revised version shown in Fig 4. This comprises a welded combined inlet and exhaust duct assembly with a transverse connecting passage, close to the ports, controlled by a plate valve. The ducts incorporate

tappings for temperature and oxygen metering to enable the EGR ratio to be determined.

Fig 4 Exhaust gas recirculation mechanism.

Results obtained with the new EGR mechanism show that the rate of exhaust gas recirculation affects the performance of the engine. Fig 5 shows the effect of two different settings of the exhaust gas recirculation valve on the brake specific fuel consumption at a constant engine speed of 1500 rev/min. Increase in the rate of EGR results in increase in brake specific fuel consumption. Possibly the fuel injection timing needs to be reoptimised to allow for the change in flame propagation with EGR as the standard diesel oil setting was used for these tests. Performance can be made to drop to the point where it is as bad as running the engine at the higher compression ratio without EGR.

Initial trials of this system, but using ethanol as a fuel without any additional ignition source, have been undertaken. The performance of the engine at a compression ratio 20:1 is shown in Fig 6. This test was carried out at 1500 rev/min and with the same EGR setting as for gasoil. The gasoil results with no EGR are also shown for comparison. It can be seen that the best bsfc point for ethanol lies close to the gasoil line but that the efficiency deteriorates markedly at higher bmep values. This has been a limited test so far and no attempt has yet been made to optimise the timing. The start of fuel injection and the injection pressure were left as for the gasoil tests. Previous attempts to run the engine on ethanol unassisted were unsuccessful at cylinder head surface temperatures below 230 °C whereas, in this test, temperatures as low as 165 °C were observed during successful operation. Exhaust temperatures with alcohol were comparable with those for gasoil at the same loads.

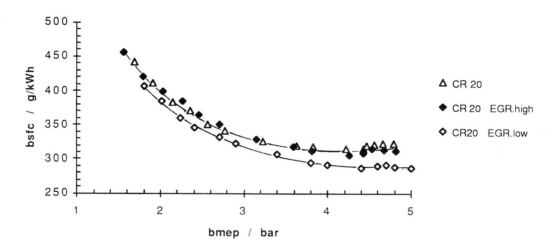

Fig 5   Comparison of engine performance on gasoil at CR 20:1 with high, low and no EGR. Engine speed is constant at 1500 rev/min.

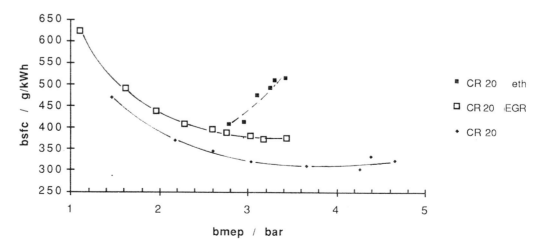

Fig 6   Engine performance at CR 20:1 using alcohol fuel with EGR compared with gasoil at the same EGR setting and with gasoil without EGR. Engine speed is constant at 1500 rev/min and bsfc for alcohol is shown in terms of gasoil equivalent.

6   CONCLUSIONS

It has been demonstrated that the single-cylinder, direct-injection engine will run satisfactorily at the higher compression ratio (20:1) on gasoil. The introduction of a modest amount of exhaust gas recirculation has a beneficial effect on brake specific fuel consumption throughout the load range although this is more marked at loads towards the rated output. Limited trials have been completed showing that this compression ratio with EGR enables unassisted combustion of ethanol to be achieved. The range of loads over which this was shown to be possible at 1500 rev/min was from 2.7 to 3.5 bar bmep.

The next stage of this research is to explore more fully the combustion characteristics of the same oxygenate fuels used in previous, spark-assisted, tests under the high-compression EGR system. This work is expected to show a much greater improvement in the combustion characteristics of alcohol fuels with their high auto-ignition temperatures once timing and injection characteristics have been optimised. The long-term aim is to demonstrate the potential of alcohols and other low cetane number fuels in diesels. The high efficiency of this form of prime mover, with the elimination of particulates in the exhaust products, combine to offer an environmentally favourable power plant using biologically-sustainable resources.

REFERENCES

(1) HENHAM, A W E and JOHNS, R A. Experience with alternative fuels for small stationary diesel engines. Proceedings of the Seminar on Fuels for Automotive and Industrial Diesel Engines, London, 1990, pp 117-122, Institution of Mechanical Engineers.

(2) JOHNS, R A, HENHAM, A W E and NEWNHAM, S. Alcohol fuels for diesel engines. *International Journal of Ambient Energy*, 1988, vol 9, no 1 pp 31-35.

(3) JOHNS, R A, HENHAM, A W E and NEWNHAM, S K C. The combustion of alcohol fuels in a stationary spark-assisted diesel engine. Proceedings of the Conference on Internal Combustion Engine Research in Universities, Polytechnics and Colleges, London, 1991, pp 61 - 66, Institution of Mechanical Engineers.

(4) DOUTHWAITE, C, HENHAM, A W E and CRAYSTON, J. An application of three-dimensional CAD to the analysis of internal combustion engine cycles. Effective CADCAM '91, Coventry, 1991, pp 35 - 38, Institution of Mechanical Engineers.

(5) TAKADA, Y, SASAKI, M and WATANABE, Y. Improved performance and durability of a heavy-duty methanol DI diesel engine. IX International Symposium on Alcohol Fuels, Florence, 1991, Vol. II, pp 455 - 460.

(6) HIKINO, K, SUZUKI, T, UEMATSU, S. Development of autoignition methanol engine system for low exhaust emissions and better fuel economy of city bus, IX International Symposium on Alcohol Fuels, Florence, 1991, Vol. II, pp 485 - 490.

(7) NEWNHAM, S K C. The combustion of ethanol in a spark-assisted diesel engine. PhD Thesis, 1990, Department of Mechanical Engineering, University of Surrey.

C448/014

Development of a high fuel economy and high performance four-valve lean burn engine

K HORIE, MSAE, MJSAE and K NISHIZAWA, MJSAE
Honda R & D Company Limited, Saitama, Japan

SYNOPSIS This paper describes the development of a four-valve lean-burn engine, especially the improvement of the combustion, the development of an engine management system, and the achievement of vehicle test results.

Major themes discussed in this paper are (1) the improvement of brake-specific fuel consumption under partial load conditions and the achievement of high output power by adopting an optimized swirl ratio and a variable-swirl system with a specially designed variable valve timing and lift mechanism, (2) the development of an air-fuel ratio control system, (3) the improvement of fuel economy as a vehicle and (4) an approach to satisfy the NOx emission standard.

1 INTRODUCTION

In recent years, the emission of CO_2 from motor vehicles has become an increasingly serious problem to the greenhouse effect. The reduction of motor vehicle fuel consumption is one of the most effective means for automobile manufacturers to lessen this problem. Among the methods to reduce fuel consumption, lean burn is well known as an effective approach. However, operation of an engine with a lean mixture requires swirl or other means to improve and stabilize the combustion which consequently decreases the rated output power due to the low flow coefficient of the intake port. Furthermore, the exhaust gases from lean-burn engines produce an oxygenated condition in which a three-way catalyst does not work effectively, thereby making it difficult to reduce NOx emission to a satisfactory level.

For these reasons, the lean-burn concept has not been well accepted by automobile manufacturers.

The authors believe that there should be ways to overcome these matters. A variable swirl system would be one way to obtain high output power. Also, NOx emission could be reduced to a sufficient level by improving the combustion to expand the lean limit and by developing a highly accurate air-fuel ratio control system.

The authors have succeeded in developing a four-valve lean-burn engine as the result of two developments: (1) a

Table 1:VTEC-E engine specifications

Cylinder Configuration		In-line 4 Cylinder
Bore × Stroke (mm)		75 × 84.5
Displacement (cc)		1493
Compression Ratio		9.3
Valve Mechanism		SOHC VTEC-E
Number of Valves		4
Valve Diameter	In.	ϕ 27.5
	Exh.	ϕ 23.5
Fuel System		PGM-FI
Ignition System		PGM-IG

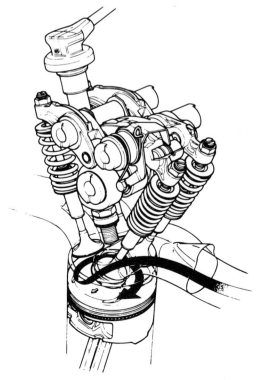

special variable-swirl system incorporating a mechanism to vary the valve timing and valve lift (hereinafter referred to as the "VTEC mechanism") which was first marketed in 1989 as a means to obtain high output power and (2) an engine management system to implement a highly accurate air-fuel ratio control with a wide range air-fuel ratio sensor using zirconia (ZrO_2).

This engine has three specific features; it (1) prevents the decrease of the volumetric efficiency by two-intake-valve operation using straight ports at high engine speeds, (2) generates swirl in the cylinder by means of the one-intake-valve deactivation mechanism at low engine speeds, and (3) produces the axially stratified charge by controlling the fuel injection timing in the intake stroke. This engine was developed with the following three targets.

● To improve brake-specific fuel consumption by 10% or more.

● To satisfy the current U.S.A. NOx emission standard.

● To ensure an optimum output power equivalent to that of conventional four-valve engines.

2 SWIRL CONTROL SYSTEM

2.1 REQUIREMENTS

As mentioned above, this engine is required to have a wider lean limit, better brake-specific fuel consumption, and a higher output power performance. Swirl motion is adopted as a means to obtain stable combustion for a lean mixture. To apply the swirl-motion method to a conventional engine, it is rather difficult to adopt a two-intake-valve system.

Operation with a single intake valve can produce a torque level equivalent to that of two-intake-valve operation at low engine speeds.However output power is very low at high speeds due to a decrease in the volumetric efficiency.

In order to satisfy these apparently incompatible requirements, i.e., a good brake-specific fuel consumption and a high output power, a valve mechanism which changes the number of intake valves actuated in either the high or low speed range was employed. For this purpose, a special valve deactivation mechanism (hereinafter called the "VTEC-E mechanism") using the mass-produced VTEC mechanism was developed. The objections

in developing the VTEC-E mechanism were as follows.

● Generate sufficient swirl by reducing radically the lift of one intake valve.

● Prevent the mechanism from decreasing the flow coefficient of the intake ports and valves.

● Make the mechanism structure as simple as possible.

2.2 VTEC-E MECHANISM

As shown in Fig.1, the VTEC-E mechanism consists of a camshaft with two profiles, one as the primary cam for intake-valve operation and the other as the secondary cam for one-intake-valve deactivation. The primary rocker arm has a roller follower with a built-in hydraulic piston.

At low engine speeds, each rocker arm works independently according to the corresponding cam profile to fully lift

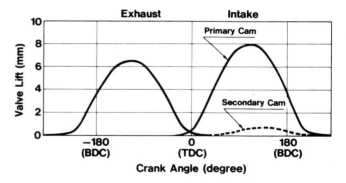

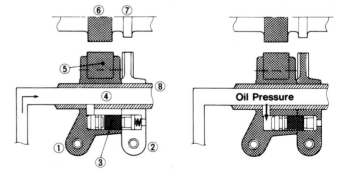

① **Primary Rocker Arm**
② **Secondary Rocker Arm**
③ **Hydraulic Piston**
④ **Oil Passage**
⑤ **Roller Follower**
⑥ **Primary Cam**
⑦ **Secondary Cam**
⑧ **Rocker Shaft**

Fig.1:Actuation of VTEC-E mechanism

the primary-side valve and to reduce the secondary-side valve to a peak lift of 0.65 mm.

The nominal value of the reduced lift, 0.65mm, is intended to flow the fuel that would otherwise remain in the secondary port, since the fuel is supplied equally to the port from the injector placed at the center of siamese port at all times.

When the engine is operating at high speeds, on the other hand, the signal from ECU opens the hydraulic circuit to remove the hydraulic piston, which engages the rocker arms to actuate the same lift for both the primary and the secondary valves.

3 LEAN-COMBUSTION CONTROL

3.1 OPTIMIZATION OF SWIRL

If a higher swirl ratio is generated by means of a helical port, then the flow coefficient of the intake port is smaller and subsequently the output power would become lower even though a wider lean limit is obtained. Therefore, the lean limit must be improved not only by increasing the swirl ratio but also by improving the combustion with modification of the combustion chamber shape. Thus securing a sufficiently high lean limit at a relatively low swirl ratio.

The combustion chamber shape of this lean-burn engine is shown in Fig.2. This compact combustion chamber with a wide squish area 0.75mm thick and 55mm in diameter compared to the bore of 75mm realizes fast burn as the result of a high turbulence generated by squish and a shortened flame-propagation distance.

This combustion chamber with a straight port has significantly improved the lean limit over the conventional pent-roof-type combustion chamber at the same swirl ratios (Fig.2).

The port specification was optimized after confirming the effects of the swirl ratio on the output power at full load, brake-specific fuel consumption, pumping loss at part load, and intake-valve flow coefficient under two-intake-valve operation. The swirl ratio is zero when two intake valves are operated and one-intake-valve operation with a straight port gives swirl ratios up to 2.8. To obtain higher swirl ratios, a helical port must be used for one-intake-valve operation. However, brake-specific fuel consumption tends to saturate at swirl ratios over 2.5 due to an increase in pumping loss (Fig.2).

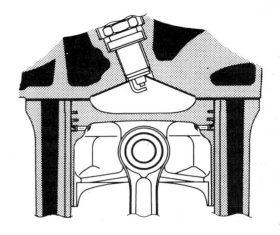

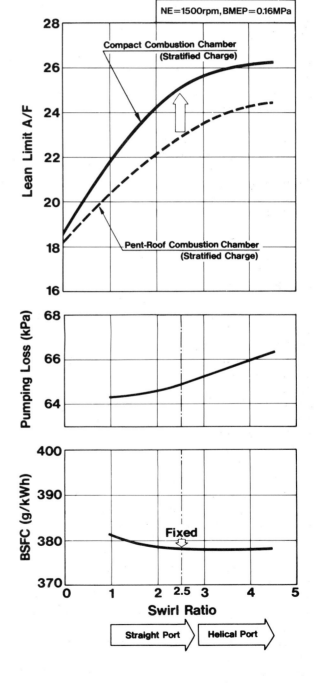

Fig.2:Influence of swirl ratio on lean limit and BSFC at part load

When a helical port is used, the intake-valve flow coefficient becomes lower in two-intake-valve operation and the rated power decreases about 8%.

Based on these test results, the swirl ratio of 2.5 was chosen. This is a value that can be achieved with a straight port, in order to achieve the two purposes of this engine; that is, power output increase and reduction of fuel consumption.

3.2 MIXTURE FORMATION

Regarding the fuel injection timing of the port-injection swirl engine, it has been reported that, when fuel is injected at a given timing during the intake stroke, an axially stratified charge is generated with a rich mixture at the upper portion in the cylinder and a lean mixture at the lower portion (ref.(1)). The rich mixture in the vicinity of the spark plug stabilizes ignition and consequently enhances the lean limit.

In Fig.3, this phenomenon is expressed as the effect of injection timing when the air-fuel ratio is maintained at 22. The optimum fuel injection timing during the intake stroke enables combustion to stabilize significantly. Although the lean limit is enhanced, NOx emission increases if fuel is injected at the most appropriate timing from the viewpoint of combustion stability by generating the axially stratified charge. This is presumably caused by the combustion of rich mixture at some portion in the cylinder. As previously mentioned, to make matters worse, the oxygenated condition of the exhaust gas from the lean-burn engine make it difficult to use a three-way-catalyst to reduce NOx. For this reason, it is necessary to control the generation of NOx in the combustion process.

Thus, we have set the fuel injection timing at a point in Fig.3 where the NOx emission does not increase even though the lean limit is slightly lowered. This injection timing is applied to the entire one-intake valve operating range.

3.3 COMBUSTION IMPROVEMENT

The axially stratified charge generated by means of swirl stabilizes ignition by enriching the air-fuel ratio around the spark plug. At the same time, combustion is enhanced through the turbulence generated by the swirl and the squish.

Fig.4 shows the comparison of combustion between the VTEC-E engine and a conventional four-valve engine. The swirl and the axially stratified charge generated by fuel injection during the intake stroke shorten the period from ignition to a point where the mass-burn fraction reaches 10% (hereinafter called the "ignition delay") and the period over which the mass-burn fraction increases from 10 to 90% (hereinafter called the "combustion period"). The compactness of the combustion chamber further shortens the combustion period, thus improving the lean limit to an air-fuel ratio of 25 (as mentioned above).

The improvement of the combustion characteristics and lean limit gives the reduction of brake-specific fuel consumption by 12% and the decrease of NOx emission to a sufficiently low level over an air-fuel ratio of 22 (Fig.4).

As previously described, the operation of a lean-burn engine requires the suppression of NOx by controlling the air-fuel mixture as lean a value as possible. If the air-fuel ratio becomes too lean, however, the combustion gradually becomes unstable. The desirable range for air-fuel ratio is extremely narrow, from 22 to 25. In this range, the U.S.A. standard for NOx emission can be satisfied and brake-specific fuel consumption and combustion stabilities can be maintained at satisfactory levels. To complete a lean-burn engine as mounted on a vehicle, it is essential to establish a highly accurate and quick-response air-fuel ratio control technology that will be described later.

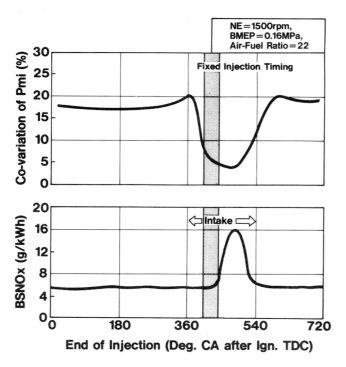

Fig.3:Influence of injection timing on co-variation of pmi and NOx emission

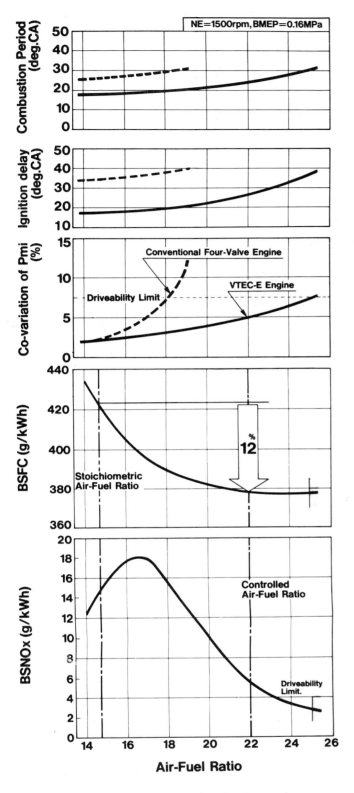

Fig.4: Influence of air-fuel ratio on
BSFC, BSNOx and combustion
characteristics

4 ENGINE MANAGEMENT SYSTEM

An engine management system was developed to reproduce the effects of good fuel consumption and stable combustion while simultaneously satisfying the NOx standard when the engine is installed on a vehicle.

The newly developed system features four sub-systems that undertake (1) swirl control by the VTEC-E mechanism, (2) fine air-fuel ratio control, to which a wide range of air fuel ratios are fed back by a linear air-fuel ratio sensor (hereinafter called the "LAF sensor"), (3) three-way catalyst, by which NOx emission is controlled at stoichiometric air-fuel ratio operations and (4) EGR control.

4.1 SWIRL CONTROL

The swirl generating mechanism in this system sets a swirl value of SR = 2.5 or SR = 0 by the VTEC-E mechanism. The change-over

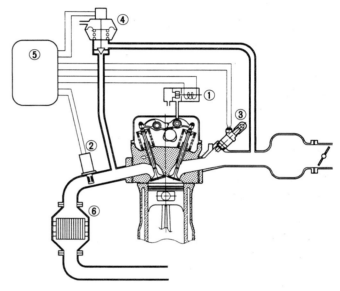

① **VTEC-E Control Solenoid**
② **LAF Sensor**
③ **Fuel Injector (Sequential Injection)**
④ **EGR Valve**
⑤ **Electronic Control Unit**
⑥ **Three-Way Catalyst**

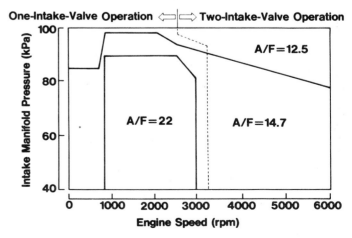

Fig.5: Engine management system and
control map

point for full load is set at 2500rpm where the torque is the same whether one or two intake valves are operating, thus allowing the hysteresis to be such that the driver feels no change-over shocks. For partial load operation, the change-over point is set at 3200rpm to improve the vehicle fuel economy as much as possible (Fig.5). The system does not change to the two-intake-valve operation mode when oil viscosity is too high to implement a quick response change-over at extremely low oil temperatures.

4.2 AIR-FUEL RATIO CONTROL

During one-intake-valve operation at part load, the air-fuel ratio is controlled at 22, at which point NOx standards, fuel consumption, and combustion stability criteria are all satisfied. During two-intake-valve operation, NOx is controlled by a three-way catalyst using EGR at stoichiometric air-fuel ratio, since NOx emissions increase at high engine speeds even under a partial load condition. For full-load operation in this case, the air-fuel ratio is controlled at 12.5 to obtain maximum output power (Fig.5).

Under the condition of a transient operation, however, it is extremely difficult to control the air-fuel ratio precisely at a required value. This requires the development of a highly accurate air-fuel ratio control system in order to reproduce good brake-specific fuel consumption, high combustion stability and sufficiently low NOx level. The most important subject for the air-fuel ratio control system is the development of a LAF sensor that can detect the air-fuel ratio over a wide range with high precision and quick response.

5 VEHICLE TEST RESULT

Emissions and fuel economy were evaluated in the LA-4 and highway-mode test using a vehicle incorporating the above-mentioned lean-burn technology and the engine management system. In this test, the fuel economy in the highway mode was 12% higher than that of an engine with stoichiometric air-fuel ratio installed on the same vehicle, but 8% higher in the LA-4 mode (Fig.6). This effect in the LA-4 mode was slightly worse than that of the steady-state engine bench test, which showed an improvement of 12%.

This result was caused by the difference in which air-fuel ratio in the LA-4 mode test was not maintained at the lean mixture over the entire operating range, instead the stoichiometric value was adopted for idling, at low coolant temperatures, and under acceleration. Thus good fuel economy was obtained and simultaneously the U.S.A NOx standard, 1.0g/mile was satisfied. However, one approach to satisfy the California state standard, 0.4g/mile is to adopt EGR at the stoichiometric air-fuel ratio and to use a three-way catalyst, which will also reduce brake-specific fuel consumption.

The most important matter with respect to the future of lean-burn engines is therefor to reduce NOx emissions by improving combustion and to promote research and development of an effective catalyst for NOx reduction.

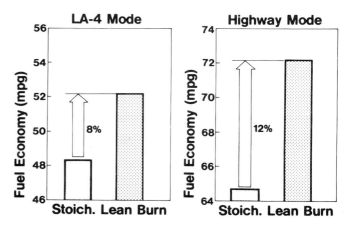

Fig.6:Effect of lean burn on vehicle fuel economy (Inertia Weight=2375lbs)

6 SUMMARY

The authors have demonstrated the possibility of using a straight port having a high flow coefficient by two kinds of technology; (1) realization of the axially stratified charge with a variable swirl system using a special valve deactivation mechanism (VTEC-E) and fuel injection in the intake stroke, and (2) further improvement of the lean limit with a compact combustion chamber shape.

As for the partial load operation, it was found that the straight port stabilizes combustion at an air-fuel ratio of 22, improves brake-specific fuel consumption by 12%, and is effective for reducing NOx emissions.

At full load operation, a high output power (69kW) can be obtained by maintaining a volumetric efficiency in the two-intake-valve operation mode at the same level as that of conventional four-valve engines.

© IMechE 1992 C448/014

At the vehicle test, the authors have also succeeded in improving the fuel economy by 8% in the LA-4 mode, 12% in the highway mode and in satisfying the current U.S.A. NOx standard by precisely controlling the air-fuel ratio under transient conditions by using a new air-fuel ratio control system with a high-response linear air-fuel ratio sensor.

REFERENCES

(1) A.A.Quader.The Axially-Stratified-Charge Engine. SAE Paper 820131, 1982

(2) G.C.Davis et al.,.Modeling the Effect of Swirl on Turbulence Intensity and Burn Rate in S.I.Engines and Comparison with Experiment, SAE Paper 860325, 1986

(3) S.Matsushita et al.,.Effects of Port with Swirl Control Valve on the Combustion and Performance of S.I.Engines. SAE Paper 850046, 1985

(4) S.Matsushita et al.,.Development of Toyota Lean Combustion System. SAE Paper 850044, 1985

(5) B.Khalighi .Intake-Generated Swirl and Tumble Motions in a 4-Valve Engine with Various Intake Configurations-Flow Visualization and Particle Tracking Velocimetry. SAE Paper 900059, 1990

(6) S.Furuno et al.,. The Effects of Inclination Angle of Swirl Axis on Turbulence Characteristics in a 4-Valve Lean-Burn Engine with SCV. COMODIA 90:437-442

(7) L.Mikulic et al.,. Development of Low Emission High Performance Four Valve Engines. SAE Paper 900227, 1990

(8) K.Inoue et al.,. A High Power, Wide Torque Range, Efficient Engine with a Newly Developed Variable-Valve-Lift and -Timing Mechanism. SAE Paper 890675, 1989

C448/015

Effects of exhaust emission control devices and fuel composition on speciated emissions of SI engines

K KOJIMA, T HIROTA, T INOUE and K YAKUSHIJI
Nissan Motor Company Limited, Yokohama, Japan

SYNOPSIS   Hydrocarbons and other organic materials emitted from S.I. engines cause ozone to form in the air. Since each species of organic materials has a different reactivity, exhaust components affect ozone formation in different ways. The effects of exhaust emission control devices and fuel properties on speciated emissions and ozone formation were examined by measuring speciated emissions with a gas chromatograph and a high-performance liquid chromatograph. In the case of gasoline fuels, catalyst systems with higher conversion rates such as close-coupled catalyst systems are effective in reducing alkenes and aromatics which show high ozone forming potential (OFP). With deterioration of the catalyst, non-methane organic gas (NMOG) emission increases, but the OFP decreases slightly because of the increase in the rate of alkanes having low OFP. With regard to the fuel composition, reducing the aromatic content results in a lower aromatic level in the exhaust emissions, and is therefore an effective way of reducing ozone formation. The NMOG level is higher with M85 than with gasoline. However, since the OFP is low, using M85 results in less ozone formation than with gasoline. The OFP of M85 vehicles is influenced mostly by formaldehyde.

1. INTRODUCTION

Speciated emissions of organic materials and their ozone reactivities were evaluated with the aim of finding an effective way to reduce the amount of ozone in the air.

Since increased ozone concectration in the air has been found to generate photochemical smog, enviromental standards have been established to regulate ozone levels in the U.S.. Ozone formation in the air is related to NOx and reactive organic gas (ROG). Each species of ROG has a different influence on ozone formation. Therefore, even with the same mass emissions of ROG, it is possible for its influence on ozone formation to differ according to the type of exhaust emission control device or fuel used. The regulations proposed by the California Air Resources Board for low-emission vehicles call for an adjustment of the standards using a Reactivity Adjustment Factor.[1] This factor is based on ozone forming potential (OFP), which indicates the contribution of an emission component to ozone formation. However, to estimate the influence on ozone formation accurately, it is necesaly to evaluate the respective effect of various factors. These include an analysis of reactivity mechanisms, air modeling taking into account air and weather conditions, an evaluation of emission inventories, including ROG sources and traffic conditions, and an accurate measurement of speciated emissions from vehicles.

Carter proposed the Maximum Incremental Reactivity (MIR) as an index for ozone formation.[2] This index shows the maximum value of the increase in ozone formation. He estimated the effect that changing from gasoline to alternative fuels, such as methanol and CNG, would have on reducing ozone formation. Russell estimated the influence of alternative fuels on ozone formation using a 3-dimensional airshed model, which they applied to different groupings of speciated emissions.[3] However, many uncertainties still remain with regard to reactivity mechanisms and air modeling, and effective methods have yet to be devised for analyzing them.

Various measurements and analyses of speciated emissions have been made. Weaver et al. at Sierra Researc h[4] compared speciated emissions from a gaso-line vehicle with those from a methanol vehicle, and Gething et al. at Chevron Research[5] compared speciated emissions from a gasoline vehicle with those from a reformulated gasoline vehicle. Dempster and Shore at Ricardo Consulting Engineers[6] identified the influence of aromatics in gasoline by measuring 23 hydrocarbons emitted from a gasoline engine. Although studies have evaluated the effects of alternative fuels on reducing the formation of ozone, no conclusive theory has been established yet.

This study examined the influence of different exhaust emission devices and fuels on speciated emissions control in relation to ozone formation by measuring the speciated emissions from vehicles equipped with spark-ignition engines. While some uncertainties with regard to the ozone formation index still remain, the ozone forming potential was evaluated on the basis of the MIR published by the Caifornia Air Resources Board in January 1992.

2. TEST VEHICLES AND EXPERIMENTAL METHODS

2.1 TEST VEHICLES AND ENGINES

The test vehicles were based on 1991 production NX Coupes for California. A flexible fuel vehicle (FFV) was used in the methanol (M85) tests. Table 1 shows the main specifications of the test vehicles. They were equipped with a GA16DE engine, a four-cylinder in-line 1.6-liter spark-ignition powerplant with dual overhead camshafts and four valves per cylinder. The fuel supply system was a fully sequential electronically controlled injection system and the catalysts were Pt/Rh three-way catalysts. The FFV modifications were made by changing certain fuel system parts and adjusting the control system calibrations.

2.2 TEST FUELS

The vehicles were tested using both gasoline and methanol (M85). Table 2 shows the main specifications of the fuels. Two types of gasoline were used, a baseline gasoline with specifications equivalent to the U.S. industry average and a reformulated gasoline, created by modifying T90 and the contents of aromatics,

		BASELINE GAS.	REF. GAS.
DENSITY (kg/l)		0.7471	0.7406
RPV (psi)		0.610	0.512
DISTILATION TEMPERATURE(℃)	(10%)	56.0	60.0
	(30%)	76.0	75.5
	(50%)	94.0	94.0
	(90%)	154.0	145.0
	EP	201.0	186.0
ALKANES (vol%)		58.4	51.7
ALKENES,ALKYNES(vol%)		8.4	11.7
AROMATICS (vol%)		33.2	21.6
MTBE (vol%)		0.0	15.0
SULFUR (%)		0.03	0.003

* ALKANES, ALKENES, ALKYNES, AROMATICS,
PROPORTIONS WERE MEASURED BY FIA

Table. 2   Fuel Specifications

Table. 1   Specifications of Test Vehicles

VEHICLE	1991MY NX Coupe	CALIFORNIA MODEL, 2 - DOOR Coupe
	TRANSMISSION	4 - SPEED AUTOMATIC
ENGINE	MODIFIED - GA16DE	4CYL., IN LINE
	BORE x STROKE	76.0 x 88.0 mm
	DISPLACEMENT	1597 cc
	COMPRESSION RATIO	9.5 to 1
	FUEL SUPPLY	SEQUENTIAL PORT FUEL INJECTION
CATALYST	CLOSE - COUPLED	MONOLITH Pt / Rh THREE WAY CATALYST
	UNDER - FLOOR	MONOLITH Pt / Rh THREE WAY CATALYST

	C1	C2 - C4	C5 - C12
APPARATUS	HP 5890	HP 5890	HP 5890
COLUMN	MOLECULAR-SIEVE 5A	PLOT 0.53MMφ x 25M	DB-1 0.32MMφ x 60M
COLUMN TEMPERATURE	50℃	-20℃ - 200℃	-60℃ - 250℃
DETECTOR	FID	FID	FID
DETECTOR TEMPERATURE	100℃	150℃	200℃
CARRIER	30ML/MIN	7ML/MIN	3ML/MIN
SAMPLE SIZE	IML	2ML	* CONCENTRATE (TCT)

*CHROMPACK TCT-CONCENTRATOR

Table. 3   HC Analysis Conditions

SAMPLE LINE TEMPERATURE	120℃
SAMPLING FLOW RATE	1L / MIN
SAMPLE VOLUME	5 μ L
COLUMN	ODS 4.5MMφ × 300MM
COLUMN TEMPERATURE	40℃
MOBILE PHASE	70%CH3CN / 30%H2O-(15MIN) CH3CN-(5MIN)
FLOW RATE	1ML / MIN
DETECTOR	UV (343NM)

Table. 4   Aldehyde and Ketone Sampling and Analysis Conditions

SAMPLE LINE TEMPERATURE	120℃
SAMPLING FLOW RATE	1L / MIN
SAMPLE VOLUME	3 μ L
VOLUME OF WATER USED FOR TRAP	15ML
APPARATUS	SHIMAZU 7A
COLUMN	DB-WAX 0.53MMφ x 30M
COLUMN TEMPERATURE	50℃ - 100℃
DETECTOR	FID
DETECTOR TEMPERATURE	220℃
CARRIER	5ML / MIN

Table. 5   Alcohol Sampling and Analysis Conditions

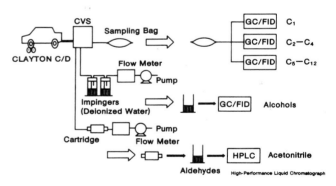

Fig. 1   Schematic of Exhaust Gas Analysis System

alkenes, sulfur and other components.

2.3 EXHAUST GAS MEASUREMENT

Exhaust gas samples were collected using three different devices for each Federal Test Procedure (FTP) mode employed in the constant volume sampler (CVS) test. Figure 1 shows the measurement system used to analyze speciated emissions. The measurement conditions of a gas chromatograph (GC) and a high-performance liquid chromatograph (HPLC) are shown in Tables 3 through 5.

For the purpose of accurately classifying hydrocarbon species, the hydrocarbon samples were collected using tedlar bags and analyzed using three methods based on their carbon number. Methane was analyzed using a molecular sieve-GC, C2 to C4 hydrocarbons were analyzed using a megabore column-GC, and C5 to C12 hydrocarbons were analyzed using a capillary column-GC with a thermal desorption cold trap injector. A dry cartridge method was used for analyzing aldehydes and ketones.[7] With this method, CVS diluted exhaust gases are drawn through a silica gel cartridge containing 2,4-dinitrophenylhydrazine (DNPH), which reacts with aldehydes and ketones. The aldehydes absorbed in the cartridge are washed out by acetonitrile injection and then measured by a high performance liquid chromatograph (HPLC). Unburned methanol was measured by drawing CVS diluted exhaust gases through impingers containing deionized water. The dissolved methanol was then analyzed using a megabore column-GC.

3. RESULTS AND DISCUSSION

3.1 INFLUENCE OF EXHAUST EMISSION CONTROL DEVICES WITH BASELINE GASOLINE

An investigation was made of how exhaust emission control devices affected speciated emissions and OFP in the Federal Test Procedure. Speciated emissions were measured under three different conditions: without a catalyst (engine-out), with an under-floor catalyst (UF catalyst), and with a close-coupled catalyst plus an

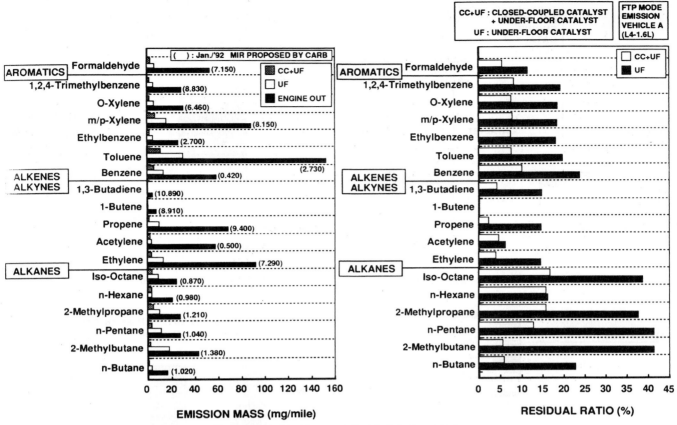

Fig. 2   Emission Mass and Residual Ratio of Main
Hydrocarbon Species for Baseline Gasoline

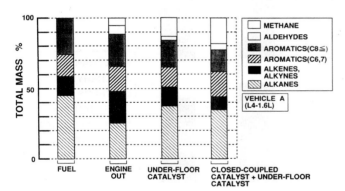

Fig. 3   Breakdown of Fuel and Exhaust Mass into
Hydrocarbon Types for Baseline Gasoline

under-floor catalyst (CC/UF catalyst system). Figure 2 shows the main components of speciated emissions and their residual ratios in the catalysts. Figure 3 shows breakdowns of the fuel and exhaust masses into their respective components.

The main components of speciated emissions from a gasoline engine that contribute to ozone formation are aromatics and alkenes. Aromatics were mainly represented by toluene and xylene, and alkenes were mainly represented by ethene and propene. Many alkanes are emitted from vehicles, but since their OFP is generally low, they do not contribute much to ozone formation. As an index for ozone reactivity, MIR, proposed by Carter, was adopted. The specific MIR values used were those published by CARB in January 1992. The amount of ozone formation was assumed to be the total amount of each type of speciated emissions multiplied by the MIR value. OFP was defined as the value obtained by dividing the amount of ozone formation by NMOG.

As Figures 2 and 3 show, a comparison of the engine-out emission components with the fuel components shows that the percentage of alkanes with a low

MIR value decreased and the percentage of alkenes and aromatics with a high MIR value increased, indicating that the OFP of engine-out emissions was higher than that of the fuel. The engine-out emissions measured with the UF catalyst and with the CC/UF catalyst system showed a much larger reduction in alkenes than in other components owing to catalytic reactions. Aromatics were the next most reactive component after alkenes, and alkanes were the least reactive. This tendency was more pronounced with the CC/UF catalyst system, because a catalyst light-off time, i.e. the period from engine start-up to catalyst light-off, becomes shorter to add a CC catalyst. Therefore the OFP with the CC/UF catalyst system is lower than that with other systems. With the use of a catalyst, the NMOG emissions are reduced and OFP is also reduced. This is because alkenes and aromatics tend to react more readily than alkanes on the catalyst. In short, higher catalyst light-off performance has a much greater influence on ozone reduction than does NMOG.

3.2 INFLUENCE OF CATALYST DETERIORATION WITH BASELINE GASOLINE

Figure 4 shows the NMOG emissions and OFP for an aged catalyst in comparison with the results for a fresh catalyst when the CC/UF catalyst system was used, and Figure 5 shows the residual ratios of an each emission species. The test data show that the NMOG emissions increase as the catalyst performance deteriorated, but the OFP declined. Catalyst reactivities showed noticeable deterioration in alkanes such as ethane, butane and hexane. With catalyst deterioration, the relative percentages of alkenes and heavy aromatics with high MIR value decreased, resulting in a lower OFP, and since ozone formation was calculated by multiplying NMOG emissions by the OFP, the rate of ozone formation increase tended to be less than that of NMOG increase.

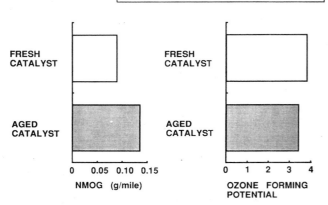

TEST VEHICLE· FFV NX COUPE
FTP MODE EMISSION
ENGINE   GA16DE (L4-1.6L)
CATALYST   CC+UF
AGING   AMA 50,000 MILES
FUEL   BASELINE GASOLINE

Fig. 4   NMOG and OFP in Durability Test

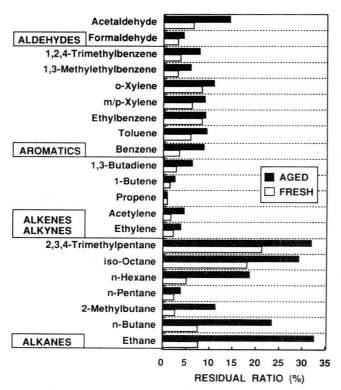

Fig. 5   Residual Ratio of Main Hydrocarbon Species
for Baseline Gasoline

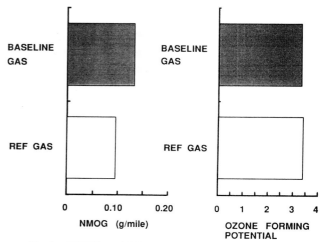

TEST VEHICLE FFV NX COUPE
FTP MODE EMISSION
ENGINE   GA16DE (L4-1.6L)
CATALYST   CC+UF
AGING   AMA 50,000 MILES

Fig. 6   NMOG and OFP of Baseline Gasoline and
Reformulated Gasoline

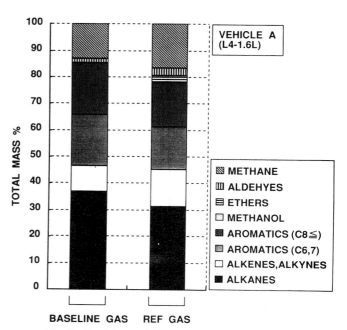

Fig. 7   Breakdown of Exhaust Mass into Hydrocarbon
Types of Baseline Gas and Reformulated Gas

3.3 INFLUENCE OF REFORMULATED GASO-LINE ON SPECIATED EMISSIONS AND OZONE FORMING POTENTIAL

The influence of reformulated gasoline on speciated emissions was evaluated. In comparison with the baseline gasoline, the reformulated gasoline had the following specifications ; T90 was lowered to 145 ℃ from 154 ℃, aromatics were reduced to 22% from 32%, the sulfur content was reduced to 30 ppm from 300 ppm and 15 % MTBE was added. The influence of the reformulated gasoline on the NMOG emissions and OFP is shown in Figure 6. Reformulated gasoline was found to be effective in reducing NMOG by about 20%, but it was not effective in reducing OFP.

The percentages of each chemical class are compared in Figure 7 for the baseline and the reformulated gasolines. The reformulated gasoline used in this ex-periment was formulated with a lower aromatic content, and it was shown to be effective in reducing aromatics in speciated emissions. In contrast, however, alkenes increased and consequently OFP was nearly equal to that of the baseline gasoline. The reason is that the reduction of aromatics in the fuel brings about the increase of alkanes in the fuel and an alkane reaction produces alkenes with a high MIR value. It is difficult to reduce the OFP through modifications to gasoline specifications. However, such modifications can reduce NMOG emissions and are effective in lowering ozone formation in the air.

3.4 INFLUENCE OF METHANOL (M85) FUEL ON SPECIATED EMISSIONS AND OZONE FORMING POTENTIAL

An investigation was made of the influence of M85 fuel on speciated emissions and the OFP. An experiment was conducted using a flexible fuel test vehicle equip-

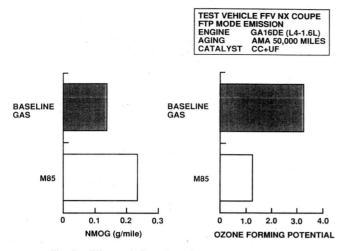

Fig. 8   Effect of Baseline Gasoline and M85 on Ozone Reduction

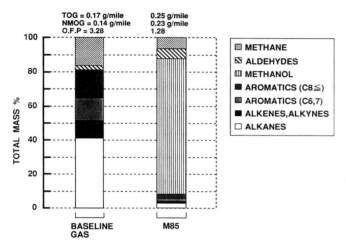

Fig. 9   Exhaust Mass into Hydrocarbon Types of Baseline Gasoline and M85

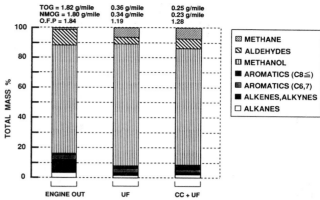

Fig. 10   Exhaust Mass into Hydrocarbon Types and Methanol for M85

ped with the CC/UF catalyst system. NMOG emissions and OFP obtained with M85 are shown in Figure 8 in comparison with the results for the baseline gasoline. Since a MIR value of methanol is a low 0.56, the OFP of the M85 fuel was only about 30% of that of gasoline. As regards exhaust emissions, NMOG emissions were 1.5 times higher with M85 than with gasoline. However, since the OFP of M85 was only about 40% of that of gasoline, using M85 would reduce ozone formation by approximately 40 to 50% compared with the use of the baseline gasoline. Figure 9 shows the proportions of chemical species. The methanol percentage in the exhaust gas was between 80 to 90%. Formaldehyde emissions with M85 were higher than with gasoline. Formadehyde has a high MIR value and it accounts for about 30% of the total ozone formed. The OFP of M85 vehicles is influenced primarily by formaldehyde.

Figure 10 shows the effect of the exhaust emission control system on the proportions of chemical species. In this experiment, when NMOG was reduced by using higher performance exhaust emission control devices, the percentage of formaldehyde increased. The lower reduction rate of formaldehyde is attributed to the formation of formaldehyde on the catalyst.

4.   CONCLUSIONS

The following conclusions can be drawn from the results and discussion presented above.

1) With gasoline vehicles, speciated emissions are affected by the exhaust emission control device used. Catalyst systems with higher light-off performance are effective in reducing alkenes and aromatics, because catalyst reactivities to such components tend to be high. Such a system can reduce both NMOG emissions and the OFP, and it is effective in reducing the atmospheric formation of ozone.

2) With gasoline, NMOG emissions increase with deterioration of the exhaust catalyst. However, the OFP tends to decrease because of the increase in the rate of alkanes with a low OFP.

3) Reducing the aromatic and alkene contents of the fuel is not effective in lowering the OFP of exhaust gas. Heavy alkanes in the fuel must be increased to maintain an acceptable octane number. Since heavy alkanes dissociate to form alkenes, the result tends to cancel out the OFP reduction obtained by reducing aromatics and alkenes. However, reducing aromatics and T90 has a large effect on lowering NMOG emissions and is effective in reducing ozone formation in the air.

4) NMOG is higher with M85 than with gasoline. However, since the OFP is low, using M85 results in less ozone formation than with gasoline. The OFP of M85 vehicles is influenced primarily by formaldehyde.

REFERENCES

1) California Air Resources Board, "Proposed Regulations for Low-Emission Vehicles and Clean Fuels," Technical Support Document, August 13,1990.

2) W. P. L. Carter, "A Method for Evaluating the Atomospheric Ozone Impact of Actual Vehicle Emissions, " SAE Paper 900710, 1990.

3) A. G. Russell, D. St. Pierre and J. B. Milford, "Ozone Control and Methanol Fuel Use," Science, January, 1990.

4) C. S. Weaver, T. C. Austin and G. S. Rubenstein, "Ozone Benefits of Alternative Fuels: a Reevaluation Based on Actual Emissions Data and Updated Reactivity Factors," Sierra Research, April 13, 1990.

5) Jeff A. Gething, S. Kent Hoekman, Annette R. Guerrero and James M. Lyons, "The Effect of Gasoline Aromatics Content on Exhaust Emissions: A Cooperative Test Program," SAE Paper 902073, 1990.

6) Nicola M. Dempster and Philip R. Shore, "An Investigation into the Production of Hydrocarbon Emissions from a Gasoline Engine Tested on Chemically Defined Fuels," SAE Paper 900354, 1990.

7) Y. Akutu, H. kanno, T. Hasegawa and M. ota, "Development of Dry Analytical Method for Formaldehyde in Exhaust Emissions," Pre-prints of JSAE, Spring 1990.

C448/044

Improving the spark ignition process in an internal combustion engine using a new spark plug concept

J BEN-YA'ISH, MSc, E SHER, PhD, CEng, MIMechE, MSAE and T KRAVCHICK, MSc
The Pearlstone Center for Aeronautical Engineering Studies, Ben-Gurion University of the Negev, Israel

SYNOPSIS   A new spark plug for SI engines has been developed. This plug provides an efficient means of improving cyclic variation and facilities the ignition of lean mixtures. In the new spark plug design the spark discharge is preceded by an ionic wind which occurs whenever a high voltage (lower than the breakdown voltage) is applied. Once the spark is introduced, the ignited mixture is convected away from the electrodes by the ionic wind. This is in order to reduce heat loss from the ignited mixture by heat transfer to the electrodes. The new spark plug has been tested in a constant volume bomb. The apparatus had two diametrically opposed quartz windows to facilitate optical observation. High speed video-photographs of the ignition process coupled with electrical current and voltage recording were taken in several stoichiometric fuel-air mixtures under high pressure conditions. Flame front velocity during the very early stages (~1 ms) was found to be about twice than that obtained without an applied electrical field. Preliminary tests with a fired conventional 4-cylinder engine have shown promising results. Owing to the improvement in the cyclic variation, a significant reduction in HC emission and fuel consumption has been observed in particular under low load conditions.

1   INTRODUCTION

Increasing fuel economy and improving the efficency of converting the energy stored in an air-fuel mixture into thermal energy under the constraint of strict smission standards have become of the most important goals in an internal combustion engine design. One of the promising solution is the use of lean fuel-air mixtures (1). However, the decrease in the fuel to air ratio is associated with a lower flame temperature which makes the flame more sensitive to the cooling effects of intense turbulence and the surrounding walls, especially during the early stages of combustion. The use of lean mixture is also limited by the increasing cyclic variability, which results in some cycles either not burning conpletely - partial burn limit, or not igniting at all - misfire limit (2). The breakdown, which initiates the ignition process, is associated with numerous kinds of phenomena which include very fast transformation of energy from one mode to another, electron avalache, streamer formation, radiation, photoionzation, thermal ionization, dissociation, fast diffusion, chemical reactions, formation and emission of shock wave, etc. Later stages of spark ignition processes are also complicated owing to the effect of many interactive factors such as the spark energy, the spark duration, the heat loss to the electrodes, etc.

Much effort has been concentrated on the examining of the influence of electrodes shape and design on the cycle variability and misfire occurence (3-6). It has been observed that operation with smaller electrodes under lean conditions is associated with a faster initial flame growth and an extension of the lean limit (5-7). It was, therfore, concluded that heat losses to the electrodes are of the most important factors in lean mixture operation. These were found to be dependent primarily on the contact area between the spark channel and the electrodes. A number of studies have investigated mechanisms of misfire in a quiescent and flowing mixtures (7-9). It was observed that there is an optimum flow velocity which is associated with a minimum required spark energy (4). This was attributed to the reduction in heat losses to the electrodes due to the convection away of the spark channel and flame from the

vicinity of the quenching electrodes. Based on these studies a simple mathematical criterion for misfire conditions has been proposed by Sher et al. (10). The cycle variability not explained by their model was attributed to turbulence or the non-homogeneity in the mixture (2).

It seems that a flow velocity of moderate magnitude (3 to 6 m/s according to Ref. 7) in the spark plug vicinity at the time of ignition is desirable for the extension of engine's operation limits and the combustion stabilization under lean conditions. The velocity should be large enough to convect the flame away from the electrodes but not too large to eliminate flame quenching in the flow. A number of investigations have indicated that an application of electric fields (either dc or ac) to combustion systems can produce a potentially useful effects (11). The electric field may not only control the chemical reactions which involve ions, but also may introduce a significant and directed momentum to the molecules, thus producing ion wind (12) in the vicinity of the electrodes. This wind may be used to convect away the spark channel and flame from the electrodes, thus minimizing the contact area and heat transfer to the electrodes.

2   THEORETICAL BACKGROUND

2.1   *Breakdown and conduction of electricity in gases*

Corona discharge plays a vital role in the conduction process of electricity in gases. However, the corona is only one of many forms of electrical discharge in gases. Therefore, the basic phenomena and principles common to all forms of gaseous conduction are an essential prerequisite to an understanding of the corona discharge itself.

In contrast to solids, gases under normal conditions contain practically no free electrons or ions, and are therefore almost perfect electrical insulators. However, when the potential between the electrodes in a typical spark plug is increased, a point is reached where the ionization and conductivity of the gas increase drastically. This transition from the insulating to the conducting state is called electric breakdown or gas discharge. There are two primary types of ionization processes in gases: (a) ionization of gas molecules or atoms themselves, in which electrons are removed from a molcule or atom; (b) release of electrons from the electrode surfaces. The dominant ion production mechanism is ionization by electron impact, in which electrons in the gas acquire energy from the applied electric field and collide with gas molecules. For the process to occur, the colliding electrons must posses a certain minimum energy which is characteristic of the molecule or atom bombarded. As long as the ionizing processes produce less electrons than required for rendering the discharge self-sustained this phase is called the prebreakdown phase. When the electrons rate production is higher, an overexponential increase in discharge currents occurs. The time required for the occurrence of a breakdown of the intervening gas, which is termed "the formative time lag of breakdown", depends on the magnitude of the over-voltage, that is, the difference between applied voltage and threshold. In practical systems the over-voltage is high and the formative time lag is around 10^{-8} to 10^{-7} s. During this period two breakdown mechanisms may take place (13) : the Townsend mechanism and the streamer formation. The "Townsend mechanism" is related to the successive development of electron avalanches between the electrodes, in which each new electron produced generates new electrons by ionization in overexponential rate, until the channel conductivity has reached a value high enough to make the current theoretically infinite and practically limited by the outer circuit. Experimental observations of the period of time required for this irreversible transition revealed that the formative times are much shorter than the Townsend mechanism (with any secondary action) can predict. By applying voltages well above the breakdown threshold, extremely short formative time lags of the order of 10^{-8} s were recorded. These short formative times do not allow for motion of the positive and negative ions, that can be regarded as stationary in time intervals of some 10^{-8} s. One more difficulty lies in the interpretation of the mechanism of spark formation at high pressures where, in many instances, the spark channel was found to be both branched and zig-zagged (13). The Townsend mechanism, in all its versions, cannot supply an explanation for such behaviour.

Early studies of the breakdown formation in the uniform field have revealed that, besides the occurrence of avalanches, another distinct mechanism of ionization also develops. Filamentary and heavily branched ionized channels, called streamers, have been observed. Streamer processes develop in very short time intervals, of the order of 10^{-8} s, in practical systems. During the build-up of this primary avalanche, excitation of gas atoms takes place at the same time as ionizations occur. Excited states have lifetimes that can be as short as 10^{-13} s, thus before the primary avalanche has reached its full size, photons will be emitted from these excited states as they return to the ground state. Following Nasser (13), these photons will be heading in all directions and will be absorbed at various distances from their origin. When a photon is absorbed, photoionization may occur. If the avalanche reaches a critical size, i.e., the space-charge field will be in the same order of magnitude as the original field, a second-generation auxiliary avalanche may be started. The photons emitted from the auxiliary avalanche create new photoelectrons which give rise to new third-generation auxiliary avalanches. Since the space-charge field distorts the original field, the electrons do not follow the original lines. Many avalanches may be created almost simultaneously, thus explaining the observed branching. When a continuous streamer exists between the two electrodes, a sudden increase in the intensity of ionization in the channel develops and breakdown takes place.

Once the electrode gap has been bridged, the gap impedance drastically drops, the voltage across it collapses and a high electrical current is developed for a short period (20 to 60 ns). This period is associated with very fast transformation of energy from one mode to another, ionization of several kinds, fast diffusion, thermalization, chemical reactions and the emission of a shock wave. Since this period is characterized by an extremely high temperature (40,000 to 60,000 K), the intervening gas may be considered to be fully dissociated and ionized to a large extent under these conditions.

2.2 *corona discharge*

One of the major fundamental differences between discharge in the uniform electric field and that in the nonuniform field is that the onset of an ionization in the uniform field usually leads to the completion of the transition and the establishment of the high-energy plasma channel. In a nonuniform field, such as exists between a sharply curved and a blunt surface, the case is entirely different and various visual manifestations of locally confined ionization and excitation processes can be viewed and measured long before the complete voltage breakdown between the electrodes occures. These manifestations are called coronae.

Under some conditions, the corona may create an effect called "electric (or corona) wind" which is characterized by a gas movement in the electrode gap which is directed toward the blunt curved electrode irrespective of the polarity of the voltage applied. A simple theoretical analysis (14), which was validated against experimental observations (15), has shown that the velocity of the gas is proportional to the square root of the corona current. A corona discharge system consists of an active sharp electrode, an ionization region, a drift region and a blunt electrode (acting as a charge collector). Charged particles travelling in the drift region, transfer their momentum to the neutral gas. The rate of the momentum exchange depends very much on the mobility of the charge carriers.

In negative corona the charge carrying particles in the drift zone are mainly free electrons. However, when the medium is an electronegative gas (air or SF6) the charge carrying particles include also negative ions which are created when electrons are caught by neutral molecules having electron affinity (oxygen molecules in air). In this case the electrons travel some distance before being caught. The travelling distance of the free electrons is characterized by the attachment coefficient which depends on a number of parameters such as the field strength, humidity and pressure. In the drift region the negative ions exchange their momentum with the neutral moleculs by collisions thus producing electric wind. In positive corona, the positive ions are created in the ionization region and are accelerated toward the negative (blunt) electrode while exchanging their momentum with neutral moleculs in the drift region.

It seems that negative corona is apparently inferior to the positive corona mainly because of the greater size of the ionization region. However, for positive arrangement it has been observed (16) that under some conditions corona wind can not be established. This was attributed to the increase with pressure of the voltage at which corona occurs which approaches the breakdown voltage at pressures relevant to those prevailing in IC engines. The threshold of this transition from corona to breakdown discharge depends on the medium type and electrode geometry. For air it varies from 0.2-0.4 MPa for a dull electrode to 1-1.5 MPa for a very sharp electrode. For practical ignition devices the lower bound should obviously be considered. In negative corona the voltage at which corona occurs never approaches (in pressure range typical to IC engines) the breakdown voltage and corona occurs at any pressure. Furthermore, the maximum current of negative corona is typically higher than in positive corona (about twice) and that means a higher wind velocity (the wind velocity is proportional to the square root of the electrical current).

3 IGNITION SYSTEM

Several possible configurations to the spark plug were suggested, in which an affective corona wind is produced; the chosen plug is shown in Fig.1. The proposed plug is a modification of a simple commonly used spark plug. It consists of the main body of a conventional plug, a high voltage central electrode (the sharp electrode) and a ground electrode. A pair of windows are situated diametrically-opposite sides to allow the fresh mixture to enter the electrodes gap. The mixture is then drifted toward the open end of the plug.

In order to operate the new plug, a new ignition system was designed (see Fig.2), which included a high-voltage source for applying high-voltage pulses to ignite the combustible mixture, and an additional voltage source which applies constant current continuously. The magnitude of the current supplied is sufficient to produce a corona discharge and thereby a corona wind. The so-produced corona wind tends to flow through the two windows in the ground electrode toward the open end and thereby directs the ignited mixture, or spark kernel, away from both electrodes. The effect is to reduce the residence time of the spark kernel near both of the electrodes, therefore reduce the heat loss from the spark kernel by heat transfer to the electrodes.

The timing diagrams and the operation of the electrical circuit are illustrated in Fig. 2.

4 EXPERIENTAL RESULTS

In order to assess the possibility of the implementation of the proposed concept to internal combustion engines, a series of experiments have been planned as follows:

- High pressure chamber tests in order to explore the relationship between the velocity of the corona wind vs. the pressure, voltage and electrodes polarity under steady-state conditions.
- Constant volume bomb tests in order to observe visually the effect of the corona wind on the ignition process in a quiescent medium.
- Fired engine tests to examine the effect of the new ignition system on the overall engine performances.

4.1 High Pressure Chamber Tests

The velocity of the corona wind was measured with the aid of a thin pitot tube (1 mm in diameter). The pitot tube was situated at a distance of 1 mm from the open side of the plug, along the sharp electrode axis. The pressure difference was measured by using a workshop made apparatus - Sher et al.(7). A careful study revealed that for a negative corona (negative sharp electrode) a wind is formed only when the gap size exceeds a certain value (in this case 2 mm), while for a positive corona, wind is obtained for any gap size. This is attributed to the travelling distance of the electrons before being captured by the oxygen molecules (see the section on theoretical background). For a gap size of 0.7 mm, which is a typical value for commonly used spark plugs, a corona wind of 1-2m/s is obtained for positive corona while no wind has been observed for negative corona. A similar behavior was observed for a wide range of relevant electrical currents. However it should be mentioned here that for higher pressures, a different figure is expected, i.e., the negative corona might yield a corona wind for a gap size of 0.7 mm or even less. The current-voltage characteristics suggest that the electrical current depends very strongly on the voltage, which means that the power supply has to be a constant current source. The resulting corona wind is depicted in Fig. 3. It seems that a wind velocity of 3-6 m/s can easily be achieved; a speed which was found (7) to be optimal for spark ignition processes.

The maximum electrical current and the resulting wind velocity for positive corona are much lower than for negative corona and the superiority of the latter over its counterpart is prominent.

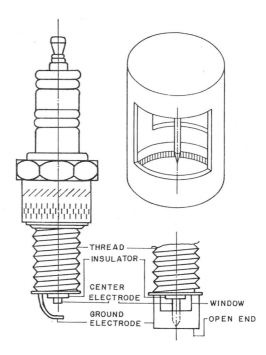

STANDARD DESING PLUG   NEW DESIGN PLUG

Fig. 1   A side view illustrating one form of spark plug constructed in accordance with the present concept and an enlarged view of the spark plug structure.

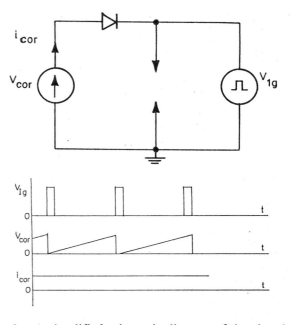

Fig. 2   A simplified schematic diagram of the electrical circuit for producing a corona wind in two-electrode system with a timing diagram of the electrical circuit.

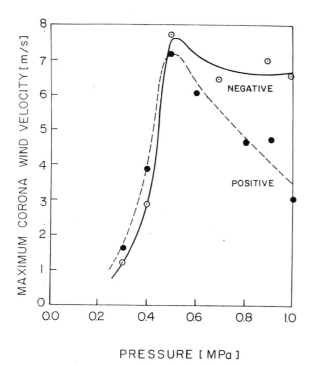

Fig. 3   Corona wind velocity vs. electrical voltage and pressure for negative and positive coronae.

4.2   Constant Volume Bomb

In order to examine the effect of the corona wind on the ignition process in quiescent medium, the proposed spark has been installed in a constant volume bomb. The apparatus has two diametrically opposed quartz windows to facilitate optical observation. A high speed Kodak video camera of up to 6,000 frames per second was employed to record the image of the flame front. Two spark plugs were examined: a conventional Champion spark plug W6D and the present proposed plug. For the latter, an additional voltage source which applies constant current continuously was employed to form a corona wind (negative corona), as shown in Fig. 2.

Figs. 4, 5 and 6 show some results of the flame front velocity in the downward direction (see Fig. 1) for stoichiometric and lean methane-air mixtures and for stoichiometric butane-air mixture at the same initial pressure. The flame front velocity was derived from the contour of the flame image as recorded by the high speed video camera, in the downward direction. In each experiment the first 10 frames were analyzed (the first 5 ms). An uncertainty of ±15% was observed in the series of 5 consecutive records. It seems that whenever a corona wind is applied, the flame front tends to travel, in its early stage, much faster than without it. Although the corona (the drive force of the corona wind) ceases at the moment when the spark is introduced, it seems that the effect is imprinted on the early stages of the spark development. In fact the flame acceleration is more than doubled when a corona is preceded the spark onset. This effect was observed for all cases. The 0 kV curve represents the flame front velocity in the absence of a corona wind. It is interesting to note that for the conventional spark plug the flame front velocity follows the 0 kV curve very closely.

It was also found that lean methane-air mixture has a higher flame front velocity than the stoichiometric one. It is, probably, due to the higher concentration of the oxigen atoms in the former.

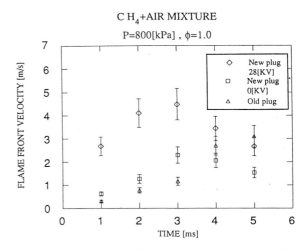

Fig. 4   Flame front velocity vs. time as photographed by a high speed video camera in a constant volume bomb at a pressure of 800 kPa in a stoichiometric methane-air mixture (new ignition system with negative corona).

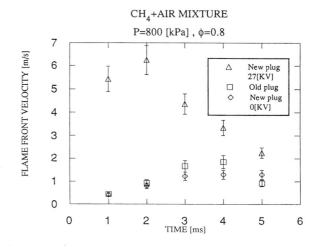

Fig. 5   Flame front velocity vs. time as photographed by a high speed video camera in a constant volume bomb at a pressure of 800 kPa in a lean methane-air mixture (new ignition system with negative corona).

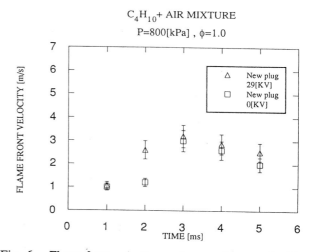

Fig. 6   Flame front velocity vs. time as photographed by a high speed video camera in a constant volume bomb at a pressure of 800 kPa in a stoichiometric butane-air mixture (new ignition system with negative corona).

4.3 Fired Engine Tests

In order to examine the effect of the proposed ignition system on the performance of a SI engine, three sets of tests were performed. In the first, a complete performance map of a VW 1297 cc engine (Table I) was measured. This includes power, break specific fuel consumption, HC emission, CO emission and NO_x emission. A pressure-transducer was installed in the first cylinder to trace the pressure inside the cylinder - this is in order to analyze the indicator diagram and the cyclic variability. Each set of tests comprised 1000 cycles which were analyzed by a Nicolet engine analyzer system.

Table I - Engine Data

Manufacturer:	Volkswagen
No. of cylinders:	4
Bore:	75 mm
Stroke:	73.4 mm
Capacity:	1297 cc
Compression ratio:	8.5 : 1

In the second test, the four conventional spark plugs were replaced by the new designed plugs and the engine complete performance map was measured again. In the third test, an additional electrical system was installed to provide a corona discharge and thereby a corona wind. The complete performance map was measured and analyzed. Prelimenary tests showed that for low engine load (5%), the HC emission has been cut down by 25% with the new ignition system, for both engine speeds of 950 and 1800 rpm. However, as the engine load increases (15%), the effect decreases in fact, for medium engine load the new system had practically no noticeable advantage over the conventional system. The superiority of the new ignition under low engine loads is attributed to the improvement in the cycle variability, which has been achieved by reducing the heat loss from the early born ignited mixture to the electrodes.

At this stage, it should be emphasized that although much attention was given to assure reliability and repeatability of the data measured with the fired engine, the accuracy of the conditions at which they were obtained has to be yet ascertained in order to treat them with confidence. The present study is therefore perforce preliminary in nature.

Based on the promising results of the constant volume bomb a further experimental investigation will be held in future, especially with lean mixtures.

5 CONCLUSIONS

A new spark plug design to improve ignition and combustion processes has been developed and examined. In the new plug the spark discharge is preceded by a corona wind. Tests in a quiescent fuel-air mixtures have indicated that a flow velocity directed from the active sharp electrode downward exists before the breakdown as a result of the corona wind. Based on a systematic study in a high pressure chamber, the spark geometry has been refined and an additional electrical circuit to provide the necessary power for the formation of a corona wind, has been designed. A negative corona has been found to be superior to its counterpart. Tests in a constant volume bomb have shown that whenever a corona wind is applied, the flame front tends to travel in its early stage, much faster than without it - the flame acceleration is doubled. Although the corona ceases at the moment when the spark is introduced, it seems that the effect is imprinted on the early sages of the spark development. Preliminary studies with a 1297 cc passenger car engine have shown a reduction of 25% in the CH emission under partial load and idle conditions.

ACKNOWLEDGEMENTS

This research was partially supported by grant no. 88-00158 from the United States-Israel Binational Science Foundation (BSF), Jerusalem, Israel, and by the Basic Research Foundation administered by the Israel Academy of Sciences and Humanities

REFERENCES

1. Heywood, J.B., Internal Combustion Engine Fundamentals, McGraw-Hill, 1988.
2. Sher, E. and Keck, J.C., "Spark Ignition of Combustible Gas Mixtures," Combustion and Flame 66, pp. 17-25., 1986.
3. Ballal, D.R. and Lefebvre, A.H., "The Influence of Spark Discharge Characteristics on Minimum Ignition Energy in Flowing Gases," Combustion and Flame, Vol. 24, pp. 99-108, 1975.
4. Kono, M., Hatori, K., and Iinuma, K., "Investigation on Ignition Ability of Composite Sparks in Flowing Mixtures," 20th Symposium (Intl.) on Combustion, The Combustion Institute, 1984.
5. De Soete, G.C., "Propagation Behavior of Spark Ignited Flames in Early Stages," IMechE. Conference Publication C59/83, 1983.
6. Douaud, D., De Soete, G., and Henault, C., "Experimental Analysis of the Initiation and Development of Part-Load Combustions in Spark-Ignition Engines," SAE Paper 830338, SAE Trans., Vol. 92, 1983.
7. Pischinger, I. and Heywood, J.B., "How Heat Losses to the Spark Plug Electrodes Affect Flame Kernel Development in an SI-Engine," SAE Paper 900021, 1990.
8. Matekunas, F.A., "Ignition Studies in a Rapid Compression Machine," Technical Meeting, Central States Section of the Combustion Institute, West Lafayette, Ind, 1978.
9. Pischinger, S. and Heywood, J.B., "A Model for Flame Kernel Developed in a Spark-Ignition Engine," 23rd Symposium (International) on Combustion, The Combustion Institute, 1990.
10. Sher, E., Heywood, J.B. and Hacohen, J., "Heat Transfer to the Electrodes: A Possible Explanation of Misfire in SI-Engines," Combs. Sci. and Tech., 1991.
11. Weinberg, F.J., Advanced Combustion Methods, Chapter 6, Academic Press, 1986.
12. Lawton, L. and Mayo, P.J., "Factors Influencing Maximum Ionic Wind Velocities," Combustion and Flame, pp. 253-263, 1971.
13. Nasser, E., Fundamentals of Gaseous Ionization and Plasma Electronics, Wiley-Interscience, 1971.
14. Levitov, V.I., Electrostatic Precipitators, Moscow, 1980.
15. Sigmond, R.S., "Mass Transfer in Corona Discharges", Rev. Int. Hautes Temper. Refrect. Fr., Vol. 25, pp. 201-206, 1989.
16. Meek, J.M. and Craggs, J.D., Electrical Breakdown of Gases, Oxford, 1953.
17. Sher, E., Ben-Ya'ish, J., Pokryvailo, A. and Spector, Y., "A Corona Spark Plug System for Spark-Ignition Engines". SAE Paper No. 920810, 1992.

C448/026

Correlation between in-cylinder flow, performance and emissions characteristics of a Rover pentroof four-valve engine

Z HU, BSc, DIC, PhD, C VAFIDIS, Dipl-Ing, DIC, PhD and
J H WHITELAW, BSc, PhD, DSc(Eng, CEng, FIMechE
Department of Mechanical Engineering, Imperial College
J CHAPMAN, BSc and R A HEAD, BSc
Rover Group Limited, Gaydon Test Centre, Lighthorne, Warwick

SYNOPSIS Measurements of in-cylinder flow, performance and emissions characteristics are reported of a four-valve, pentroof chamber, S.I. engine equipped with two different inlet duct configurations, generating two levels of induction barrel swirl ratios, (BSR). The high BSR configuration is shown to result in 50% higher in-cylinder turbulence in the pre-ignition flow field and leads to a 5% reduction of specific fuel consumption, 30% increase in burn rate and 50% reduction in cyclic variability of the indicated mean effective pressure at air/fuel ratios in excess of 20:1. This performance improvement, however, is accompanied by increased HC and NOx emissions.

NOTATION

AFR   Air Fuel Ratio
BMEP Brake Mean Effective Pressure
BSFC Brake Specific Fuel Consumption
BSR   Barrel Swirl Ratio (steady flow)
CAD   Crank Angle Degrees
COV   Coefficient of Variation
IMEP Indicated Mean Effective Pressure
IVC   Inlet Valves Close
MBT   Mean Best Torque
PMEP Pumping Mean Effective Pressure
TVR   Tumbling Vortex Ratio (engine flow)
u_t Rms of the tumbling velocity component fluctuations
U_t Tumbling mean velocity component
V_p Mean piston speed
TDC   Top Dead Centre
WOT   Wide Open Throttle

1   INTRODUCTION

Current trends in the design of medium size spark ignition (S.I.) engines favour the use of multi-valve induction and exhaust systems. These configurations increase engine volumetric efficiency and allow control of the induction-generated flow over the engine operating regime, thus improving the engine torque and drivability characteristics, (1-3). Of particular interest, however, is the extension of the engine lean operating limit and the increased tolerance in exhaust gas recirculation (EGR) usually achieved with these combustion chamber geometries. Stable engine operation with air/fuel ratios in excess of 23:1 and EGR ratios up to 30% is reported as feasible (1,4-6).

Lean, or highly diluted, mixture combustion has the inherent advantages of good fuel economy and reduced CO and NOx emissions but the associated slow burning rates may lead to increased cyclic combustion variability, partial burn events and misfires, (7). Burn rates can be increased by means of turbulence enhancement prior to combustion and an efficient way to achieve this is through the break-down of an induction-generated tumbling (barrel swirl) vortex, (8,9). As shown by numerous investigations, (1,4-6,10,11), the inlet port geometry of multi-valve, pentroof chamber, S.I. engines favours the formation of such vortices and their strength, which is a function of geometric port/valve details, largely determines the pre-combustion turbulence levels. Although increased pre-combustion turbulence compensates for the low laminar burning velocity of lean mixtures and improves burning rates, excessive turbulence levels may lead to ignition difficulties and quenching due to aerodynamic flame straining, (12). This, together with the fact that the optimum burning rate varies with engine load, suggests that the induction system design is critical in overall engine performance.

This paper summarises some findings of an extensive study of the correlation between in-cylinder flow and performance and emissions characteristics of a prototype four-valve, pentroof chamber S.I. engine. The motored in-cylinder flow field for two identical combustion chambers, differing only in the details of the inlet port geometry, is characterised by means of laser Doppler velocimetry(LDV). The results are used to interpret the differences observed in the corresponding, conventional, performance and emissions tests.

2 EXPERIMENTAL SYSTEM

2.1 The research engine

The single-cylinder research engine employed in this study is a Ricardo Hydra MKIII, adapted to accommodate the Rover cylinder head/liner assemblies. The two cylinder heads used were identical in every respect except for the fact that the inlet ports were partially filled in their lower or upper part in order to, accordingly, enhance or decrease the strength of the induction-generated tumbling vortex. The discharge and tumble-generating characteristics of the two cylinder head configurations, which in the following will be termed Head #1 and Head #2, were studied under steady flow conditions and the inlet ports were trimmed to produce identical discharge coefficients. The Barrel Swirl Ratio evaluated according to (5) was 1.74 and 0.4 for Head #1 and #2, respectively. The pumping mean effective pressure (PMEP), measured with the two cylinder heads under firing engine conditions, exhibited differences of less than 2%.

Following the performance tests, the two cylinder heads were modified to provide optical access into the combustion chamber for the motored LDV flow studies. These modifications involved the installation of two small quartz windows in the spark plug and pressure transducer openings of the combustion chamber and of a thick acrylic annulus between the cylinder head and liner sealing faces. The piston was also elongated to preserve the engine compression ratio. A diagram of the optically accessible engine is shown in Fig. 1 and the engine operating characteristics are given in Table 1.

2.2 Measurement system

The engine performance tests were conducted in a standard Cussons test bed which included a Datac pressure analysis system. The emissions measurements were conducted with a Horiba Mexa 9100/EGR analyser. The results presented here concern data obtained at 2,000 rpm and part load (2 bar BMEP) operation.

The in-cylinder flow field was studied under motoring conditions at 1,000 and 2,000 rpm with velocity measurements obtained with laser Doppler velocimetry at various locations of the combustion chamber, as indicated in Fig. 2. The laser velocimeter was operated in the off-axis forward, 90° and back scatter

modes depending on measurement location. Despite the limited access through the cylinder head windows, the in-cylinder flow field was characterised throughout the induction and compression strokes in terms of cycle-resolved ensemble mean and rms of three velocity components with a resolution of 1 CAD (crank angle degree). Particular emphasis was given to the velocity field in the vicinity of the spark plug gap and its evolution near the ignition angle. The results are presented in terms of temporal (crank angle) evolution or spatial distribution of individual mean velocity components and of the rms of the corresponding turbulent fluctuations according to the conventions of Fig. 2.

Table 1 Operating Characteristics of the Research Engine

Bore	84.45	mm
Stroke	89.00	mm
Mean piston speed (V_p)		
@ 1,000 rpm	2.97	m/s
@ 2,000 rpm	5.93	m/s
Compression ratio	10.5:1	
Connecting rod length	155.4	mm
Inlet valves:		
Diameter	31.95	mm
Open	19	CAD BTDC
Close	41	CAD ABDC
Peak lift	9.5	mm
Peak lift angle	79	CAD BBDC
BSR: Head #1	1.74	
Head #2	0.40	
Exhaust valves:		
Diameter	27.90	mm
Open	42	CAD BBDC
Close	18	CAD ATDC
Peak lift	9.5	mm
Peak lift angle	78	CAD ABDC
Engine speed	1,000 & 2,000	rpm

3 RESULTS

3.1 Velocity measurements

Both cylinder head configurations were found to produce a tumbling vortex structure during induction which was stronger and better organised in the case of the high BSR cylinder head #1. Although the in-cylinder flow field was studied in great detail, only a characteristic sample of the results will be presented and discussed in the following paragraphs. These results concern measurements of the tumbling velocity component (U_t) obtained along the symmetry plane of the combustion chamber.

© IMechE 1992 C448/026

The temporal evolution of the mean and rms of the tumbling velocity (U_t) component during the 400 CAD measurement window and at location L3 in the middle of the combustion chamber is shown in Fig. 3 for the two cylinder head configurations. The characteristic jet flow from the two inlet valves merging along the engine's symmetry plane is clearly evident from 10 to 170 CAD, reaching velocities of $12\text{-}13V_p$ for both cylinder heads. After BDC of induction the mean velocity magnitudes with cylinder head #1 remain nearly constant at around $2.5V_p$ directed towards the exhaust valve side of the combustion chamber and suggesting that the induction-generated tumbling vortex is sustained during the early part of the compression stroke. The corresponding results with the low BSR cylinder head #2 reflect the lower velocity magnitudes associated with the weaker tumbling vortex. At 310 CAD, and with head #1, the mean velocity magnitudes start decaying quite abruptly as the centre of rotation of the vortex approaches the measurement location and within 20 CAD the U_t-velocity component reaches a maximum of $2V_p$ towards the opposite direction as the measurement location L3 is now situated in the lower part of the vortex. This flow structure persists well after TDC of compression and decays at, approximately, 390 CAD. At the same time, the nearly constant rms magnitudes of the turbulent velocity fluctuations during early compression increase by 60% after 300 CAD to reach levels of the order of $1.7V_p$ near the crank angle of flow reversal which, coincidentally, is very close to the typical MBT ignition angle of lean-burn engines (30-40 CAD BTDC). Contrary to the above, the results with cylinder head #2 suggest that the tumbling vortex decays by TDC of compression and its decay is associated with a moderate increase of turbulence magnitudes to levels of the order of $0.75V_p$.

The spatial distribution of the mean and rms of the tumbling velocity component along the spark plug and cylinder axes is shown in Fig. 4 for both cylinder head configurations and at three instances of the compression stroke, at 290, 315 and 325 CAD. The above mentioned differences between the two configurations are evident with the high BSR configuration exhibiting higher velocity magnitudes. The development of the tumbling vortex from 290 CAD with cylinder head #1 shows that, as the aspect ratio of the combustion chamber decreases, it is transformed into a near solid-body type rotation for both cylinder heads. This leads to increasingly steeper velocity gradients with associated high shear around the point of flow reversal and locally increased turbulence production. The maximum U_t-velocity magnitudes increase from $3.3V_p$ at 290 CAD to $4.3V_p$ at 315-325 CAD before they start decreasing towards TDC of compression. At the same time, and as mentioned above, the vortex centre of rotation shifts towards the apex of the combustion chamber. Turbulence also increases from $0.7V_p$ at 290 CAD to $1.3V_p$ at 325 CAD, being generated at the centre of the vortex and diffused along the cylinder axis. In fact, turbulence generated by the shear in the U_t-velocity component was found to be redistributed to the other velocity components. During the same period, between 290-325 CAD, the development of the tumbling vortex with cylinder head #2 follows similar stages to the ones described above. The maximum U_t-velocity magnitudes, however, do not exceed $1.5V_p$ and decay to $1.1V_p$ at 330 CAD, near the spark plug location. Turbulence at 330 CAD is around $0.7V_p$, which is 50% lower than that with cylinder head #1. Still this is higher than the $0.5V_p$ turbulence levels encountered in a typical quiescent combustion chamber.

The above results show that, in both cases, the induction-generated tumbling vortex "spins-up" during the compression stroke as a result of its tendency to conserve its angular momentum within an increasingly smaller space. This spin-up process, which was first identified in (8), is clearly illustrated for the high BSR configuration in Fig. 5. The data of this figure represent the instantaneous equivalent rotational speed of the vortex during the compression stroke in terms of Tumbling Vortex Ratio (TVR) and were obtained by evaluating the angular momentum of the in-cylinder charge based on all available axial and tumbling velocity component measurements according to the procedure outlined in (9). They show that the average rotational speed of the air charge, which gradually decays after inlet valve closure, increases after 280 CAD to reach a maximum at 310 CAD before its final decay 25 CAD after TDC of compression. As shown above, and according to the analysis of (8), this vortex spin-up results in steep velocity gradients, with associated high turbulence production leading to overall turbulence enhancement inside the combustion chamber. It is interesting to note that the maximum TVR occurs at 315 CAD which corresponds to the crank angle of maximum turbulence levels with cylinder head #1 (Fig. 3) and that the corresponding numerical value (TVR=2) is similar to the BSR estimate obtained with the steady flow tests (1.74). Further to that, the crank angle of maximum turbulence production in the present combustion chamber is the same as that observed in the disc-type combustion chamber geometries of (8,9). The tumbling vortex in the present high BSR configuration, however, was longer-lived than those of similar strength in disc-type combustion chambers and this was shown to be due to its near-two-dimensional development into the favourably-shaped pentroof combustion chamber.

A final observation concerns the effect of engine speed on the in-cylinder flow. Our detailed investigation showed that, for the particular induction system used in this engine, the mean and turbulence velocity magnitudes scaled very well with engine speed in the range between 1,000 and 2,000 rpm. This implies that, in the high BSR configuration under WOT conditions, the absolute magnitude of the convective and turbulent velocities near the spark plug reach very high values at 2,000 rpm (11 and 8 m/s, respectively) and, according to (12,13) they may lead to ignition difficulties or flame kernel quenching.

3.2   Performance and emissions tests

The results of the performance tests discussed here were obtained for part load (2 bar BMEP) and 2,000 rpm conditions which are generally considered as the most critical for lean burn engine operation, (6). They are summarised in the graph of Fig. 6 and are presented as a function of air/fuel ratio.

The optimum (MBT) ignition advance provides a measure of the early flame development period. Early ignition, usually required by lean mixtures due to their low laminar burning velocity, may lead to ignition difficulties due to the associated low pressure and temperature in the combustion chamber. The corresponding results of Fig. 6 show that the low BSR configuration required 10-15 CAD earlier ignition for best torque operation throughout the 13.5-21 AFR range examined demonstrating that, as expected, the increased turbulence intensity around the ignition angle with the high BSR combustion chamber enhanced the early flame development rate, (14). This is also illustrated by the ignition delay (flame development) graph of the same figure which shows that the 0-5% mass fraction burned period for the high BSR head was 20-30 CAD (33%) shorter than that for the low BSR configuration. The rapid burning period (5-95% mass fraction burned) is also shorter by 25-30% for the high BSR configuration suggesting that the increased turbulence enhanced the overall flame propagation rate, as also reported in (15).

An important parameter related to lean burn engines is the combustion cyclic variability which is directly related to burn rate (7). Two measures are used here in order to quantify the relative performance of the two cylinder head configurations, the COV of maximum pressure and of indicated mean effective pressure. The first, which directly reflects the combustion variability and is shown in Figure 6 suggests that the high BSR configuration exhibits nearly 50% lower cyclic variations than the low BSR configuration in the AFR range below 19 with this difference decreasing to 20% at AFR=21. Although this shows a significant improvement of engine performance, a better indication of the engine drivability is given by the COV of IMEP which, in essence, characterises the engine in terms of variation of actual work done on the piston by the combustion gases from one cycle to the next. From the corresponding results of Fig. 6 it is clear that the high BSR configuration extends the practical lean limit of engine operation well above AFR=21, with a conventional stability limit taken as COV IMEP=5%, (6).

Finally, the comparison of the two configurations in terms of specific fuel consumption is also shown in Fig. 6 and suggests that the high BSR configuration leads to an average BSFC reduction of the order of 5% throughout the AFR range, consistent with the trends reported in (6).

In conclusion, the above results clearly indicate that the two-fold increase of turbulence intensity magnitudes in the engine combustion chamber, which was achieved with a simple re-shaping of the inlet ducts, resulted in significant improvements in engine performance and extension of its lean operating limit. The effect of this modification on the engine emissions will be discussed in the following.

The results of the specific HC and NOx emissions for the two combustion chamber configurations are also shown in Fig. 6 and suggest that the high BSR configuration led to a relative increase in both pollutants. The specific NOx emissions are shown to increase with air/fuel ratio up to AFR=16 and then, as expected, to decrease with increasingly leaner mixtures. A relative rise in NOx emissions with the high BSR configuration was expected as a result of the associated higher combustion temperature and earlier pressure rise due to its faster burn rate (16) and no attempt was made to adjust the ignition characteristics for their reduction. However, it should be noted that the specific NOx emissions at AFR=21 are nearly identical, and indeed very low, for both configurations. In this context, the attribute of the high BSR engine is the fact that it exhibits stable operation at this AFR range and, therefore, the inherent advantage of low NOx emissions with lean mixture operation can be exploited. Further to that, the increased EGR tolerance of this configuration allows further significant reduction of NOx emission levels, (3,16).

The fast burning characteristics of the high BSR configuration were expected to lead to reduced HC emissions (16). This was also suggested by the results of similar tests conducted with combustion chamber configurations exhibiting BSR levels intermediate to those considered here. However, as also reported in (5), a two-fold increase in specific HC emissions was observed throughout the AFR range with the increase of BSR from 0.4 to 1.7. The source of this phenomenon is difficult to identify since a number of operating parameters have a significant effect on HC emissions. Given the care taken during this study and the results of the previous tests, it is postulated that the observed rise in HC emissions is due to occasional early flame quenching, and therefore incomplete combustion, due to excessive flame front straining (12) in the highly turbulent combustion chamber of cylinder head #1. An additional factor may be the high mixture velocities near the spark plug at the ignition angle and persistence of the tumbling vortex long after ignition which, according to (13), may detach the flame kernel from the spark plug electrodes. An investigation is under way aiming to clarify this point.

4   CONCLUSIONS

The experimental study of the relationship between induction-generated mixture flow, performance and emissions characteristics of a Rover research S.I. engine equipped with a pentroof 4-valve chamber revealed the following:

1) The directional flow characteristics of the engine inlet ducts generated a tumbling vortex during induction the strength of which could be varied widely through small modifications to the ducts. Barrel Swirl Ratios of 0.4 and 1.7 were achieved with insignificant effects on engine breathing characteristics.

2) Turbulence was generated during compression through the increase of the velocity gradients during the tumbling vortex spin-up process and was diffused throughout the combustion chamber. This generation was larger with the high BSR configuration due to the associated steeper velocity gradients and peaked at around 315 CAD which is close to the MBT ignition angle. On average, an increase of mixture turbulence from 0.7 to $1.3V_p$ was achieved in the specific combustion chamber geometry by increasing the BSR from 0.4 to 1.7. In the high BSR case, the tumbling vortex persisted long after the ignition angle, partly due to the favourable combustion chamber shape, resulting in high convective velocities near the spark plug gap.

3) The increased turbulence intensity associated with the higher BSR configuration led to a reduction of specific fuel consumption under part load operation by 5% and an increase in burn rate by 30% as compared to those of the low BSR configuration. This increase of burn rate was instrumental in the reduction of cyclic combustion variations which, effectively, extended the lean operating limit of the engine to air/fuel ratios above 21:1 with associated variations in terms of IMEP of less than 4%.

4) The specific HC and, to a lesser extent NOx, emissions increased as the BSR was raised from 0.4 to 1.7. This was attributed to the excessive turbulence and high convective velocities prevailing in the combustion chamber of the high BSR configuration and appears to support the arguments regarding the limitations in burning rate enhancement through turbulence in lean burn engines.

ACKNOWLEDGEMENTS

The authors wish to thank the Directors of Rover Group Ltd for funding this work and permitting publication of the results and the SERC for supporting one of the authors (C.V.) under Advanced Research Fellowship B/AF/1202.

REFERENCES

(1)   Inoue, T., Iguchi, S. and Yamada, T. In cylinder gas motion, mixture formation and combustion of 4 valve lean burn engine. Proc. 9th Vienna Motor Symposium, 1988.

(2)   Mikulic, L. A., Quissek, F. and Fraidl, G. K. Development of low emission high performance four valve engines. SAE Paper 900227, 1990.

(3)   Endres, H., Schulte, H. and Krebs, R. Combustion system development trends for multi-valve gasoline engines. SAE Paper 900652, 1990.

(4) Benjamin, S. F. The development of the GTL "barrel swirl" combustion system with application to four-valve spark ignition engines. Paper C54/88, Proc. Int. Conf. on Combustion in Engines - Technology and Applications, 203-212, I.Mech.E., London, 1988.

(5) Chapman, J., Draper, A., Fairhead, G. S. and Wallace, S. Optimization of combustion chamber design. I.Mech.E. Paper C382/030, presented at 2nd Int. Conf. on "New Developments in Powertrain and Chassis Engineering", European Automobile Engineers Co-Operation, Strasbourg, 1989.

(6) LE COZ, J. F., HENRIOT, S., HERRIER, D., MARIE, J. J. AND BARRET, P. Development of a lean burn multi-valve S.I. engine: A new approach based on flow field optimization. Proc. 3rd Int. Conf. on Vehicle Dynamics and Powertrain Engineering, EAEC Paper 91012, 233-242, Strasbourg, 1991.

(7) Matekunas, F.A. Modes and measures of cyclic combustion variability. SAE Paper 830337, 1983.

(8) Gosman, A. D., Tsui, Y. Y. and Vafidis, C. Flow in a model engine with a shrouded valve - A combined experimental and computational study. SAE Paper 850498, 1985.

(9) Arcoumanis, C., Hu, Z., Vafidis, C. and Whitelaw, J. H. Tumbling motion: A mechanism for turbulence enhancement in spark-ignition engines, SAE Paper 900060, 1990.

(10) Kent, J. C., Mikulec, A., Rimai, L., Adamczyk, A. A., Mueller, S. R., Stein, R. A. and Warren, C. C. Observations on the effects of intake-generated swirl and tumble on combustion duration. SAE Paper 890296, 1989.

(11) Omori, S. Iwashido, M. Motomochi, M. and Hirako, O. Effect of intake port flow pattern on the in-cylinder tumbling air flow in multi-valve S.I. engines. SAE Paper 910477, 1991.

(12) Bradley, D., Hynes, J., Lawes, M. and Sheppard C. G. W. Limitations to turbulence-enhanced burning rates in lean burn engines. Paper C46/88, Proc. Int. Conf. on Combustion in Engines - Technology and Applications, 17-24, IMechE, London, 1988.

(13) Pischinger, S. and Heywood, J. B. How heat losses to the spark plug electrodes affect flame kernel development in an SI-engine. SAE Paper 900021, 1990.

(14) HADDED, O. AND DENBRATT, I. Turbulence characteristics of tumbling air motion in four-valve S.I. engines and their correlation with combustion parameters. SAE Paper 910478, 1991.

(15) KYRIAKIDES, S. C. AND GLOVER, A. R. A study of the correlation between in-cylinder air motion and combustion in gasoline engines. Paper C55/88, Proc. Int. Conf. on Combustion in Engines - Technology and Applications, 195-202, I.Mech.E., London, 1988.

(16) HEYWOOD, J. B. Internal combustion engine fundamentals. 1988, McGraw-Hill.

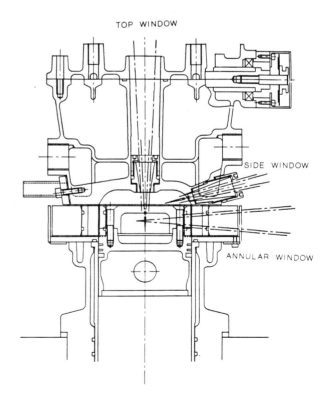

Figure 1 Optically accessible engine.

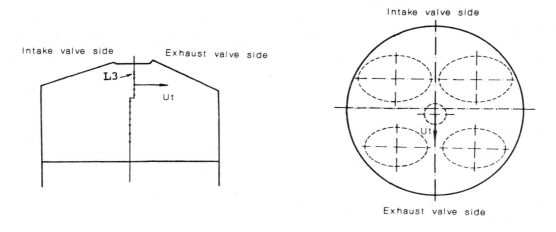

Figure 2 Velocity measurement locations and sign conventions.

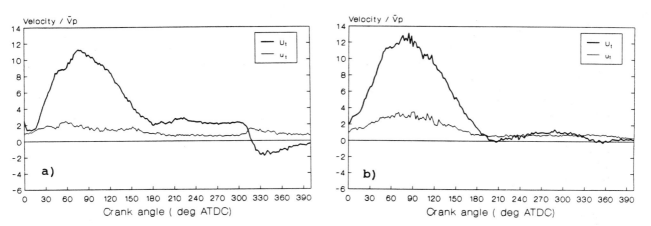

Figure 3 Crank angle-resolved ensemble average mean and rms of the tumbling (U_t) velocity component at location L3. a) Cylinder head #1 and b) cylinder head #2.

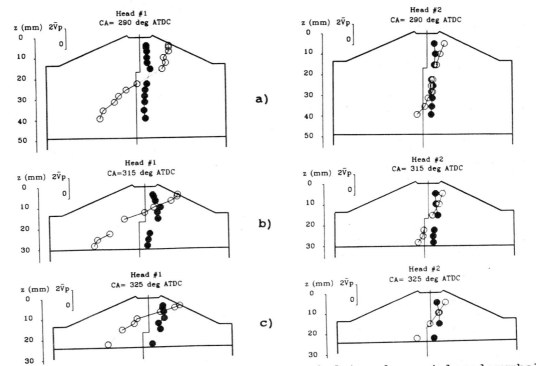

Figure 4 Spatial distribution of the mean (open symbols) and rms (closed symbols) tumbling velocity component along the spark plug and cylinder axes with the two cylinder heads. a) 290 CAD, b) 315 CAD, c) 325 CAD.

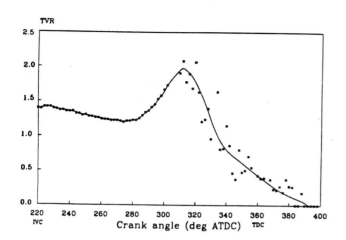

Figure 5 Variation of Tumbling Vortex Ratio with crank angle. Cylinder head #1.

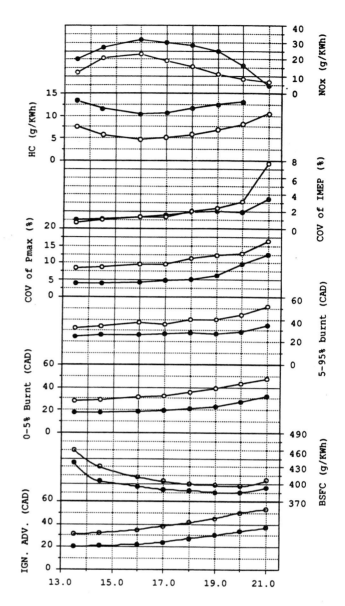

Figure 6 Performance and emissions characteristics of cylinder head #1 (closed symbols) and #2 (open symbols) as a function air/fuel ratio.

Port throttles applied to a high performance four-valve S.I. engine

T G BAKER, BEng and C J E NIGHTINGALE, MA, PhD, CEng, MIMechE, FIMarE
Department of Mechanical Engineering, University College, London
D J MYERS, BSc, CEng, MIMechE
Ford Motor Company, Laindon, Basildon, Essex

SYNOPSIS

The application of port throttles in series with a single upstream throttle has been investigated both through the use of a simple model and through engine testing on a modern 4-valve s.i. engine. Port throttling was found to limit the back flow of exhaust gases into the inlet manifold at part load, thus allowing the use of a camshaft with a relatively large valve overlap period. Other possible benefits of port throttling were better mixture preparation and a reduction of pumping loop work.

1 INTRODUCTION

The relatively high volumetric efficiency at high speeds of modern 4-valve engines brings with it a tendency towards less stable combustion at low-speed, part-load conditions due to reduced inlet gas velocities. One solution that has been the subject of a number of investigations (1) (2) involves blocking one of the two inlet ports at low-speed, part-load conditions using a throttle disk. This has the effect of increasing the gas velocity in the open port which can be oriented to establish axial swirling within the cylinder when it is the sole supplier of fresh charge. This technique is usually referred to as 'port de-activation'.

Another method of obtaining both satisfactory low and high speed performance is to adopt a system of variable valve timing, for which many designs (3) exist. The simplest manifestation of variable valve timing is where camshaft phasing alone is altered. Altering the phasing allows an improvement in low-load combustion stability through reducing the proportion of exhaust gases in the cylinder as a result of having a shorter valve overlap period. It can be argued that this approach is not tackling the fundamental problem of low intake velocities which leads to insufficient in-cylinder charge motion and hence low turbulence levels. The more complicated designs, which allow changes to valve lift and valve opening period, do lead to increased charge velocities at part load, and therefore deliver a more fundamental solution.

A third solution to the problem of establishing a satisfactory compromise between low and high speed performance is to use port throttles. In this approach throttle disks are fitted to individual ports instead of, or as well as, having a single throttle disk upstream of the plenum chamber. A schematic layout of a port throttle design is shown in figure 1. The obvious advantage of using port throttles is the improved control of exhaust gas back flow into the inlet tract during the valve overlap period at low speed and load. This enables the camshafts to be designed with a considerable degree of overlap to give good volumetric efficiency at high speed and load, yet with no greater exhaust gas dilution of the charge at low speed and load operation. In this respect, port throttles achieve similar effects to those of variable camshaft phasing.

An investigation by Newman et al. (4) confirmed that port throttles do indeed serve to control exhaust gas back flow, and also bring a further advantage of reduced pumping loop work. The investigation covered both experimental work on a 4-cylinder, 1.6-litre research engine, and modelling using a computer simulation which is proprietary to the Ford Motor Company. This model has received several years of development and is very comprehensive covering both gas exchange and in-cylinder combustion processes throughout the engine cycle. In particular, it accounts for the effects of turbulence, swirl and exhaust gas fraction on combustion rate, and it predicted only a small difference in burn rate for the port throttled version compared with its standard form. However, engine testing indicated that the burn rate with port throttles was significantly slower than with the standard arrangement. It is important at this stage to appreciate the details of the engine then under investigation: one of its intake valves was 'de-activated' so as to promote axial swirl, and the throttle was positioned in the remaining active port. Fuel was supplied by an injector situated *downstream* of the port throttle and there was no initial throttle upstream of the plenum chamber. The investigators argued that the slower burn rate was caused by the port throttle tending to reduce gas velocities at the inlet valve simply through reducing back flow and hence the overall gas flow past the valve; this, in turn, would lead to poorer mixing between exhaust gas and fresh charge as well as lower turbulence intensities. The model did not account for poor mixing and hence was likely to over-estimate the burn rates with port throttles. Newman et al. also noted that the port throttles would have to be manufactured very precisely in order to obtain an even distribution of air between the cylinders.

The investigation of Newman et al. was comprehensive in that it explored the effect of several important variables, such as valve overlap and port volume, on the performance of their port-throttled engine. However, some important questions still remained unanswered:

(a) If the concept of port throttles had been implemented on an engine with intake porting designed to promote tumble rather than axial swirl at low speed and part load, would the burn rate still have been slower than for the standard arrangement?

(b) Would any advantage be gained by utilising the excellent mixture preparation that could be obtained at part load by injecting *upstream* of the port throttle disk?

(c) Could the problem of air flow balance between the cylinders be significantly reduced by having a further throttle upstream of the plenum chamber, and dividing the pressure drop between upstream and port throttles?

The objective of the work reported here was to build on the earlier investigation by searching for answers to the above questions. A combination of computer modelling, flow rig work and engine testing was used for the investigation.

2 COMPUTER MODELLING

The number of variables with which a designer has to contend on a port throttled engine are numerous enough to require a computer model; engine testing alone would be very tedious, and also would probably not provide a basic understanding. A comprehensive model such as that used by Newman et al. (discussed in the preceding section) was not readily available for this investigation, so a relatively simple one was constructed. It was based on applying the 'filling and emptying' equations to a 3-volume system (cylinder volume, port volume and plenum volume). This approach is well described by Watson and Janota (5) for a 2-volume system, and

it was extended to cover the extra port volume. A number of simplifying assumptions were made:

(a) The effect of fuel on the gas flow processes was ignored. That is to say, the gas was assumed to be exhaust gas, or intake air, or a mixture of the two.

(b) 'Perfect' mixing was assumed between the exhaust gas and the intake air whenever the two came into contact.

(c) No pressure recovery was allowed for across the throttle disks as the downstream passage geometry is unfavourable to this.

Heat exchange between the gas and the port walls was estimated using equations developed from the Reynolds analogy applied to pipe flow. The Annand (6) correlation was used to calculate heat exchange between the gas and cylinder walls. The discharge coefficients for forward and reverse flow past the inlet and exhaust valves were initially taken from values quoted by Annand and Roe (7) and then modified slightly after flow testing a cylinder head of the engine type under investigation.

A decision was taken not to model the burning part of the cycle but to utilise instead available engine data on values of cylinder pressure and temperature at exhaust valve opening for various engine configurations and operating conditions. The main reason for adopting this approach was to save time, but it could be argued that relatively little was lost since the more comprehensive model (4) failed to compute burn rate accurately, yet still required engine data itself for 'calibration'.

A flow chart for the simple model is shown in Figure 2 and a complete description including considerably more detailed flow charting is given by Baker in (8).

2.1 Flow rig and engine testing

The pressure loss tests to be described in Section 3.4 were performed with air supplied under pressure from a Howden 8A3 screw-type compressor, while the remaining flow tests (cf. Sections 3.1 & 3.6) were carried out using the vacuum pump of the UCL Fuel Systems Test Facility (9) to induce flow. Pressures were measured by manometer and flow rates by nozzle pressure drop, orifice pressure drop and hot hire anemometer, as appropriate.

The engine used for this investigation was a 1.8-litre, 4-cylinder, in-line engine with a 4-valve head and multi-point fuel injection system. The salient details are given below:

Bore 80.60 mm
Stroke 88.00 mm
Compression ratio 10:1
Inlet valve head dia. 32.00 mm
Exhaust valve head dia. 28.00 mm
Spark plug location central

A schematic layout of the head, porting and port throttle for one cylinder is shown in Figure 3. The engine tests reported here were carried out during a period of time when the base engine was being developed such that three different versions were used (termed Vers1, Vers2 and Vers3 in this report). These changes did not affect the conclusions as the tests were performed as back-to-back comparisons with the base configuration of the appropriate period, but it does mean that the engine data quoted is not representative of the engine's current performance level.

Cylinder pressure was measured using a Kistler 6121 A1 piezoelectric transducer together with a charge amplifier and an Apple Macintosh II computer for data acquisition and post-processing. LABVIEW software by National Instruments was used, together with associated interface cards including an on-board A/D converter. Further details of the system and the software developed for cylinder pressure data acquisition are given in (9). It should be noted that the only drawback of the system was the relatively long time required for post-processing, so that the results quoted in the following sections were computed from averages ranging from 40 cycles for speed of processing up to 225 cycles for accuracy. It should be noted that the pressure of only one of the four cylinders was measured as there were not sufficient transducers and amplifiers available to instrument all the cylinders.

3 MODELLING AND TEST RESULTS

A review of the potential of port throttles suggests that they should show benefits in the areas of mixture preparation, reduced part-load pumping work and control over charge dilution by exhaust gas. On the other hand, problem areas could be full-load pressure drop, inter-cylinder air distribution, and in-cylinder air motion. Each of these areas will be discussed in turn in the following sub-sections.

3.1 Mixture Preparation
The extremely good atomisation produced when fuel is introduced into the air stream passing a throttle disk is illustrated by the comparison between the part-load carburettor spray (throttle disk atomisation) and a normal pintle-type injector spray in Figure 4. The results are given in the form of equivalent diameter rather than Sauter mean diameter for reasons discussed in (10). The equivalent diameter, $d_{e,0[y]}$, is defined as the diameter of a mono-size spray that would take the same computed time to evaporate from 0% to y% as the real spray, assuming a similar d^2/time gradient throughout. The use of port throttles with suitable injection timing allows the production of a carburettor-like spray without the external air supply required by an air-assisted atomiser.

A brief series of engine tests was carried out with a port-throttled, Vers1, engine fitted with both upstream and downstream injectors. The engine was run at 800 r/min idle, 1200 and 1500 r/min road load, with a multi-way switch arrangement so that the drive current could be fed to either the upstream or the downstream injectors without change to any other settings. Figure 5 indicates that the cycle-to-cycle variability was reduced at light road load conditions by switching to upstream injection. The results are averages of several readings taken at each condition. Further study of the traces suggested that burning was more rapid with upstream injection, presumably resulting from better mixture preparation. The results shown in Figure 5 were obtained with just under half of the pressure drop across the upstream throttle and just over half across the port throttles. It is possible that a further improvement could have been obtained with the upstream injection configuration by changing the injection timing to allow for the different fuel transport time, rather than accepting the timing of the base engine.

The conclusion of this short series of tests was that, in the case of the engine tested, improved mixture preparation brought about a significant reduction in cyclic variability. Care must be taken in extrapolating this result to other engine types, as different engine designs are known to exhibit different sensitivities to mixture preparation. If upstream injection were to be adopted, there would need to be a reassessment of the programming of the fuel system controller in the area of fuelling strategy during transients.

3.2 Reduction in pumping loop work
On a port throttled engine, there is flow past the port throttle during the period of the cycle when the inlet valve is closed. This flow causes the port pressure to recover towards the value of pressure that exists in the part of the manifold that includes the plenum. Newman et al. (4) pointed out that the port volume was critical in determining whether full pressure recovery is obtained during the available time. Their model results showed that a port volume had to be 22% of the swept volume of a single cylinder or less for full pressure recovery to be obtained at the 800 r/min, idle condition. A similar prediction was obtained from the model used in this work; estimations of port and cylinder pressure characteristics from this model are shown in Figure 6 for the 20% volume ratio case. It can be seen that the benefit of the port pressure recovering to plenum pressure is that, during the initial induction period, the cylinder pressure starts from near the value of plenum pressure rather than the value required at the end of the induction stroke, with a consequent reduction in the area of the pumping loop. Figure 6 refers to the situation where all the pressure drop is set across the port throttles. The benefit is, of course, less if the pressure drop is divided between

port and upstream throttles.

Figure 7 shows the order of reduction in indicated specific fuel consumption that has been observed in practice. The results were calculated by obtaining an average value of work per cycle (ie 2 engine revolutions) for the instrumented cylinder. The calculations used the fuel flow value obtained when the total fuel flow was divided by 4, since only one cylinder was instrumented. The reduction in indicated specific fuel consumption with port throttles was 9% at idle, 8% at 1200 r/min, and 10% at 1500 r/min road load. The tests were performed on a Vers1 engine and the results quoted for each configuration were an average of at least 4 separate tests. It should be noted that the engine configuration was far from ideal in terms of reducing pumping loop work, as the volume ratio was too high (@ 35%) and only just over half of the pressure drop was set across the port throttles. It is believed that the majority of the observed improvements could be attributed to reductions in the pumping loop work, but it is also likely that other factors, such as faster burning resulting from reduced charge dilution and better mixture preparation, played their part as well.

The conclusion is that port throttling can lead to significant reductions in fuel consumption at idle and part load conditions. However, as will be argued in a later section, it may be necessary to sacrifice some of this potential improvement in order to reduce the inter-cylinder air distribution problem.

3.3 Improved control over charge dilution
The original raison d'être for port throttles was to control the back flow of exhaust gas into the inlet manifold at part load during the valve overlap period. In fact, it is advantageous to have a certain proportion of exhaust gas in the charge in order to reduce NO_x emissions, but there is a stage reached for high performance engine versions with relatively long valve overlap periods, where the charge dilution becomes excessive and causes unstable combustion. Figure 8 shows the predicted flow past the inlet valve, with positive values representing flow into the cylinder and negative values reverse flow. It demonstrates the effectiveness of port throttling in reducing the back flow into the port, particularly when it is realised that the conventionally throttled case was computed for a 20° valve overlap, while the port throttled case used 30° overlap.

Figure 9 contains a comparison of engine stability at idle between a base Vers2 engine and the same engine fitted with port throttles. The comparison is made at the standard overlap (20°) and at the simulated overlaps of 30° and 50°. The increased overlaps were obtained by advancing the inlet camshaft through half the required angle and retarding the exhaust camshaft through the remaining angle. The engine was very uneven at 50° overlap without port throttles such that the engine speed was far from constant, making any reading of standard deviation in imep rather inaccurate; however, a single reading of 1.4 bar was recorded just to obtain some measure of the roughness. Efforts were made to optimize fuel injection and spark timing for the various configurations listed in Figure 9. From the tests performed, it was not possible to identify for certain that the control of charge dilution was the main cause for the improvement evident with port throttles as mixture preparation and other factors might also have been significant. However, it seemed likely that charge dilution was one of the most significant factors even if it was not the most significant.

3.4 Full-load pressure drop
One potential drawback of a port throttle is the likely increase in manifold pressure drop at full load that it creates due to the obstruction to the flow caused by the throttle spindle. Flow testing was performed on a port throttle unit designed for the Vers3 engine and the results showed (Figure 10) that this increase could be kept quite small.

3.5 Inter-cylinder air distribution
Newman et al. recognised that the main problem with port throttles from the production point of view is the precise machining and setting operations that would be required to obtain a uniform distribution of air between the cylinders. However, this difficulty could be reduced if the pressure drop is divided between the port throttles and the upstream throttle. The likely benefit of such a split can be estimated from the computer model. Figure 11 shows the computed relationship between change in overall mass flow to a cylinder for a small change in air flow to an individual cylinder when the flow area around the port throttle is decreased by 1 mm². It can be seen that there is a significant reduction (29.3%) in air flow below the nominal idle value of 0.61 gm/s in the case where 100% of the pressure drop is set across the port throttle, but the decrease is much less (4.8%) when the pressure drop is divided 25/75% between port/upstream throttles respectively.

Having less of the pressure drop across the port throttle obviously loses some of the pumping loop work benefit referred to in section 3.2, but how will it affect mixture preparation and exhaust gas dilution gains? The indication from throttle disk atomisation measurements (11) is that mean droplet sizes of 10 µm can be obtained over the range from 100% to 30% of idle pressure drop across the port throttle, and so atomisation benefits are not likely to be sensitive to changing the split of pressure drop within this range. Results from the computer model indicate that a port/upstream throttle pressure drop split of 25/75% still leads to a relatively low level of exhaust back flow. Further strength is given to the argument by the realisation that the port throttle results in Figure 9 were obtained with 25/75% pressure drop split between port/upstream throttles.

The applications of port throttle could well be limited to in-line engines due to difficulties in synchronising throttles situated in different banks of cylinders. An obvious remark concerning the application to in-line engines is that the fewer the cylinders, the easier the task. In situations where port throttles are used in conjunction with an upstream throttle, it is envisaged that the upstream throttle would be linked to the accelerator pedal, while the port throttles would be spring-loaded open and held in their closed postion at part load by a vacuum motor. The vacuum motor would operate using plenum depression controlled through a solenoid valve.

To summarise, the production problems of obtaining a uniform and consistent inter-cylinder distribution of air can be reduced by splitting the pressure drop between port and upstream throttle disks at the expense of increasing pumping loop work and, to a lesser extent, increasing the dilution of the charge by exhaust gas.

3.6 In-cylinder air motion

The base engine used for the investigation incorporated inlet ports oriented to promote in-cylinder tumble. De Boer et al. (12) indicated the importance of having high charge velocity in the port and a suitable port approach angle if an active tumbling motion is to be established. The presence of port throttles will disrupt the port flow of the base engine, and the danger is that the subsequent in-cylinder flow patterns will not be so favourable to the generation of the high turbulence levels around the time of flame propagation, leading to the slow burn rate problems experienced by Newman et al. The indications from engine tests performed on the Vers1 engine were that port throttles actually increased the burn rate, whereas the effects on Vers2 engine burn rate were that burn rate was increased at some engine conditions and marginally worse at others.

Work is proceeding on Vers3 engine hardware to obtain a more fundamental understanding of the influence of port throttles on the flow processes. Figure 12 illustrates some results obtained so far showing the distribution of flow around the inlet valve periphery at two different valve lifts. A rigorous analysis would involve the use of laser Doppler anemometry backed by computational fluid dynamics. However, some indication of the likely in-cylinder behaviour can be obtained from quite straightforward testing (12). The approach used to obtain the results shown in Figure 12 involved manufacturing an inlet valve with a hollow stem. The passage in the stem was connected to a pitot tube mounted in the flat end of the valve and bent in a loop to face the flow

issuing between the valve and valve seat. A manometer was connected to the outer end of the valve passage and was referenced to the pressure in a plenum chamber situated between a dummy cylinder and the vacuum source. Steady flow tests were performed using flow rates calculated from the model as being appropriate for the valve lift and engine configuration under investigation. The points on the polar plots in Figure 12 joined by a solid line are for the case where the port volume is 45% of the cylinder volume, while the points joined by the dashed line are for the 20% volume case. It would appear that the 45% case, which is tending towards a conventionally-throttled arrangement, has the flow quite evenly distributed around the periphery of the valve, but the 20% case has an imposed bias which would have a significant effect on in-cylinder flow patterns. As pointed out at the beginning of this section, engine testing has indicated that the changes to in-cylinder flows have not decreased burn rates significantly. However, further work is necessary to understand and quantify the mechanisms involved.

4 CONCLUSIONS

The application of port throttles is an option available to engine designers in situations where they wish to extract a high power output from an engine by using camshaft timing with a large valve overlap. In this situation, port throttles will control the back flow of exhaust gas into the inlet manifold and so prevent the charge dilution from becoming excessive.

Port throttles bring with them the two further advantages of reducing pumping loop work and better mixture preparation if upstream injection is used. In the case of the former, some of the benefit may have to be sacrificed by using port throttles in series with an upstream throttle to reduce the difficulties of obtaining an even distribution of air between the cylinders. In the case of the latter, improvements in steady-state engine performance through better mixture preparation must be weighed against a possible deterioration in performance during transients.

The additional full-load pressure drop incurred by port throttles is unlikely to lose a significant amount of engine power.

Port throttles do affect in-cylinder air motion at part-load conditions. It appears that the effect is detrimental to burn rate on some engines but beneficial on others. Further work is required to understand this complex topic.

5 ACKNOWLEDGEMENTS

The authors would like to thank the Ford Motor Company for permission to publish this paper and for their support of the investigation. The authors are also indebted to the SERC for the provision of a CASE award which is currently helping to support the work. The significant contributions of Mr P A Williams and Mr M J Miller in performing the Vers1 and Vers2 engine testing is very gratefully acknowledged. Thanks are also due to Mr T H Curtis and other technicians of the UCL Thermodynamics Laboratory for their work on the engine test beds and flow rigs.

REFERENCES

(1) MIKULEC, L.A., QUISSEK, F., and FRAIDL, G.K. Development of low emission high performance four valve engines, SAE Paper 900227.

(2) ENDRES, H., SCHULTE, H., and KREBS R. Combustion system development trends for multi-valve gasoline engines, SAE Paper 900652.

(3) DEMMELBAUER-EBNER, W., DACHS, A., and LENZ, H.P. Variable valve actuation systems for the optimization of engine torque, SAE Paper 910447.

(4) NEWMAN, C.E., STEIN, R.A., WARREN, C.C., and DAVIS, G.C. The effects of load control with port throttling at idle - measurements and analyses, SAE Paper 890679.

(5) WATSON, N, and JANOTA, M S. Turbocharging the Internal Combustion Engine, pp 517-594, The Macmillan Press, 1982.

(6) ANNAND, W J D, Heat transfer in the cylinders of reciprocating internal combustion engines, Proc. IMechE, Vol. 177, No. 36, pp 973-990, (1963).

(7) ANNAND, W J D, and ROE, G E, Gas Flow in the Internal Combustion Engine, pp 52-84, Foulis, 1974.

(8) BAKER, T G, Computer modelling of a port-throttled engine, UCL Report for Ford Motor Company, 1992.

(9) MILLER, M J, NIGHTINGALE, C J E, and WILLIAMS, P A, Measurement of spark ignition engine mixture preparation and assessment of its effects on engine performance, IMechE Conf. on Internal Combustion Research in Universities, Polytechnics and Colleges, IMechE HQ, 30-31 January 1991, IMechE Publication No. C433/022, pp 73-84.

(10) WILLIAMS, P A, NIGHTINGALE, C J E, and BECKWITH, P, An alternative to Sauter Mean Diameter (Smd) for characterizing fuel sprays in the manifold of a spark ignition (si) engine, Sprays and Aerosols Conference, University of Surrey, 30 Sept - 2 Oct 1991, Paper No. D6.

(11) FINLAY, I C, McMILLAN, T, BANNELL J L K, and NIGHTINGALE, C, The measurement of the sizes of droplets leaving the throttle plate of an air-valve carburettor, J.Phys.D:Appl.Phys., 1985, 18, pp 1213 - 1222.

(12) DE BOER, C D, JOHNS, R J R, GRIGG, D W, TRAIN, B M, DENBRATT, I, and LINNA, J-R, Refinement with performance and economy for four-valve automotive engines, IMechE International Conf. on Automotive Power Systems, Chester UK, 10 -12 Sept 1990, IMechE Publication No. C394/053, pp 147 - 155.

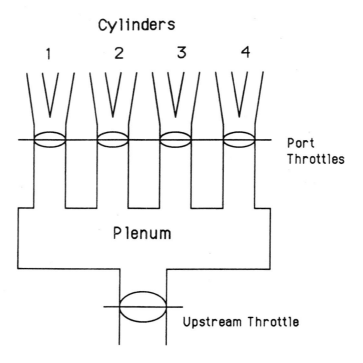

Fig 1   Schematic layout of combined port and upstream throttle arrangement

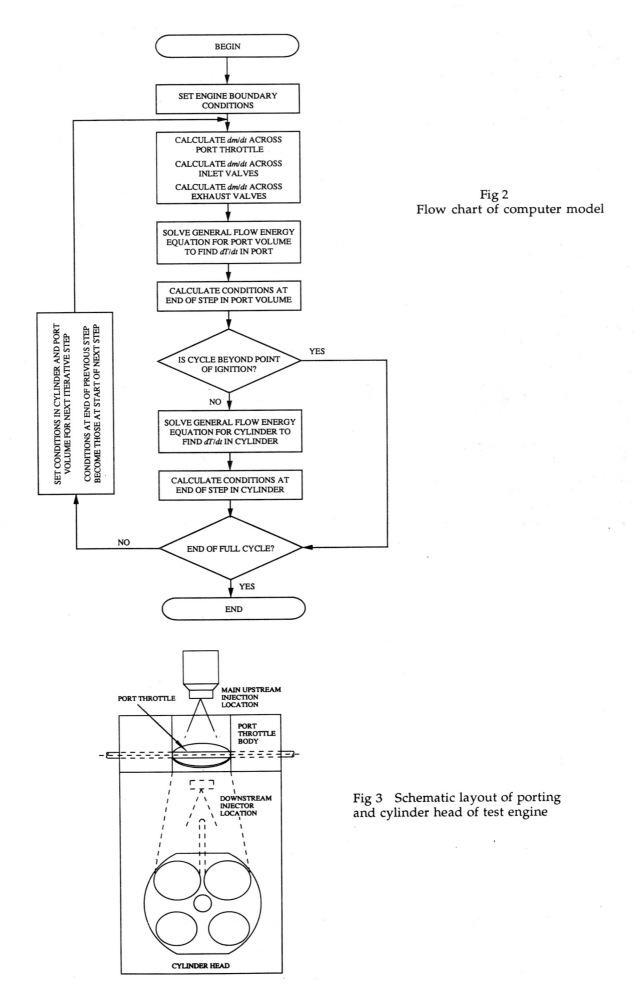

Fig 2
Flow chart of computer model

Fig 3   Schematic layout of porting and cylinder head of test engine

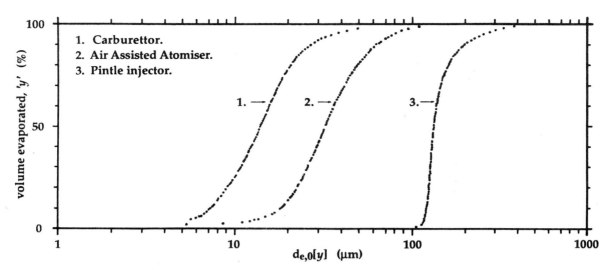

Fig 4 Comparison of si engine fuel sprays

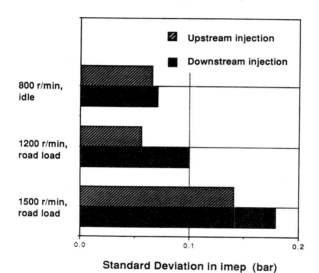

Fig 5 Effect of injection position on cyclic variability (Vers1 engine)

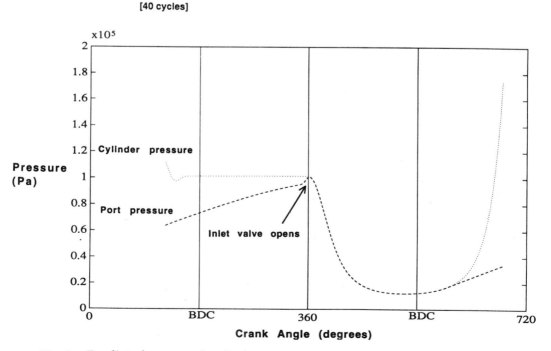

Fig 6 Predicted port and cylinder pressure characteristics at 800 r/min idle, illustrating port pressure recovery

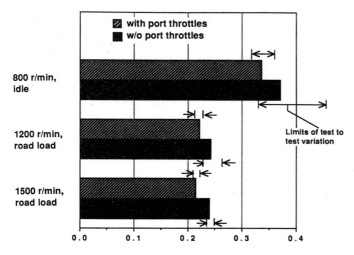

Fig 7   Reduction in fuel consumption achieved with port throttles (Vers2 engine)

Ind. specific fuel consumption (kg/kWh)

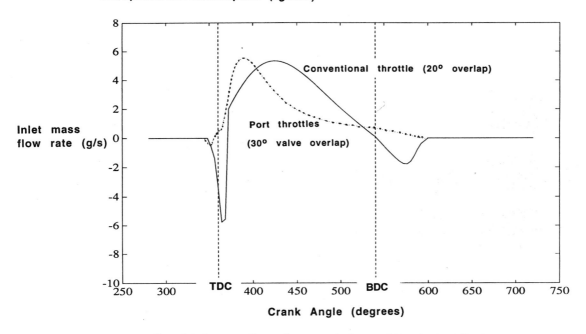

Fig 8   Predicted inlet gas flow characteristics at 800 r/min idle (100% pressure drop across port throttles)

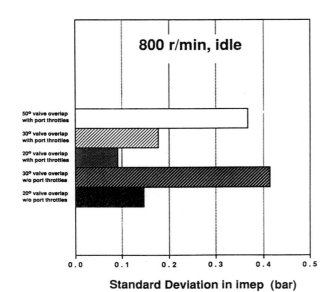

Standard Deviation in imep (bar)

[225 cycles]

Fig 9   Effect of port throttles on cyclic variability for different valve overlaps (Vers2 engine with 25% pressure drop across port throttles)

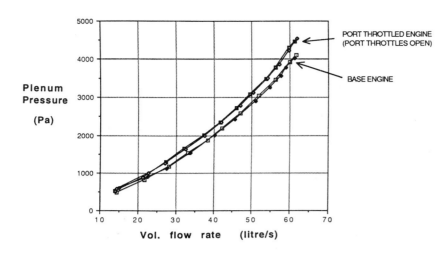

Fig 10 Pressure drop vs flow characteristics for Vers3 engine with and without port throttles

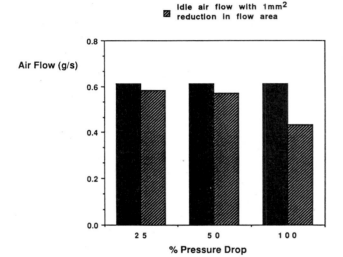

Fig 11 Computed change in air flow rate when the effective throttle disk flow area is reduced by 1.0 mm^2

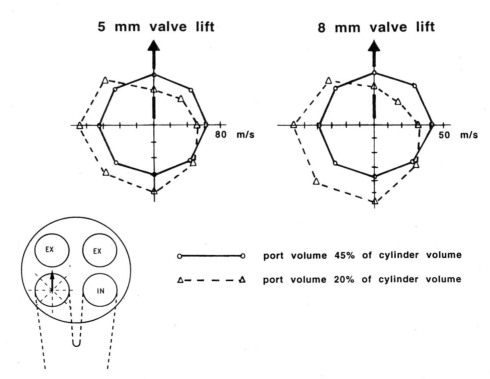

Fig 12 Polar plots of radial velocity in the inlet valve curtain area at 5 and 8 mm valve lifts for simulated 1500 r/min road load condition

C448/058

A basis for the predictive control of cyclic dispersion in a spark ignition engine

S P STEVENS, MA and P J SHAYLER, MSc, PhD, CEng, MIMechE, MSAE
Department of Mechanical Engineering, University of Nottingham
T MA, BSc, MSc, PhD
Ford Motor Company Limited, Basildon, Essex

SYNOPSIS

Cycle-to-cycle variability in combustion quality adversely affects engine roughness, exhaust emission levels and fuel economy. Here, the possibility of feed forward control to minimise variability is investigated. A method based on time series analysis has been developed to predict next-cycle combustion data for a spark ignition engine. The prediction technique constructs a sequence of parameter values from the most recent cycles and compares the pattern to other patterns stored in a library array. Knowledge of subsequent cycles for the most similar library patterns is used in the prediction of the next-cycle event. The prediction method has been tested on cycle-to-cycle cylinder pressure and burn rate data. Results indicate that it is possible to accurately predict some next-cycle parameters, particularly peak cylinder pressure and IMEP. A system to control cycle-to-cycle variations based on this predictive technique is discussed.

INTRODUCTION

A number of previous studies, reviewed by Young [1], indicate that cycle-to-cycle variations in combustion are created in the early part of the combustion cycle. This results in cyclic variations in burning rate which, in turn, affects the energy release rate. With respect to the work output, the effect of this variation is amplified by variations in the phasing between the energy release rate profile and piston motion. It has been concluded that the most effective way to minimise cycle-to-cycle variations is to create fast burning cycles. This would permit a substantial fraction of the combustion energy to be released around top dead centre (TDC) when piston motion, and therefore the rate of change in cylinder volume, is at a minimum. Several methods of achieving this have been proposed including swirl and/or turbulence inducing devices such as shrouded valves and alternative port designs [2] or ignition system modifications such as a dual spark plug system [3].

The majority of work related to cycle-by-cycle variations in the 1980s centred on gaining an improved understanding of the early stages of the combustion process and of the influence of in-cylinder gas motion in particular. The application of techniques such as laser Doppler anemometry (LDA) and hot wire anemometry (HWA) have led to detailed analyses of in-cylinder flow [4].

By the very nature of the internal combustion process, cycle-to-cycle combustion variations appear to be unavoidable to some degree. Cyclic variations occur throughout the range of engine operating conditions,

both for stoichiometric and lean mixture cases. For engines running at stoichiometric fuelling with a three way catalytic converter, the gains associated with reduced cyclic variability are significant but probably small compared to the potential benefits for lean burn engines, when cyclic variability is a limiting characteristic of lean operation. Matekunas [5] and Martin et al [6] suggested that there are two separate regimes which characterise cyclic variability depending on the air fuel ratio (AFR) and exhaust gas recirculation (EGR).

The aim of the work described in this paper has been to assess a technique which can be used in the prediction of cycle-by-cycle engine data. The method uses previous-cycle information to predict the parameter for the next-cycle combustion event. It is believed that cyclic variations can not be totally suppressed by passive design improvements alone since the causes of the variations can never be eliminated. The motivation for the work has been the possibility of applying feed-forward control of engine parameters to minimise cyclic variations. The paper focuses the quality of the predictions the technique produces together with an outline of the application of the prediction procedure to the feed-forward control of spark timing.

THE DATA PREDICTION PROCESS

A summary of the prediction process is given here and the authors direct the reader to a number of other publications for a more detailed description of the type of procedure used [7,8,9]. The principal elements of the prediction process are shown in Figure 1. The forecasting method does not require a knowledge of the mechanisms which determine the behaviour of the system. Instead, the technique bases predictions on patterns repeated within the data set. The time series used in this work consisted of 500 consecutive cycles which were transformed by first differencing and divided into two halves. Patterns were then constructed using sequences of consecutive differenced values. The length of the patterns was defined by the "embedding dimension" (denoted by E). Differenced parameter values from the second 250 cycles were predicted by comparing the pattern constructed from the most recent values (the "sample" pattern) to patterns from the first 250 cycles (referred to subsequently as "library" patterns). The selection process is based on a type of technique known as the "K-Nearest Neighbours" approach.

TEST CONDITIONS

One of the primary objectives of this work was to determine the feasibility of predicting cycle-to-cycle variations in a combustion system representative of a production engine. For this reason the test work was carried out using a standard four-cylinder, eight-valve engine. This was to ensure that factors which may influence the mechanisms of cyclic dispersion such as the flow conditions in the intake ports and the fuelling characteristics of the injection system were present. The engine used was a Ford 2.0 litre DOHC coupled to a Froude EC38TA eddy-current dynamometer via an Ford MT75 5-speed gearbox. A version of the Ford EEC IV engine management system was used to control sequential fuel injection and ignition with modifications to the strategy made possible using a calibration console. AFR measurement was made using an NTK MO-1000 AFR meter, the sensor of which was located in the exhaust downpipe approximately one metre from the exhaust ports. Throttle position and engine oil and coolant temperatures were monitored and controlled using an IBM PC-based, rig control system developed at Nottingham.

At each steady state test condition, data were recorded in batches of 500 consecutive cycles. A Hewlett Packard 9000 Series 300 computer was loaded with "Quikburn" data acquisition software developed at Nottingham and which is described elsewhere [10]. For each cycle, the parameters recorded are listed in Table 1 together with corresponding symbols which are used in subsequent figures and tables.

Table 1: Parameters Recorded and Symbols Used

Symbol	Description
WMEP	Work Mean Effective Pressure = IMEP calculated between 180° BTDC to 180° ATDC
PP	Peak cylinder pressure
PPL	Location of peak cylinder pressure
PRPR	Peak rate of cylinder pressure rise
PRPRL	Location of peak cylinder pressure rise
10% BL	Location of 10% mass burnt
50% BL	Location of 50% mass burnt
90% BL	Location of 90% mass burnt

In order to calculate these parameters, the cylinder pressure for number one cylinder was acquired at one degree intervals using a Hohner Series 3000 shaft encoder. Cylinder pressure was measured using a Kistler 6121 pressure transducer connected to a Kistler 501 charge amplifier. The pressure data was read into the computer via a 12-bit analogue-to-digital converter board.

The nature of cycle-to-cycle variations has been found to be different under different operating conditions. Martin et al [6] suggested two, superimposed types of variation. At a relatively rich AFR, the dominant type of variation is caused by current-cycle effects such as variations in in-cylinder gas flow and mixture fluctuations. It was found that these were largely random in nature and resulted in relatively low cyclic variations. However, close to the lean limit, where relatively large variations in burn duration exist, the residual exhaust gas composition may vary significantly. This may then be influential in the next-cycle combustion event. At lean fuelling conditions, this prior-cycle effect is the main cause of cyclic dispersion. The implication of these conclusions is that a prediction process described above may successfully predict parameter values at lean AFR or high EGR conditions but not at stoichiometric fuelling conditions with low EGR. In order to test this hypothesis and to assess the possibility of predicting cycle-to-cycle variations at different operating conditions, the engine test conditions defined in Table 2 were used.

Table 2: Test Conditions

Test Number	Speed (rev/min)	Load (Nm)	AFR	Spark Timing (°BTDC)
1	1000	0	15.5:1	10
2	1500	20	20.5:1	10
3	2000	44	14.7:1	31

RESULTS

The prediction accuracy of the technique has been quantified using the correlation coefficient between the predicted and actual differences in cycle-by-cycle data. The correlation coefficient is a measure of the degree of correlation between two variables. The larger the absolute value of the coefficient, the closer the observations lie on a straight line. If there is no significant relationship between the variables, the points are widely scattered and the value of the correlation coefficient will be low (of the order of 0.1). In this work, the correlation coefficient has been calculated for 250 points. According to Belmont et al [11], 2000 data points are necessary in order to ensure stability in cycle-to-cycle data statistics. Although this criterion is not satisfied, the quoted statistics are appropriate measures of the prediction ability of the technique.

Before a detailed analysis of the prediction method was performed, an investigation was carried out to determine the minimum number of library patterns necessary to achieve the best possible prediction performance. Clearly, from a practical point of view, the fewer the number of library patterns required, the faster the next-cycle prediction can be made and the quicker library data may be acquired for a particular operating condition. To quantify this requirement, the WMEP cycle-by-cycle data for the three tests were used. The data were split into two sections and the second 250 cycles were predicted using libraries based on part or all of the first 250 cycles. The library section was always taken from the first cycle recorded to a given number of cycles later, leaving a gap as necessary between the end of the library and the start of the sample data. The results shown in Figure 2 indicate that prediction performance improves rapidly as the number of library patterns is increased up to typically 150, after which further increases in library size have a small effect. The correlation coefficients in this figure have been normalised with the correlation coefficient obtained using the full library array of 250 cycles.

The effect of library size has a similar effect with a smaller pattern size or embedding dimension (Figures 3a and 3b). Again, WMEP values were used in this prediction analysis from the test at a lean AFR and light load (Test 2). In this figure, the effects of library size and when the library data were recorded relative to the sample data (cycles 250 to 500) are shown. At steady operating conditions and using the optimum embedding dimension (pattern size), any part of the first half of data set can be used to construct the library without affecting the general prediction accuracy. Also, with the smaller embedding dimension, the necessary library size to obtain the best accuracy is lower (100 cycles for E=2 compared with 200 cycles for E=5). With regard to a real-time application, the results indicate that updating the library would not have to be done continuously under steady operating conditions. It may also be possible to improve the speed of the prediction process by storing only a small set of library patterns which corresponded to particular abnormal operating features which were to be identified in the sample data.

Figures 4a, 4b and 4c show the prediction results using each of the parameters recorded at each test condition. The maximum possible library size of 250 cycles was used to produce all of the results shown. The labels in the figures refer to the parameters listed in the previous section in Table 1. As expected, the best predictions were made when the engine operating conditions were least stable, that is the test at fast idle and at light load and lean mixture (Tests 1 and 2). The coefficient of variation or COV (the standard deviation divided by the mean) in WMEP for each test was 19.81 per cent, 24.76 per cent and 1.30 per cent for Tests 1, 2 and 3 respectively. Test 2 was particularly unstable due to the combination of lean fuelling and retarded ignition, chosen specifically as a condition likely to produce predictable cyclic variations. The results for the three tests generally follow the trends indicated by previous investigations [5,6]. However, the results for Test 3 show that cyclic variations are not predominantly random, as expected, at conditions of relatively rich fuelling. Although the prediction results are generally poorer, particularly for the parameters which characterise cylinder pressure, the correlation coefficient values of 0.4 and higher indicate with a high degree of certainty that some form of pattern exists within the data. If the cyclic variation was predominantly random, the correlation coefficient between the predicted and actual data should be significantly lower. A value of 0.16 corresponds to the 1 per cent level of significance for the size of data set used.

The parameter which is predicted most accurately in the two unstable tests is WMEP, probably due to the high degree of variation this exhibited. The effect of a high COV can also be observed in the results for peak cylinder pressure (PP). For the test at fast idle (Test 1), the COV of peak cylinder pressure was 20.03 per cent compared to 1.58 per cent for Test 2 and 4.16 per cent for the part-load, stoichiometric operation in Test 3. The low value for Test 2 can be attributed to an effect of slow combustion, which gives poor phasing of heat release on most cycles. For these cycles, charge compression has the major effect in determining the peak cylinder pressure and its location, which are then quite repeatable. The relative stability of the peak cylinder pressure results in a comparatively low prediction accuracy as shown in Figure 4b. Sampling at a fixed crank angle location might be more appropriate under these operating conditions. Douaud et al [12] suggested using the cylinder pressure recorded at 40° ATDC, which will be assessed using the prediction process as part of future work. In general, the other parameters examined were poorly predicted. In the case of peak rate of pressure rise and its location (PRPR and PRPRL respectively), the effect of signal noise may have a particularly significant influence which can create erroneous patterns in the time series data. This view was supported by Matekunas [5]. For the burn rate parameters, the error in determining the end of combustion point accurately can alter the burn rate values. The errors may be averaged out over a large number of cycles but it is suggested that the parameters are less suited to cycle-by-cycle analysis.

The data acquired during the test at fast idle (Test 1) contained a number of misfiring cycles, evident from the points corresponding to large negative differences in actual WMEP in Figure 5. It is clear that the misfire event itself has not been anticipated. Once a misfire cycle is incorporated within the sample pattern, the next cycle, which has a large positive difference, is very accurately predicted. The positive difference for the cycle exceeds the negative difference of the misfire cycle (+2.3 compared to -2.0) because of the effect of unburnt residuals carried over from the misfire cycle. The inability to predict misfire cycles from prior-cycle WMEP data has also been observed by Martin et al [6]. Surprisingly, the prediction of extreme peak pressure differences (including misfiring cycle cases) is substantially more accurate, as illustrated by Figure 6. Analysis of the data on a cycle-to-cycle basis indicates that the extremes of the peak cylinder pressure differences are not uniquely associated with misfires, however, and so can not be used to predict a misfire event.

An alternative approach that can be taken using this technique involves combinations of previous cycle parameters rather than a time series to predict a particular parameter value for the next cycle. For example, the values of WMEP, peak cylinder pressure (PP) and peak rate of cylinder pressure rise (PRPR) for cycle "A" may be treated as a pattern to predict the WMEP value for cycle "A+1". The use of auto-correlation and cross-correlation coefficients calculated between parameters has been investigated to determine possible combinations of parameters. In the case of the test at fast idle for example, correlations were found to exist between the last-cycle WMEP, peak cylinder pressure and its location, and the current-cycle WMEP value. These parameter values were used to form cycle-to-cycle patterns with an embedding dimension of 3, to predict WMEP as shown in Figure 7. The corresponding correlation coefficient of this data is 0.68, which can be compared with a value of 0.72 obtained from the WMEP differenced time series. The conclusion drawn is that there is an association between last-cycle pressure parameters and the present-cycle WMEP value. This is probably due to the relationship between the parameters which describe the cylinder pressure profile and the WMEP value for a given cycle. However, though this relationship is strong, it produces less accurate predictions than those given by using the differenced time series approach. This was found to be the case for all combinations of parameters used. It is also interesting to note that no parameter combinations were able to predict the WMEP value for a misfiring cycle even when the previous-cycle WMEP value was not included in the sample sequence.

The work completed to date indicates that the technique is most applicable to lean-burn engine operation. The WMEP prediction results for the test at light load and lean fuelling are shown in Figure 8. These predictions correspond to the correlation coefficient of 0.82 for Test 2 in Figure 4b. The results show that good predictions are consistently achieved at this test condition and are sufficiently accurate for the application of the technique to be considered.

APPLICATION TO SPARK TIMING CONTROL

The work presented indicates that, particularly under lean-fuelling operation, the technique can predict next-cycle performance with sufficient reliability for its application to be considered. It is beyond the scope of this paper to quantify the anticipated performance of such a system, but some relevant features can be defined. The feed-forward control of cyclic variability could use either WMEP or peak cylinder pressure as a suitable prediction parameter. Fuelling or spark timing changes could be made to modify the next-cycle conditions. Fuelling modifications made using current fuel injection systems create complications because of the effects of wall wetting on fuel transfer from the intake port into the cylinder. Characteristic transport delays and induced mixture excursions would tend to distort the intended control action, making accurate cycle-by-cycle control difficult. More direct control may be realised by adjusting spark timing and, at first sight, provides a relatively simple means of effecting compensation. For this application, however, the influence of adjustments to spark timing are uncertain and so a preliminary assessment has been undertaken at the operating conditions of light load and lean AFR used for Test 2. To assess the possible effect of such changes, the spark timing was changed to minimum advance for best torque (MBT) timing from MBT-5° timing. Even at such lean fuelling, the 5° advance in spark timing improved the average value for the peak cylinder pressure (PP) by 7.2 per cent and the average WMEP value by 11.3 per cent. These results suggest that modifications to spark timing of this size can affect the phasing of the combustion event and, therefore, the work transfer. Thus, it is anticipated that with accurate next-cycle predictions, appropriate adjustments made to spark timing may result in reduced cyclic variability.

The criteria for determining when action should be taken can be defined in various ways. One of the simplest strategies is illustrated in Figure 9. Modifications to the standard spark timing are made based upon the predicted deviation of WMEP falling outside a pre-determined limit. So in Figure 9, modifications would be initiated if values within the bands labelled "2" were predicted. No modifying action would be taken if the predicted value lay within band "1". Typically, such a strategy would produce actions in approximately 40 per cent of cycles.

Although open issues remain, the illustration above indicates how the component parts of a feed-forward control system may be assembled. The indications are that a practical system which takes advantage of time series based predictions to improve cyclic variability at lean-fuelling conditions can be demonstrated. At this stage, the final form of such a system has yet to be defined and performance benefits quantified. These are subject to further investigations which will determine the potential for production applications.

CONCLUSIONS

1. A versatile technique has been developed which can be used to predict next-cycle data at any engine operating condition. This is possible because the technique does not require knowledge of the mechanisms which determine the nature of the data.

2. Peak cylinder pressure and WMEP were generally found to be the most suitable parameters to use in the prediction process. This is due to the relatively high variation of the parameter values suggesting that these parameters are most sensitive to the causes of cycle-to-cycle variation.

3. Results indicate that deterministic elements exist within cycle-to-cycle engine data at conditions of high dilution created by either lean fuelling or high EGR. It is believed that this characteristic is due to the effect of variations in the residual exhaust gas composition.

4. It is apparent that misfiring cycles are caused by purely same-cycle effects. There is no evidence to suggest that prior cycles have any influence in determining the occurrence of such an event.

5. Results using the prediction technique show that it may be possible to use the technique to control cyclic dispersion at extreme operating conditions. This is of particular relevance to lean-burn combustion concepts.

ACKNOWLEDGEMENT

The authors would like to express their thanks to the Ford Motor Company for their financial support and permission to publish this work which was carried out at the Department of Mechanical Engineering of the University of Nottingham.

REFERENCES

[1] M. B. Young, "Cyclic Dispersion in the Homogeneous-Charge Spark-Ignition Engine- A Literature Survey", SAE 810020.

[2] S. Matsouka, T. Yamaguchi, Y. Umemura, "Factors Influencing the Cyclic Variation of Combustion of Spark Ignition Engine", SAE 710586.

[3] R. W. Anderson, "The Effect of Ignition System Power on Fast Burn Engine Combustion", SAE 870549.

[4] P. G. Hill, A. Kapil, "The Relationship Between Cyclic Variations in Spark-Ignition Engines and the Small Scale Structure of Turbulence", Combustion and Flame, 78, pp 237-247, 1989

[5] F. A Matekunas, "Modes and Measures of Cyclic Combustion Variability", SAE 830337.

[6] J. K. Martin, S. L. Plee, D. J. Remboski, "Burn Modes and Prior Cycle Effects on Cyclic Variations in Lean-Burn Spark Ignition Engine Combustion", SAE 880201.

[7] J. D. Farmer, J. J. Sidorowich, "Predicting Chaotic Time Series", Physical Review Letters, Vol. 59, No. 8, pp 845-848.

[8] M. Casdagli, "Non Linear Prediction of Chaotic Time Series", Physica D., 35, pp 335-356, 1989.

[9] G. Sugihara, R. M. May, "Nonlinear Forecasting as a Way of Distinguishing Chaos From Measurement Error in Time Series", Nature, Vol. 344, No. 6268, pp 734-741

[10] P. J. Shayler, M. W. Wiseman, T. Ma, "Improving the Determination of Mass Fraction Burnt", SAE 900351.

[11] M. R. Belmont, M. S. Hancock, D. J. Buckingham, "Statistical Aspects of Cyclic Variability", SAE 860324.

[12] A. Douand, G. de Soete, C. Henault, "Experimental Analysis of the Initiation and Development of Part-Load Combustions in Spark Ignition Engines", SAE 830338.

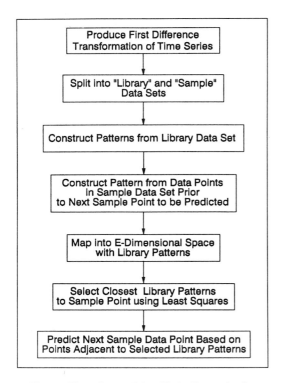

Fig. 1: Flowchart of the Main Steps in the Prediction Procedure

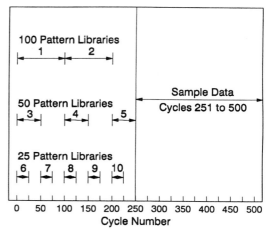

Fig. 3a: Position of Libraries Relative to Sample Data

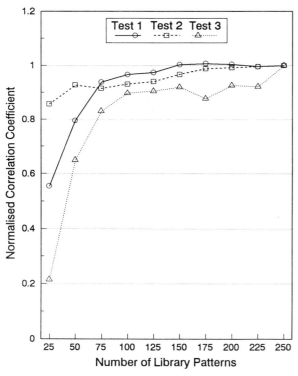

Fig. 2: Effect of Library Size on Prediction Ability of WMEP Using an Embedding Dimension of 5

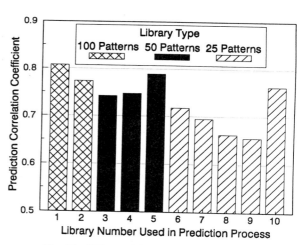

Fig. 3b: Effect of Size of Library and Position Relative to the Sample Data on Prediction Accuracy of WMEP Sample Data Using an Embedding Dimension of 2 with Test 2 Data

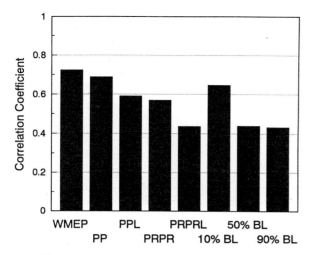

Fig. 4a: Effect of Parameter Used on Prediction
Performance Using Data from Test 1
1000 rpm, No Load, 15.5:1 AFR, 10 BTDC

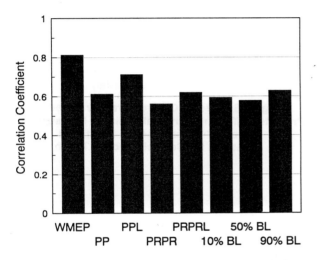

Fig. 4b: Effect of Parameter Used on Prediction
Performance Using Data from Test 2
1500 rpm, 20 Nm, 20.5:1 AFR, 10 BTDC

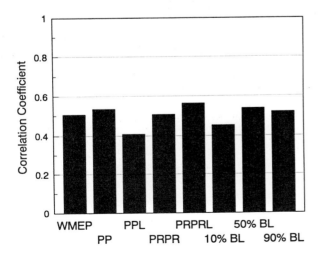

Fig. 4c: Effect of Parameter Used on Prediction
Performance Using Data from Test 3
2000 rpm, 44 Nm, 14.7:1 AFR, 31 BTDC

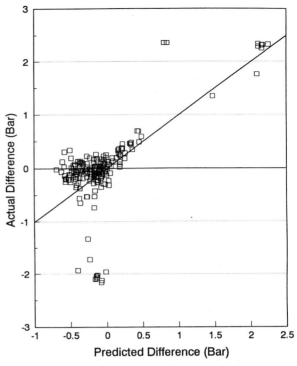

Fig. 5: Predicted Against Actual Differences in WMEP
Using an Embedding Dimension of 3 with Test 1 Data
1000 rpm, No Load, 15.5:1 AFR, 10 BTDC

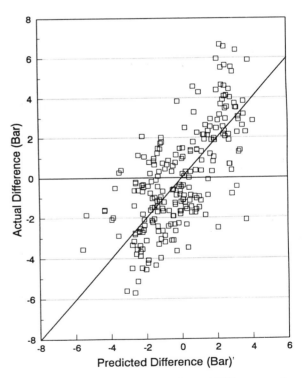

Fig. 6: Predicted Against Actual Differences in
Peak Cylinder Pressure (PP) Using an Embedding
Dimension of 2 with Test 1 Data
1000 rpm, No Load, 15.5:1 AFR, 10 BTDC

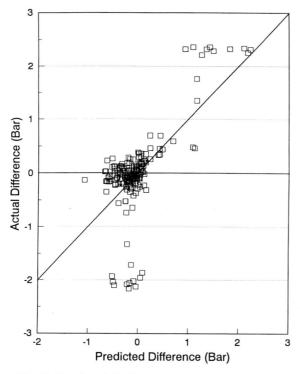

Fig. 7: Predicted Against Actual Differences in WMEP
Using Previous Cycle WMEP, PPR and PPL with Test 1
Data: 1000 rpm, No Load, 15.5:1 AFR, 10 BTDC

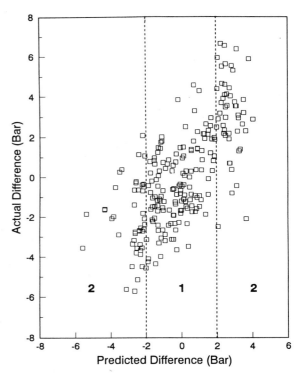

Fig. 9: Suggested Control Strategy Based on
Predicted Differences in Peak Cylinder Pressure (PP)
Area "2": Control Action is Initiated
Area "1": No Control Action is Taken

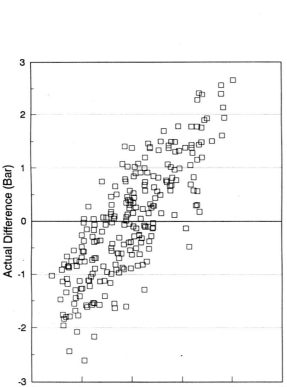

Fig. 8: Predicted Against Actual Differences in WMEP
Using an Embedding Dimension of 2 with Test 2 Data
1500 rpm, 20 Nm, 20.5:1 AFR, 10 BTDC

C448/051

Experimental studies of pre-ignition on a four valve engine

N COLLINGS, BSc, PhD, CEng, MIMechE
Cambustion Limited, Cambridge
G D GOOLD, BSc and G D MORRIS
Ford Motor Company Limited, Laindon, Basildon, Essex

ABSTRACT Miniature flame arrival (ionization) probes and the spark plug itself have been used to investigate the pre-ignition phenomena in a four valve engine with central ignition. It was found that for the conditions investigated, pre-ignition was always associated with the end gas. At 'safe' spark timings, stable pre-flame reactions were observed in the end gas region close to the exhaust valves, whereas such reactions were only observed near to the inlet valves immediately preceding pre-ignition proper. A large variability in flame arrivals at the flame ionization probes was observed, and may be indicative of cyclic variability in turbulence.

1 INTRODUCTION

At high load and speed, all gasoline engines can exhibit pre-ignition at sufficiently advanced ignition. Recent changes in fuel characteristics, and the ever present need to improve power density, mean that the pre-ignition is more than ever a limiting condition. This paper contains details of a technique for pre-ignition measurement, and data for a particular engine type.

Ionization probes have been used for many years to measure combustion related phenomena, such as flame arrival (1,2) knock (3) and misfire (4). All of these studies rely on the fact that when a hydrocarbon burns, a small but measurable proportion of the products is ionised. This study takes advantage of this phenomena, via the ionization produced by pre-flame reactions.

Pre-flame reactions and **pre-ignition** have to be carefully distinguished. Both refer to oxidation reactions occurring before the arrival of the flame originating at the sparking plug. In a sense pre-flame reactions are always occurring, as there is a finite reaction rate for air/hydrocarbon mixtures even at room temperature. However, the term is normally used to mean significant reaction rates, but not involving a complete burning, the latter being pre-ignition.

The term pre-ignition is often used to mean only those reactions that occur before the spark has actually fired. Reactions occurring before flame arrival, but after the spark, have been termed surface ignition. This demarcation is of course arbitrary - the results presented below show that there is no sudden transition when the pre-flame reactions occur before flame arrival.

The term pre-ignition could be used exclusively for the runaway damaging condition, but this would certainly not then include all pre-spark reaction conditions.

2 EXPERIMENTAL

The engine was a 1.8l 4 valve engine of conventional design, and the ionization sensors were inserted through existing pressure transducer access holes, one between the inlet valves and one between the exhaust valves. Fig 1 shows the position of the sensors, and fig.2 the probes themselves.

In order to measure any ionization present in the vicinity of the probe or sparking plug, they were electrically biased by a voltage of 180 VDC - fig 3. Note that a high voltage diode is used to block the HT spark pulse when the circuit is applied to the spark plug. The signal output was amplified and then recorded and observed utilising an AVL Indiskop.

3 RESULTS

At 6000 rpm WOT, 13.0 AFR, using 95ULG fuel, it was observed at normal ignition timing values that the flame arrivals at both probes was sharp, but varied considerably in arrival time (CA), possibly due in part to variability in tumble breakdown. At the inlet valve probe, some cycles had arrivals as late as 90°ATDC, though the majority were in the range 15 - 35°ATDC - fig 4. (Note that although the ionization current scale is arbitrary, all the plots have the same scaling.) At the exhaust valve, a similar picture was seen, except that occasionally no flame arrival whatever was recorded. The reason for this not understood. It was also observed that even at the safe timing value, pre-flame reactions occurred in the vicinity of the exhaust valves. These reactions were steady, however, showing no tendency to run away into ever increasing and earlier reaction.

The spark timing was then incremented (advanced) by 2°, the engine was allowed to stabilise, and then the data was recorded. This procedure was repeated every 2 minutes until pre-ignition occurred.

As the spark timing was advanced, the average time (in CA) between spark and first flame arrivals decreased markedly, and the level of the ionization signals both on the sparking plug and the flame arrival sensors increased significantly. These features are emphasised in fig 5, which is for a spark advance of 30°, but using 100 octane fuel. This spark advance was beyond the PI limit on 95ULG fuel, but the position and size of the flame arrivals is a good extrapolation of the non-PI relationship of flame arrivals to spark advance.

At spark timings near to those where PI was expected (from previous work), small, but distinct features appeared on the flame arrival probe before the flame arrival signal, though the flame arrival proper was still quite distinct - fig 6. This result emphasises the difference between full combustion, represented by a large and distinct signal level change, and the relatively small pre-flame reaction signals. At this stage, the pre-flame reactions were steady, that is there was no tendency for them to advance towards the spark trace during the 2 minute period at each spark advance. (The spark plug ionization signal often exhibited clear signs of heavy knock as the timing was advanced. This was seen as a characteristic high frequency modulation of the ionization trace after TDC, much the same as is seen in pressure signals.)

The ignition was then advanced by a further 2°, and the pre-flame reactions immediately advanced significantly towards the spark trace, and furthermore began a steady advance at this (fixed) spark timing. The pre-flame reaction signals were also becoming much larger, though the 'true' flame arrival could still be differentiated.

At this point, the energy release in the pre-flame reactions was becoming comparable to the heat release from the flame itself, and pre-ignition was expected. After a few moments, the pre-flame reactions advanced to a point prior to the spark event.

Fig 7 shows a typical result obtained at this time, just (a few seconds) before pre-ignition proper occurred, (i.e. sudden torque drop) requiring the engine to be shut down immediately.

It was always clear that the first activity seen at the plug was the spark itself. I.e. there was no evidence of pre-flame (pre-spark) reactions occurring at the plug. In other words there was no evidence that the spark plug was the source of PI. Fig 8 shows an artificially induced PI at the plug, caused by fitting a plug of non-standard heat rating. As well as illustrating clearly the pre-spark reactions, knock-induced gas conductivity variations are also evident, as discussed above.

The pre-flame reactions seen at the exhaust end gas sensor remained 'in check' until PI occurred, and this seemed to be associated with a simultaneous advance of both sensors' pre-flame reaction events up to and in front of the spark event.

It should be added that the PI limit was not changed by the use of ionisation probes - whether they were fitted or not, the PI limit was at exactly the same spark advance.

4 CONCLUSIONS

The use of flame arrival sensors and spark plugs for pre-ignition investigations has been successfully demonstrated.

It is concluded that the PI event in the case tested is due to reactions occurring in the end gas, rather than at the spark plug. Since the pre-flame reactions are very sensitive to pressure and temperature, it may be that although the region near the exhaust valves is hot enough to get some reaction started, it is only when the ignition is advanced, and further pressure and temperature increases occur, that full pre-ignition begins.

5 REFERENCES

1. Curry, S., 'A Three Dimensional Study of Flame Propagation in a Spark Ignition Engine', SAE Trans., Vol. 71, No. 3 (1963)

2. May, M.G., 'Flame Arrival Sensing Fast Response Double Closed Loop Engine Management'., SAE Trans. Vol 92 (1984)

3. Blauhut, R.B., Horton, M.J., Wilkinson, A.C.N., 'A Knock Detection System Using the Spark Plug Ionization Current', 4th International Conference on Automotive Electronics, IEE, London 1983.

4. Collings N., Willey J., 'Cyclically resolved HC emissions from a spark ignition engine', SAE paper No. 871691 (1987)

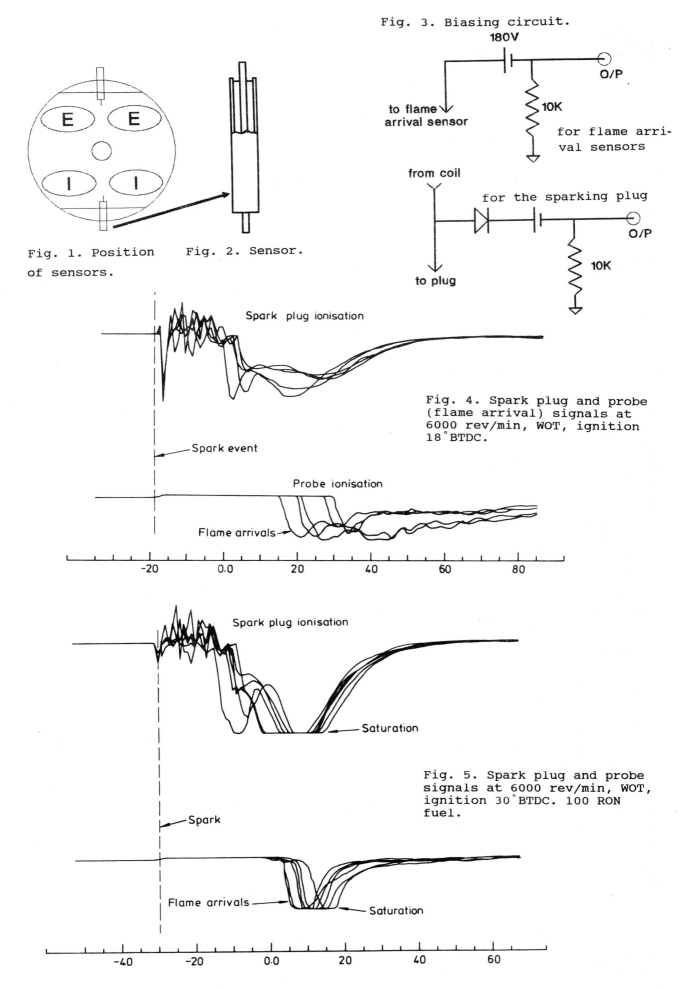

Fig. 1. Position of sensors.

Fig. 2. Sensor.

Fig. 3. Biasing circuit.

180V

to flame arrival sensor

10K

for flame arrival sensors

from coil

for the sparking plug

to plug

10K

O/P

O/P

Spark plug ionisation

Spark event

Probe ionisation

Flame arrivals

Fig. 4. Spark plug and probe (flame arrival) signals at 6000 rev/min, WOT, ignition 18°BTDC.

Spark plug ionisation

Saturation

Spark

Flame arrivals

Saturation

Fig. 5. Spark plug and probe signals at 6000 rev/min, WOT, ignition 30°BTDC. 100 RON fuel.

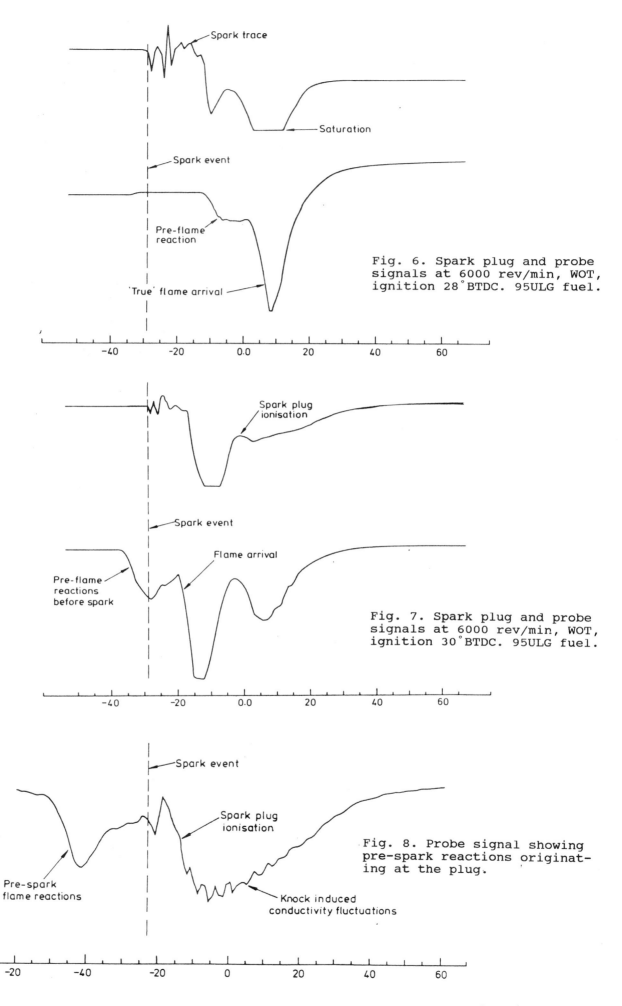

Fig. 6. Spark plug and probe signals at 6000 rev/min, WOT, ignition 28°BTDC. 95ULG fuel.

Fig. 7. Spark plug and probe signals at 6000 rev/min, WOT, ignition 30°BTDC. 95ULG fuel.

Fig. 8. Probe signal showing pre-spark reactions originating at the plug.

latter being, as a rule, sharp-edged to minimize the contact of hot gases with the walls and thus prevent the recombination of active radicals by surface collisions. Thereupon, combustion is reinstated in the cores of the large scale vortex structures in the turbulent plume created by the jet. The function of active radicals is for this purpose of essential significance, as brought out authoritatively by the extensive studies conducted at the Institute of Chemical Physics reported by Gussak and Turkish at the first IMechE Conference (6). The two step combustion process, arrest at efflux and restoration in vortex cores, is of prime importance, as confirmed by our observations (7-10). Adequate concentration of active radicals is maintained by keeping them within the stream core at a sufficiently high temperature so that, upon their distribution in the jet plume, they provide a multitude of ignition sources, rather than generating a diffusion flame, as is actually the case in the majority of stratified charge engines.

2.3 Background

In view of the plethora of aborted engine concepts, in order to be effective, the refinement offered by PJC has to be treated with sufficient care and attention, especially with respect to fluid mechanical and chemical kinetic features. In particular, its mechanism and control have to be catered to with diligence. It is for this reason that, so far, all our investigations were associated with experiments carried out in a constant volume enclosure, while the test charge was kept initially at rest (7-10).

The effect of an initially turbulent state has been studied at the same time by Wolanski and his associates at the Warsaw Institute of Technology. This included experiments using a constant volume vessel, where the turbulent field and swirl were introduced by rapid filling of an initially evacuated chamber (11), as well as a rapid compression machine (12). In both cases the superiority of the PJC system over a spark ignited FTC has been demonstrated, especially insofar as the initiation and execution of the process of combustion in extremely lean mixtures are concerned.

In a multi-stream version - a system realizable by the use of a number of orifices in a single PJC generator, or a number of generators, or both (9,10) - most of the exothermic process takes place then in the form of a fireball - a turbulent regime of combustion established in the middle of the enclosure, away from the walls. The beneficial effects of a significantly decreased contact of combustion products with the walls are of importance to both the rate of the process and its extent, primarily as a consequence of the sustenance of circulation promoted by the vortex structure of the turbulent jet plume, and the reduction of heat transfer losses, as brought out in the studies carried out recently on this subject (13,14).

On the basis of all this it should be expected that premixed charge engines equipped with a PJC system should be capable to operate at part load, down to idling, at wide open throttle - the essential plank in bridging the gap between spark-ignition and compression-ignition engines.

Application of similar principles for diesel engines was presented by us in two papers (15,16) illustrating the potential benefits obtainable from the equivalent system of pulsed air-blast atomizers. In the meantime, the originality

of the PJC concept has been established by three U.S. patents (17-19).

Finally it should be noted that PJC ushers in the ultimate in internal combustion: the environmentally benign hydrogen fueled engine (20). The beneficial consequences of introducing an air-fuel mixture via a PJC system, rather than injecting pure fuel, should be in this case of crucial importance in assuring knock-free operation.

3. MODUS OPERANDI

3.1 Principle

The proposed modus operandi for an engine of the future, based on these principles, consists of two fundamental elements:

1. dilution by excess air with *residual gas recirculation* (RGR)

2. *fluid mechanical governance* (FMG) of combustion

Dilution is the necessary first step. As a consequence of it:

- temperature peaks are reduced, impeding, according to the well known Zeldovich mechanism ($N + O_2 = NO + O$ and $O + N_2 = NO + N$) that occurs predominantly in lean mixtures (the influence of prompt NO mechanism being then relatively small), the highly temperature sensitive formation rate of NO,

- exothermic power pulse is spread out, whereby its peak is drastically reduced preventing the onset of knock.

- cycle efficiency is increased (21) - a significant factor in enhancing fuel economy.

Advantages of dilution by RGR in its extreme form have been demonstrated by a number of compression-ignition premixed charge engines with excessive recirculation of products. Prominent among them are: the *Active Thermo-Atmospheric Combustion* (ATAC) of Onishi (22), the *Toyota-Soken Combustion* (TSC) of Naguchi, et al. (23), and the *Homogeneous Charge Compression Ignition* (HCCI) four stroke engine tests of Najt and Foster (24) and, more recently Thring, et al. (25).

The application of FMG is of crucial importance to assure proper mixing of reactants and distribution of reaction sites in order to prolong the residence time of the reacting medium in the zones of essential chemical activity. As a consequence, the air-fuel mixture is processed effectively by a set of combustors within the combustion chamber, rather than being swept by a flame traversing the charge - a process causing all sorts of problems, as described in the introductory section.

The advantages accrued thereby include:

- distribution of the exothermic process of combustion in space and time associated with the entrainment of products into the reaction zones, impeding the formation of NO, as well as inhibiting the yield of HC.

- attainment of high rates of energy conversion, curtailing the concentration of CO.

- ushering in high-tech, whereby the evolution and deposition of exothermic energy is modulated by a microprocessor governed control system.

3.2 Implementation

As pointed out in the introductory section, the principles advanced here can be best implemented by a PJC system. Its mechanism is illustrated in Fig.1. Combustion is executed thereby upon entrainment of reactants into distributed, large scale vortex structures, formed within a shear-generated, turbulent jet plume. The reacting medium in the cylinder is then entrained and subjected to spiral mixing with the products of combustion issuing from the cavity of a PJC generator. This furnishes thus optimum conditions for the initiation and execution of the exothermic process in the vortex cores forming a set of concomitantly acting combustors within the turbulent jet plumes. The explosive potential of such a system is negated by the use of a diluted charge. Employing near stoichiometric mixtures, as done in today's engines, is, of course, in this case quite inappropriate. If one wishes, however, the engine can be equipped for this purpose with a conventional spark ignition system.

Conceptual features of a PJC generator for premixed charge engines, in a configuration suitable for a JPIC system described in the previous section, is depicted in Fig. 2.

Projected version of a PJC generator for non-premixed charge engines is portrayed in Fig. 3. It represents a pulsed version of the air-blast atomizer - a well established system, as brought up in the introductory section, in the technology of gas turbine engines (4). Here, spray droplets are formed by shear between the central air stream and the fuel sheet fed peripherally. For best performance, the fuel should be preheated, as close to the coke limit as possible, and the atomization should be sufficiently fine, so that the droplets are of adequately small diameters to vaporize before they get involved in the vortex structures of the turbulent plume.

Prospective features of a premixed charge engine equipped with a microprocessor controlled PJC system are presented in Fig. 4, while those of its equivalent for a non-premixed charge engine are shown in Fig. 5.

To complete the description of an engine of the future embodying a PJC system, illustrated in Fig. 6 are the salient features of its block. It is a two-stroke - the only sensible way of realizing a competitive, significantly diluted engine that the PJC system is capable to operate. Added to it are some less essential but, nonetheless, noteworthy devices to illustrate how far-reaching can be the feasible means for interactive, feedback control of combustion in engines. Included among them here are: precompression of air on the down stroke of the piston, valved intake and exhaust ports to regulate the amount of RGR, as well as an air-assisted fuel injection system. It is then, as pointed out before, that, under such circumstances, a premixed charge engine, similarly as the non-premixed, should be capable to operate efficiently at part load, down to idling, at a wide open throttle.

4. RECIPE

To sum up, the recipe for a controlled combustion engine, manifesting a definite progress of technology, consists of three ingredients:

1. preparation of the charge by dilution, using excess air combined with RGR.

2. execution of the exothermic process by a PJC system forming fireball combustors within the cylinder-piston enclosure, produced by generators provided with rich air-fuel mixtures by air-assisted injectors.

3. modulation of performance by a micro-processor control of the PJC process and its ancillary system in a two-stroke, loop-scavenged engine.

5. REFERENCES

(1) OBERT, E. F. Internal Combustion Engines and Air Pollution, 1973, xiii + 740 pp. + 5 charts, Harper and Row, Publishers, New York.

(2) HEYWOOD, J. B. Internal Combustion Engine Fundamentals, 1988, xxix + 930 pp., McGraw-Hill Book Company, 1988.

(3) BENSON, R. S. The Thermodynamics and Gas Dynamics of Internal-Combustion Engines, 1982 (edited by J. H. Horlock and D. E. Winterbone), Vol. I, 582 pp., Clarendon Press, Oxford; J. H. Horlock and D. E. Winterbone, The Thermodynamics and Gas Dynamics of Internal-Combustion Engines, 1986, pp. 583-1237, Clarendon Press, Oxford.

(4) LEFEBVRE, A. H. Gas Turbine Combustion, 1983. xvii + 531 pp. (vid. pp. 413-459), Hemisphere Publishing Corporation, New York.

(5) I.MECH.E. Conference on Stratified Charge Engines, London, 1976, 240 pp.

(6) I.MECH.E. Conference on Stratified Charge Automotive Engines, London, 1980, 116 pp.

(7) OPPENHEIM, A. K., BELTRAMO, J., FARIS, D. W. MAXSON, J. A., HOM, K., and STEWART, H. E. Combustion by Pulsed Jet Plumes - Key to Controlled Combustion Engines. SAE Congress Paper 890153, 1989, 10 pp.

(8) MAXSON, J. A., and OPPENHEIM, A. K. Pulsed Jet Combustion - Key to a Refinement of the Stratified Charge Concept. Twenty-Third Symposium (International) on Combustion, The Combustion Institute, Pittsburgh, Pa., 1991, pp. 1041-1046.

(9) MAXSON, J. A., HENSINGER, D. M., HOM, K., and OPPENHEIM, A. K. Performance of Multiple Stream Pulsed Jet Combustion Systems. SAE Congress Paper 910565, 1991, 9 pp.

(10) HENSINGER, D. M., MAXSON, J. A., HOM, K., and OPPENHEIM, A. K. Jet Plume Injection and Combustion. SAE Congress Paper 920414, 1992, 10 pp.

(11) ABDEL-MAGEED, S. I., LEZANSKI, T., and WOLANSKI, P. Comparative Performance of a Pulsed Jet Combustion System in a Swirl and a Turbulent Field. 13th International Colloquium on Dynamics of Explosions and Reactive Systems, Nagoya, Japan, 1991.

(12) GMURCZYK, G. W., LEZANSKI, T., KESLER, M., CHOMIAK, T., RYCHTER, T., and WOLANSKI, P. Single Compression Machine Study of Pulsed Jet Combustion (PJC), Twenty-Fourth Symposium (International) on Combustion, Sydney, Australia, 1992.

(13) OPPENHEIM, A. K., Thermodynamics of Combustion in an Enclosure. 13th International Colloquium on Dynamics of Explosions and Reactive Systems, Nagoya, Japan, 1991.

(14) MAXSON, J. A., EZEKOYE, O. A., HENSINGER, D. M., GREIF, R., and OPPENHEIM, A. K., Heat Transfer from Combustion in an Enclosure. Twenty Fourth Symposium (International) on Combustion, Sydney, Australia, 1992.

(15) BECK, N. J., HOM, K., MAXSON, J. A., STEWART, H. E., and OPPENHEIM, A. K., Studies of Advanced Fuel Injection Concepts for Diesel Engines. Diesel Fuel Injection Systems, The Institution of Mechanical Engineers, London, England, 1989, pp. 41-48

(16) OPPENHEIM, A. K., BECK, N. J., HOM, K., MAXSON, J. A., and STEWART, H. E., A Methodology for Inhibiting the Formation of Pollutants in Diesel Engines. SAE Congress Paper 900394, 1990, 10 pp.

(17) OPPENHEIM, A. K., Method and System for Controlled Combustion Engines. U.S. Patent No. 4,924,828, May 15, 1990.

(18) OPPENHEIM, A. K., STEWART, H. E., and HOM, K., Pulsed Jet Combustion Generator for Premixed Charge Engines. U.S. Patent No. 4,926,818, May 22, 1990.

(19) OPPENHEIM, A. K., and STEWART, H. E., Pulsed Jet Combustion Generator for Non-Premixed Charge Engines. U.S. Patent No. 4,974,571, December 4, 1990.

(20) HISHIDA, M. and HAYASHI, A. K., Numerical Simulation of Ignition and Combustion Using a Premixed Pulsed Jet, 18th International Symposium on Space Technology and Science, Kagoshima, Japan, 1992, 6 pp.

(21) DALE, J. D., and OPPENHEIM, A. K., A Rationale for Advances in the Technology of I.C. Engines. SAE Congress Paper 820047, 1982, 15 pp.

(22) ONISHI, S., JO, S. H., SHODA, K., JO, P. D., and KATO, S., Active Thermo-Atmosphere Combustion (ATAC) - A New Combustion Process for Internal Combustion Engines. SAE Congress Paper 790501, 1979, 10 pp.

(23) NOGUSHI, M., TANAKA, Y., TANAKA, T., and TAKEUCHI, Y., A Study on Gasoline Engine Combustion by Observation of Intermediate Reactive Products during Combustion. SAE Paper 790840, 1979, 13 pp.

(24) NAJT, P. M., and FOSTER, D. E., Compression-Ignited Homogeneous Charge Combustion. SAE Congress Paper 830264, 1983, 16 pp.

(25) THRING, R. H., Homogeneous-Charge Compression-Ignition (HCCI) Engines. SAE Congress Paper 892068, 1989, 9 pp.

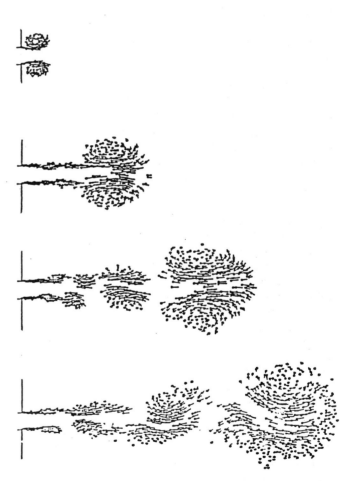

Fig. 1 Mechanism of a PJC mode of combustion
Formation and structure of a turbulent jet plume revealed by numerical solution of Navier-Stokes equations; clearly evident are the large scale vortex structures and the fjord-like paths of entrainment they induce.

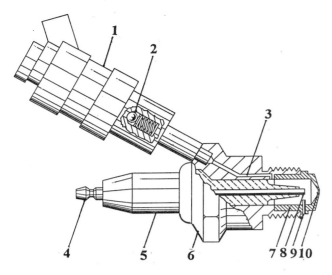

Fig. 2 PJC generator system
1-fuel injector; 2-check valve; 3-feed duct; 4-high voltage terminal; 5-insulator body; 6-plug body; 7-cavity gap; 8-high voltage electrode; 9-ground electrode; 10-exit orifice.

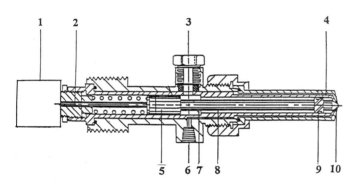

Fig. 3 Pulsed air-blast atomizer
1-actuator; 2-pintle rod; 3-fuel inlet; 4-fuel passage; 5-pintle piston; 6-air inlet; 7-air plenum chamber; 8-air passage; 9-pintle swirl guide; 10-mixing and atomizing nozzle.

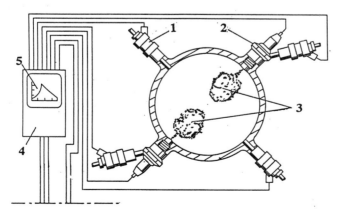

Fig. 4 Conceptual features of an advanced engine system
1-air assisted injector; 2-PJC generator; 3-turbulent jet plumes; 4-electronic control modulator; 5-indicator diagram, the primary sensing element of the control system.

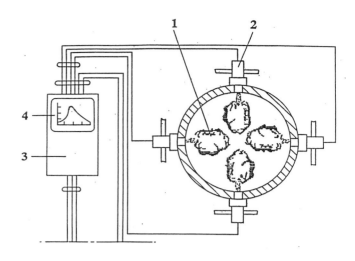

Fig. 5 Controlled combustion non-premixed charge engine
1-turbulent jet plume; 2-pulsed air-blast atomizer; 3-microprocessor controller; 4-indicator diagram, primary sensor of the closed loop control system.

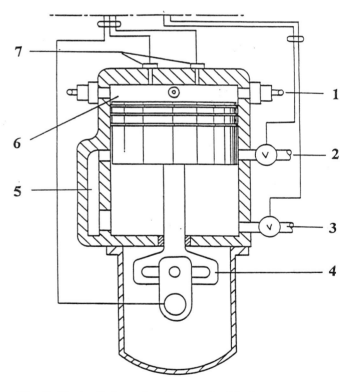

Fig. 6 Controlled combustion engine block
1-PJC (or Pulsed Air-Blast Atomizer) generator; 2-tunable exhaust duct; 3-tunable inlet duct; 4-scotch yoke transmission; 5-air transfer duct; 6-disc combustion chamber; 7-primary and secondary sensing elements of the control system.

A comparison of hydrocarbon measurements with a fast response flame ionization detector and a rapid acting sampling valve

W A WOODS, DEng, PhD, FEng, FIMechE, P G BROWN, BSc, MSc, PhD and
O S SOGUT, BSc, MSc, PhD
Department of Mechanical Engineering, Queen Mary & Westfield College, London

Synopsis The objects of the paper are, first, to compare the results of tests using a 'Cambustion' fast response flame ionization detector (FRFID) with those obtained using a 'Sokken' rapid acting sampling valve. The experiments were carried out on the exhaust system of a 2-litre OHC Ford Sierra engine and on an experimental Rover-Hydra engine.

It is well known that the unburnt hydrocarbon emissions from spark ignition engines have their origin in different locations in the combustion chamber and that they are formed and released at different crank angle times in the engine cycle.

Many investigations have been carried out using timed, rapid acting sampling valves in order to gain more understanding of the processes involved. This method relies on the engine operating at the "cyclically repeating condition". It is, therefore, not suitable for investigating a transient or warm-up condition. The FRFID is a miniature FID continuous sampling system designed to have a very short time delay, and it aims to resolve hydrocarbon measurements on individual cycles. In addition to reporting new experimental results, the paper discusses the analysis of the results. This includes a gas dynamic calculation of the delay times in the capillary tube of the FRFID.

The conclusions of the work are that, for modest engine conditions, the FRFID gives results which show reasonable agreement to those of the sampling valve and they confirm the usefulness of the new instrument. Under more severe engine conditions indicated by higher speeds and loads, the FRFID underestimates the magnitude of the peaks and smooths out other fluctuations. In view of the considerable potential of the FRFID, it merits further theoretical study and experimental investigation.

1. INTRODUCTION

Sampling techniques are widely used to investigate changes in gas composition caused by chemical reactions and mixing in engines and other gas flow systems. Rapid acting sampling valves have been employed for many years to obtain 'snatch' samples to investigate the scavenge efficiency of two stroke cycle engines [1] and, more recently, to examine unburnt hydrocarbons on modern automotive engines [2]. With this technique, a small sample of gas is extracted from the engine over a prescribed crank angle period of perhaps 5° or 10°. The valve is set to open and close over the specified period on each consecutive cycle for, typically, 800 engine cycles while a sufficient quantity of sample gas is collected for chemical analysis. Over the years, further developments have taken place in which the new technology of magnets and computer controls have been incorporated. The authors have used a Sokken valve; the system built up in their laboratories has been described recently [3]. During the 1980s, a different method was developed by Collings and

his colleagues [4] and this has become known as the 'Cambustion' fast response flame ionisation detector (FRFID). With this system, a continuous flow is drawn from the sampling location along a small diameter tube to a miniature flame ionization detector. Various pipe branches, bleed flows and constant pressure chambers are incorporated so that the distribution in time of the sample is converted to a lengthwise distribution along the tube to the flame ionization detector. As the various parts of the sample arrive at the detector, the distribution is re-converted to be a function of time in the form of an electrical signal. This signal responds well to the changes, after allowance is made for time lag corresponding to the transit time for the gas to flow along the tubes to the instrument.

Each sampling system has shortcomings particular to itself and, therefore, both are subject to measurement errors. The objects of the paper are to present and discuss the results of experiments which give direct comparisons of the two instruments. Although some differences were observed, it was concluded that the FRFID was able to respond quickly and, under moderate engine conditions, it gives results very similar to those of the sampling valve.

Under similar engine conditions, indicated by higher speeds and loads, the FRFID underestimates the magnitude of the peaks and smooths out other fluctuations. In the discussion, further theoretical study and experimental investigation are advocated.

In what follows, the experimental set up is described and the test results are presented and discussed in some detail.

2. APPARATUS

The tests reported here are from two engines; these are a two valve design, as shown in Fig. 1, and a four valve type, as shown in Fig. 2. The two valve design was a Ford $2l$ SOHC engine, connected to a Heenan and Froude DPX2 water brake. This engine had a carburettor which had been modified with a commercial conversion kit to allow the use of propane fuel. The spark timing system was modified to incorporate an Ignition Timing and Measurement System unit (ITAMS) so that either the standard setting of the distributor could be used or a timing set by the ITAMS unit. The four valve engine was a long bed Hydra single cylinder research engine fitted with a cylinder head from Rover group M16 engine. The long bed Hydra has an electric dynamometer which is used to start the engine, rotating at the desired speed before switching on the ignition. The engine was fitted with a fuel injection system and the ITAMS unit was used as the ignition system. Unleaded 95 RON gasoline was used. Other details of the engines are given in Table 1. The main difference between the two test bed arrangements is that the hydra engine has an electric dynamometer with thyristor control equipment which maintains the engine speed to within ± 3 r/min. The Hydra is fitted with a fuel injection system which makes control of the air fuel ratio easier than it would be if a carburetted engine was used. In the work reported here, the carburetted engine was run on propane and the air fuel ratio was set, in this case, by altering the gas flow with a needle valve for adjustment.

Both engines had a sampling valve, as described in [3], and the FRFID, as shown in Fig. 3, fitted in the exhaust pipe. The particular layout of each engine determined where the sampling valve could be positioned, but in both cases the valve was placed as close as possible to the appropriate exhaust valve of the engine concerned. Sectional diagrams of the cylinder head and exhaust port of both engines are given in Figs. 1 and 2. These diagrams show the location of the sampling position from the exhaust valve seat. The FRFID probe was located in the same plane as the sampling valve, and the tip of the probe was set to project into the pipe by 6 mm. Exhaust gas analysis, together with the Spindt equations [5], was used to establish the air fuel ratio at which the engine was operating. Exhaust gas carbon monoxide and carbon dioxide levels were measured with

NDIR instrument; a Sermomex paramagnetic analyser measured the amount of oxygen. The hydrocarbon mole fraction was measured with an FID as ppm equivalent propane. The same analyser was used for the sample extracted with the sampling valve. The exhaust gas for air fuel ratio measurement was drawn off with a tube placed just downstream of the sampling valve and on the axis of the exhaust pipe, as indicated in Figs. 1 and 2.

The sampling valve data was obtained from a chart record taken while the sampling valve was operated, at the chosen point in the engine cycle. Each point comprises the average, estimated from the chart recording, of some 800 cycles of operation. The FRFID trace was obtained by digitising the signal from the instrument at a resolution of one degree crank angle and storing the data in a computer. The memory management system of the computer limited the number of data points that could be conveniently stored to a maximum of 32,767. By taking 360 points over the exhaust period, it was possible to store enough data in the computer for 80 consecutive engine cycles. The FRFID traces shown are the average of these 80 cycles.

3. TESTS AND RESULTS

Tests were carried out on both the Sierra engine and the Rover Hydra engine. Diagrams were obtained in the exhaust system of each engine using both the sampling valve method and the fast response, flame ionization detector. The test conditions and the figure numbers which show the corresponding results are summarised in Table 2.

As shown in Fig 1, for the Sierra engine the sampling valve was mounted at the top of the stub pipe, which is rectangular in cross-section. The FRFID probe was mounted in the corner, close to the sampling valve position. As shown in Fig. 2, for the Rover/Hydra engine the sampling valve was mounted on the top of the exhaust pipe and the FRFID probe was mounted, horizontally, at the side of the pipe. The sampling valve and the FRFID probe were located in the same plane, where the exhaust pipe has a circular cross-section.

The main results shown in this paper were obtained with the probe mounted 6 mm into the pipe.

Bench tests were carried out on the FRFID, in which the instrument was set up at the conditions used for the test on the Sierra engine. The probe was mounted horizontally, with the capilliary tube open to the atmosphere. A small flexible tube, discharging span gas of propane and nitrogen was rapidly moved past the end of the FRFID probe and the point of passing was indicated by means of an electrical contact. The results showed that there was a time interval between the electrical contact and the start of the rise in the signal from the FRFID. The signals were displayed on a cathode ray oscillograph. The time delay varied from about 5 ms. to 8 ms, and the hydrocarbon pulse signal showed a rise time of approximately 3 ms. The amplitude reached at the peak of the hydrocarbon pulse was about half of the value reached when the jet of span gas was played steadily on to the end of the probe.

4. DISCUSSION

As this discussion is quite wide-ranging, it is appropriate to explain at the outset the objectives of the discussion and how it is set out. As far as the authors know, this is the first time a direct comparison has been made between a sampling valve system and the FRFID instrument. The aims of the work and of this discussion, in particular, are to give guidance to users of these instruments based upon the results of the present experiments, and to discuss aspects of the work which the authors consider merit further investigation.

The results of the present experiments for the Sierra engine are given in Fig. 4 and those for the Rover-Hydra engine are given in Figs. 5 and 6. Next, some comments are given on the reasons for doing the comparative experiments on firing spark ignition engines. Accounts are then given on the known errors and limitations of the sampling valve system and the FRFID instrument. The discussion is concluded with an account of the methods which have been suggested for correcting the timing results.

The sampling valve results are shown directly but the crank angles for the FRFID traces have been corrected by crank angles corresponding to the particle-transit times. These times were obtained from wave action path line calculations, carried out to simulate unsteady flow in the transfer tube of the FRFID. This calculation was based upon the methods described in ref. [6], and the key data for these calculations is given on Table 2. During the course of the experiments, comparative results were taken with the FRFID instrument, with the sampling probe mounted flush with the pipe wall and with it projecting 6 mm into the exhaust pipe on the Rover/Hydra engine. The results obtained were almost identical, as shown in Fig. 7. The results shown in Figs. 4, 5 and 6, were obtained with the probe mounted 6 mm into the pipe. Measurements taken with each instrument are, in general, similar. This result applies to those obtained from the Sierra and the Rover-Hydra engines. However, the results are not identical and the discrepancy between the results from each instrument first becomes more pronounced as the engine speed is increased and, then, as the bmep and, therefore, the release pressure are increased. Increasing the bmep also increases the velocities of the discharge gases, generated initially by the exhaust blowdown and maintained by piston displacement.

The results for the Sierra engine, at a speed of 1000 r/min and a bmep of 1 bar, are given in Fig. 4. This shows very similar trends for each instrument, indicating that the FRFID is able to respond to the changes in hydrocarbons,

but during blowdown and the initial part of the period of piston generated motion, the FRFID gives lower values of hydrocarbon mole fraction, say about 300 ppm lower than those of the sampling valve. During the distinctive second hydrocarbon peak of the two valve engine, the magnitude of the hydrocarbons measured by both instruments is remarkably similar.

The results for the Rover/Hydra engine at a speed of 1000 r/min and bmep of 5.5 bar are given in Fig. 5. Here, the hydrocarbon blowdown peak shown by the FRFID is down by 1000 ppm on that of the sampling valve, but the resolution is good. In the piston displacement phase, the hydrocarbon peaks are smoothed out but the phasing is still reasonable. The amplitude shown by the FRFID is down relative to the sampling valve by between 0-200 ppm.

The results for the Rover/Hydra engine, at a speed of 2000 r/min and a bmep of 2 bar, are given in Fig. 6. This time, the hydrocarbon blowdown peak shown by the FRFID is down by 400 ppm on that given by the sampling valve. In the subsequent part of the discharge period, the trace from the FRFID shows that the hydrocarbon peaks have been smoothed out most markedly and the magnitudes are lower than those of the sampling valve by 150 ppm.

When allowance is made for small differences in sampling position, exhaust system configuration and engine operating conditons, the FRFID results shown in Fig. 6 are similar to those obtained by Finlay et al [7].

It would appear logical to carry out a comparative experimental study on the performance of the present two different instruments in an environment in which the physical and chemical properties are changing at the truly cyclically repeating condition. This would require the construction of a special test machine. This principle has been adopted in the past, for example, when comparing the performance of various types of pressure transducer. However, the present instruments were developed, primarily, if not specifically, for use on firing engines and it was considered most relevant to compare the operation of these instruments in practical and realistic conditions.

The first complication to note is that when the spark ignition engine is operating at the nominally, cyclically repeating condition, it exhibits, cycle by cycle, fluctuations in flow and combustion. Evidence of this has been revealed by diagrams of pressure and velocity fluctuations, e.g. [8], and more recently, by FRFID diagrams [9].

The significance of this is that experimental studies made with a sampling valve system must necessarily be made on the basis of ensemble averages, taken at prescribed crank angles, and the diagrams for the whole cycle are built up of ensemble averages obtained at different times. In principle, results obtained with perfect FRFID instruments may be compared on an individual cycle basis. This is the basis for using these instruments for transient or warm up operation of an engine.

Considering now the operation of the sampling valve system in more detail, the electronic controls start to open the valve at a prescribed crank angle and it takes a small, but finite, time to open, perhaps, 0.5 ms. It remains open for a similar interval and closes in about a further 0.5 ms. The sample, is therefore, extracted over an interval of perhaps 12° crank angle at 1000 r/min through a variable area valve. The sample is set into motion by means of an expansion wave which propagates into the exhaust pipe. The shape of this is approximately spherical or toroidal for needle valve or poppet valve types, respectively. An analysis of these flows, taking account of the unsteady boundary layers, has been considered by Hicks, Probstein and Keck [10] and a spherical type of expansion wave has been studied experimentally and theoretically by Woods et al. [11]. It is, therefore, difficult to determine the precise location from which a sample originally came, particularly in the case of one drawn from a flowing system such as an exhaust pipe. The result of the valve operating with a finite open period is that during periods when the hydrocarbons are increasing, the reading will be overestimated and, likewise, the signal will be underestimated when the hydrocarbons are decreasing. The net effect of this is to advance the apparent timing of a pulse and to reduce its amplitude. Another possible problem is that with a signal having sharp peaks, the spacing of the crank angle measurements could miss the actual peak. These effects are illustrated in Fig. 8.

The next consideration concerning the sampling valve results is that although the engine was fully warmed up and was running at the nominally cyclically repeating condition (CRC), the results for each crank angle location are the average of different sets of 800 cycles.

The results using the sampling valve for the Sierra engine are shown in Fig. 4. Here, each symbol shows the average hydrocarbon mole fraction and the crank angle at which the sampling valve is set to start opening. The rectangles show the range of hydrocarbon reading and the crank angle period from the start of opening to the closure of the sampling valve. The band of variation of the results using the sampling valve almost encompasses the results obtained with the FRFID. The errors and reasons for discrepancies in the results using the FRFID are discussed next. With the FRFID, a continuous flow is drawn from the source along a capillary tube, called the transfer tube, to a sudden enlargement into the tee top tube from which a second capillary carries a fraction of the sample to the flame. The transfer tube is much longer than the others, as illustrated in Fig. 3. The tee top tube is open to the constant pressure chamber and it is included to reduce the effects of pressure fluctuations on the readings of the instrument; this has been investigated by Finlay et al [7]. It is clear from this work that if significant pressure fluctuations occur at the source, this can lead to unsteady flow in the transfer tube and cause errors due to fluctuations in pressure in the so called constant pressure chamber. In the operator's manual [12] an analysis is given of the flow in the capillary tubes of the instrument based upon steady, one-dimensional, isothermal, compressible flow. They also discuss the classical work of Taylor [13][14] to give justification for their analysis. The main purpose of these calculations is to provide time delays so that the hydrocarbon traces can be corrected on a crank angle basis to show the events in relation to engine events, such as exhaust port opening. In the present work, the unsteady flow calculation, mentioned earlier, provided a crank angle correction which varies over the engine cycle.

It is suggested that the FRFID instrument has two main types of error. The first concerns the sample transit time from the source to the flame detector. This time interval varies over the engine cycle and it depends, importantly, on the pressure ratio across the relatively long transfer tube, on the length and diameter of this transfer tube and on other parameters, such as the gas temperatures and the dimensions of the tee top and FID tube.

Three different methods were used to estimate this transit time for the results shown in Fig. 4 for the Sierra engine. The Cambustion software [12] calculated a value of 6 ms, the wave action calculation gave values which varied over the engine cycle from 4.8 ms to 5.6 ms and the simple bench experiment gave values ranging from 5 ms to 8 ms. The Cambustion software is sensitive to the temperature used for the transfer tube and the wave action calculation is sensitive to the exhaust gas temperature. The simple experiment is by no means perfect or complete but the measure of agreement between these different estimates gives some credibility to the estimated transit time.

The second type of error influences the magnitude of the signal and there are two sources of this error. The first is due to the mixing effect and the second concerns the capture of a representative sample. Mixing is introduced by the combined influence of longitudinal convection and radial diffusion and it leads to a smearing of changes in mole fractions which are, initially, very sharp. In the case of engine tests, the sharp increases and sharp decreases occur several times within an engine cycle. This effect tends to reduce the magnitude of the signal. An illustration of this type of effect may be observed in Fig. 5.

The case shown in Fig. 6 is perhaps more severe and a longer delay is more appropriate. Allowing for this, it shows that the FRFID instrument gives a lower level of mole fraction and that the hydrocarbon pulses, after that corresponding to the blowdown, are smoothed out.

The question of capturing a representative sample may be the explanation for the FRFID missing the sharp peaks showing in Figs. 5 and 6. If steep radial composition gradients are present, the peak values near the centre line of the pipe are more likely to be included in the catchment volume of the sampling valve because it will be 100 times larger than that of the FRFID.

Returning to the question of the transit time interval taken for the sample to flow from the source to the FID. A further theoretical investigation of the FRFID has been carried out [15] and this points out that the region near the probe entry contributes 100 times as much to the spreading as does the region near the FID. It also recommends the use of a short transfer tube which should be kept hot. It is now suggested that this instrument has considerable potential and it merits further study both theoretical and experimental. The goals of subsequent work should be to improve the performance of the instrument when being used under engine conditions perhaps more severe than those illustrated in Figs. 5 and 6. The authors are planning to investigate the use of the instrument for sampling gas from the cylinder of a firing engine. It is also suggested that further fundamental studies, both theoretical and experimental, should be carried out on gas dynamics and viscous flows in the capillary tubes of the instrument.

As the traces given in Figs. 5 and 6 are probably the first time sampling valve results for a four valve engine have been openly reported it is, therefore, appropriate to comment on the new aspects. The interesting feature of these results is the replacement of the pronounced second peak, which is characteristic of a two valve engine, by three smaller and distributed peaks. These effects may be observed by comparing the sampling valve results of Figs. 5 and 6 with those shown in Fig. 4. It is thought that the smaller peaks are associated with inflow from the cylinder through the smaller exhaust valves on the Rover engine rather than a wave action effect in the exhaust system.

5. CONCLUSIONS

1. Time resolved samples have been taken from the exhaust systems of a two valve SOHC engine and a four valve DOHC engine with both a sampling valve system and an FRFID instrument.

2. The engines were operated at the cyclically repeating condition and in general the results obtained with the sampling valve and the FRFID instrument show reasonable agreement for moderate engine conditions, as shown in Fig. 4.

3. As the engine conditions were made slightly more severe, as indicated by higher speed and power, the discrepancies between the results from the two instruments became larger.

4. The discrepancies, mentioned in 3, are that the FRFID instrument tends to indicate lower hydrocarbon levels and miss the peak amplitudes of the sharp hydrocarbon pulses. Other ripples are also smoothed out.

5. The time delay, caused by the sample transit time in the FRFID instrument, has been estimated by three separate methods to be about 6 ms. for the test on the Sierra engine shown in Fig. 4. The wave action calculation indicated that the delay time varied 17% over the engine cycle and this depends upon the temperatures and pressures at the sampling point.

6. It is suggested that the lower hydrocarbon peaks shown by the FRFID instrument in Figs. 5 and 6 is due to the combined effects of (a) unrepresentative samples being drawn into the probe and (b) samples from the peak being mixed in the capillaries and tee top sudden enlargement with pre-peak or post-peak samples.

7. The FRFID is a relatively new instrument and because of its considerable potential it merits further theoretical study and experimental investigation to help further development of it.

ACKNOWLEDGEMENT

The authors wish to acknowledge that this work was supported, in part, by the SERC under Grants Nos. GR/F64685 and GR/H18272. They also wish to mention the assistance received from the UK hydrocarbons P4 consortium, especially the help from Mr T Biddulph and Mr S Wallace. The authors also wish to thank Dr N Collings and Dr T Hands of Cambustion Ltd. for their help and continued interest in this work.

6. REFERENCES

1) Taylor, C F, "The internal combustion engine in theory and practice". Vol. 1, The Technology Press of MIT and John Wiley Inc. New York, 1960, pp. 1-574.

2) LoRusso, J A, Lavoie, G A, Kaiser, E W, "An electrohydraulic gas sampling valve with application to hydrocarbon emissions studies". SAE paper 800045, SAE Trans., Vol. 89, 1980.

3) Panesar, A, Woods, W A, Brown, P G, "Instrumentation for determining the time-history of hydrocarbons in spark ignition engines". Proc. of I.Mech.E. Seminar, London, Dec. 1991, pp. 19-27.

4) Collings, N and J Willey, "Cyclically resolved HC emissions from spark ignition engine". SAE paper 871691, Sept. 1979, pp. 89-115.

5) Spindt, R S, "Air/fuel ratios from exhaust gas analysis". SAE paper No. 650507, 1965.

6) Woods, W A, Sogut, O S, Rutter, C, "Application of path line streams to the exhaust system of an internal combustion engine". Int. Conf. on Computers in Engine Technology, I.Mech.E. C430/001, 1991-11, pp. 101-110, 1991.

7) Finlay, I C, Boam, D J, Bingham, J F and Clark, T A, "Fast response FID measurement of unburned hydrocarbons in the exhaust port of a firing gasoline engine". SAE paper 902165, Int. Fuels and Lubricants Meeting and Exposition, Oklahoma, Oct. 1990, pp. 125-146.

8) Heywood, J B, "Internal combustion engine fundamentals". McGraw Hill, 1988, pp. 1-930.

9) Brown, P G, Sogut, O, Woods, W A, "Fast sampling digital system for investigating engine unburnt hydrocarbons". A paper submitted for the ASME European Joint Conf. on Eng. Systems Design and Analysis. Istanbul, Turkey, July 1992, pp. 1-8.

10) Hicks, R E, Probstein, R F, Keck, J C, "A model of quench layer entrainment during blowdown and exhaust of the cylinder of an internal combustion engine". SAE paper 750009, presented at SAE Automotive Eng. Congress and Exposition, Feb. 24-8, 1975; SAE Trans. Vol. 84 (1975).

11) Woods, W A, Cleaver, J W, El-Shobouksy, M, "Unsteady flow inside a pressure vessel during discharge". 6th Thermo. & Fluid Mech. Convention, I.Mech.E., Univ. of Durham, 1976, Paper C55/76, pp. 201-210.

12) Cambustion Ltd. HFR 300 High Frequency FID. User Manual and Specs. TH 90.4, Cambridge , 1991.

13) Taylor, G I, "Dispersion of soluble matter in solvent flowing slowly through a tube". 1953, Proc. Roy. Soc, A. 219, pp. 186-203.

14) Taylor, G I, "Dispersion of soluble matter in solvent flowing slowly through a tube". 1953, Proc. Roy. Soc. A 219, pp. 186-203.

15) Smith, R, "Loss of frequency response along sampling tubes for the measurements of gaseous composition at high temperature and pressures". J. Fluid Mech. 1989, Vol. 208, pp. 25-43.

Engine Make & Model	Ford Sierra 2.0l	Ricardo Hydra MK III, Built to Rover Group M16 Specification
Engine Type	4 stroke, 4 cylinder, spark ignition	4 stroke, single cylinder, spark ignition
Engine Details	SOHC, 2valves per cylinder	DOHC, 4 valves per cylinder
Fuel System	Carburettor	Port Injection, timed 90 ° BTDC firing stroke.
Bore	90.82	84.45
Stroke	76.95	88.9
Compression Ratio	9.2	10
Inlet Valve Dia	41.8mm	31.7mm
Exhaust Valve Dia.	35.8mm	29.2mm
IVO(Inlet Valve Opens)	24°BTDC(696°)	10°BTDC(710°)
IVC (Inlet Valve Closes)	64°ABDC (244°)	55°ABDC (235°)
EVO (Exhaust Valve Opens)	70°BBDC (470°)	44°BBDC (496°)
EVC (Exhaust Valve Closes)	18°ATDC (18°)	21°ATDC (21°)

Table .1. Engine Details

Engine Type	2 Valves	4 valves	4 valves
Air Fuel Ratio	16.2	16.8	15.6
Fuel	Propane	Unleaded Petrol	Unleaded Petrol
Speed r/min	1000	1000	2000
Coolant Temperature °C	88	88	88
Lubricant Temperature °C	88	83	90
b.m.e.p. (bar)	1 (approx.)	5.4	2.1
Spark Timing	20°BTDC	12°BTDC	30°BTDC
Fig. No.	4	5	6

Engine	Sierra	M16 Hydra
CP Chamber vacuum (mm Hg)	410	300
Overall Length of tube (mm)	195	195
Friction Factor f	0.06	0.06

Table .2. Test Condtions

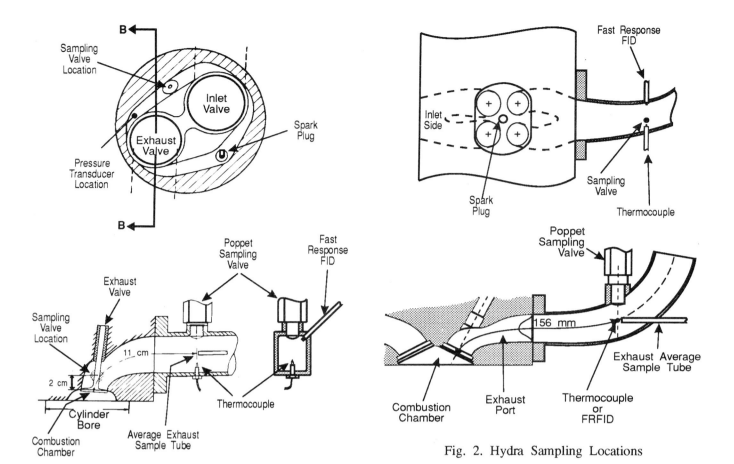

Fig. 1. Sierra Engine sampling locations

Fig. 2. Hydra Sampling Locations

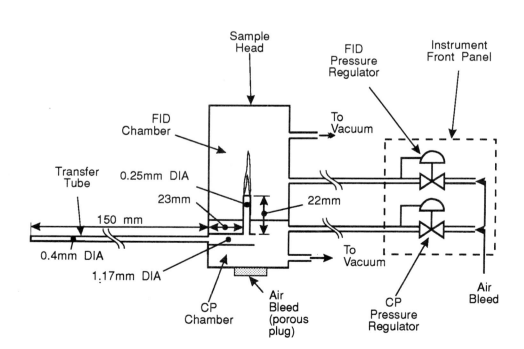

Fig. 3. Fast Response FID Schematic

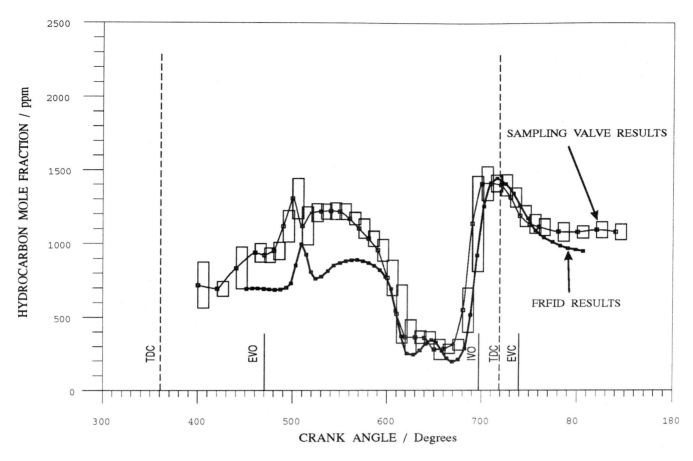

FIG. 4. Time resolved hydrocarbons taken from the exhaust
system of the Sierra engine at 1000 r/min and 1.0 bar bmep.

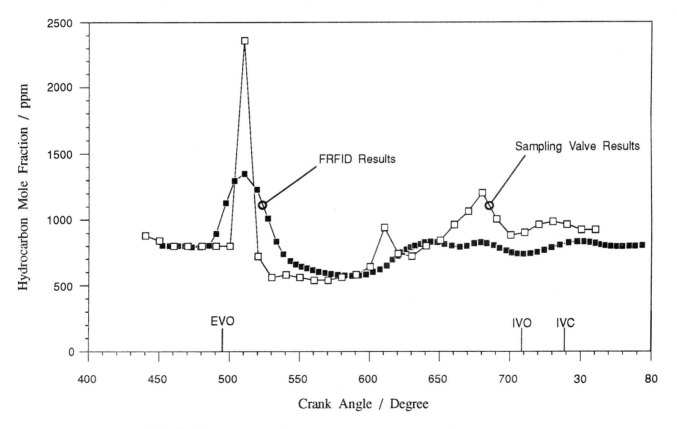

FIG. 5. Time resolved hydrocarbons taken in the exhaust system
of the Rover/Hydra engine at 1000 r/min and 5.4 bar bmep

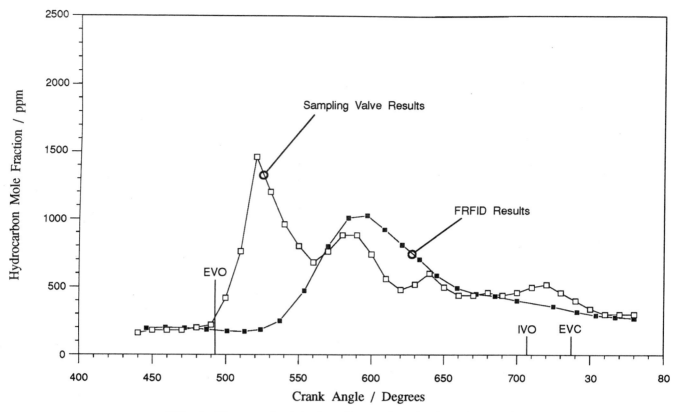

FIG. 6. Time resolved hydrocarbons taken in the exhaust system
of the Rover/Hydra engine at 2000 r/min and 2.1 bar bmep

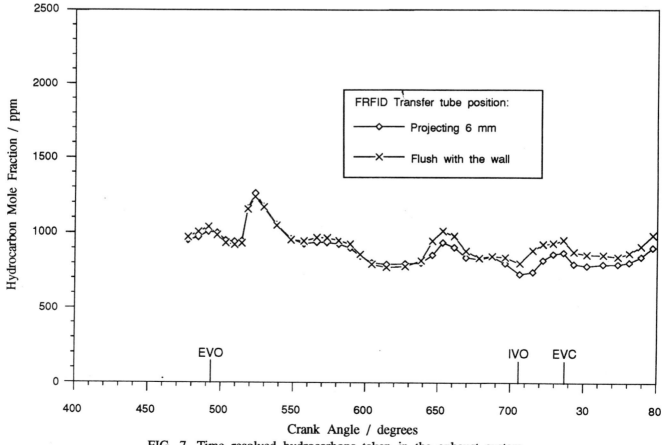

FIG. 7. Time resolved hydrocarbons taken in the exhaust system
of the Rover/Hydra engine at 1000 r/min and 5.4 bar bmep with the FRFID

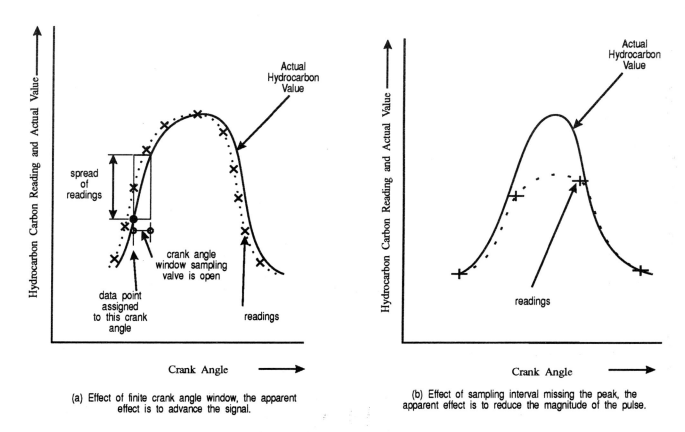

(a) Effect of finite crank angle window, the apparent effect is to advance the signal.

(b) Effect of sampling interval missing the peak, the apparent effect is to reduce the magnitude of the pulse.

Fig. 8. Possible errors in sampling valve measurements

C448/029

Flame development in an S.I. engine measured by an infra-red camera

H ZHAO, BSc, PhD and N COLLINGS, BSc, PhD
Department of Engineering, University of Cambridge
T MA, BSc, PhD, CEng, MIMechE
Ford Research & Engineering Centre, Laindon, Essex

SYNOPSIS A novel technique has been developed to measure the flame propagation in a spark ignition engine using an infra-red camera. A special CO_2 optical filter mounted in the camera allows only the absorption/emission of CO_2 to reach the infra-red detector, thus eliminating radiation from the cylinder walls. The technique offers both spatial and temporal measurements of flame development from the spark ignition up to 40° crank angle duration. Simultaneous recording of cylinder pressure and flame propagation speeds in many consecutive cycles has been performed under different engine speeds and equivalence ratios. The relation between the flame propagation pattern and mass fraction burned rate is discussed.

Introduction

Improvement of thermal efficiency and fuel economy in the spark ignition engine are among the major goals of motor industry. As a fundamental approach to this goal, studies are needed of combustion processes such as flame propagation and cycle-to-cycle variation in the engine. Many flame propagation measurements in I.C engines have been performed with different experimental techniques, such as the direct light emission photography (e.g. 1, 2), Schlieren photography (e.g. 3, 4), and laser tomography[5].

In this study, a cycle-resolved flame propagation measurement technique using an I.R camera, is developed and applied to a single cylinder spark-ignition engine. The detection of I.R emission is preferable to visible light since it has much higher intensity and is less prone to the deposits on the window due to its longer wavelength. Simultaneous recordings of cylinder pressure and the flame propagation speed of up to 70 consecutive cycles have been made. Temporal distributions of flame propagation speed and their relation with the mass fraction burned rate are to be discussed.

Experimental

Experimental Set Up– A schematic diagram of the experimental set-up is shown in Figure 1. A Ricardo Hydra single cylinder optical engine was used, which has an extented cylinder block and piston. A 45 degree front surfaced mirror placed beneath the window allows the combustion chamber to be viewed through a port in the cylinder block.

The optical window employed on the piston is made of silicon, which is infra-red transmitting and has a similar heat conductivity to that of aluminium. More details about the window and its properties can be found in reference [6].

Thermal imaging system– The results reported here were taken with an AGEMA 870 Thermovision system as shown in Figure 2. The scanner produces thermal images from a single sensing element by scanning the optical image onto the single sensing element using a system of rotating/oscillating mirrors. The scanner has a line frequency of 2500Hz. In the normal operation mode of the scanner a thermal image consisting of 70 uninterlaced lines is produced via the control unit and the PC computer and is presented at a field rate of 25 times per second. Such mode of operation has been used in surface temperature measurements reported previously [6,7].

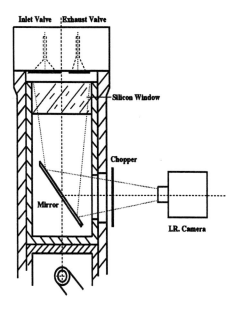

Figure 1: Schematic drawing of the experimental set-up

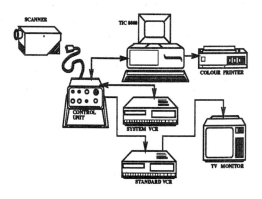

Figure 2: The Schematic Drawing of I.R Camera System

The present result, however, is obtained using LINETIC, which is an optional subsystem to AGEMA Thermovision system. In this mode of operation, the I.R scanner only has one horizontal scanning line in operation in the middle of the field while its vertical scan is disabled. The output image is therefore the time history of the single scanning line image. In order to apply this mode of operation for measuring flame propagation, a CO_2 optical filter was mounted between the sensing element and viewing optics in the I.R scanner so that only the absorption/emission of CO_2 is allowed to reach the infra-red detector, thus eliminating radiation from the cylinder walls. 101 frames were stored in the PC computer's RAM(2Mbytes) before dumping them onto the hard disk. Each frame in this mode consists of 140 consecutive lines. The number of cycles that can be recorded are 43, 57 and 70 and the crank angle increments between adjacent lines are 2.16°, 2.88° and 3.6° for engine speeds 900rpm, 1200rpm and 1500rpm respectively.

The I.R scanner was aligned so that the scanning line crossed the spark plug and was in the direction of flame propagation as shown in Figure 3. A sample image obtained from such measurement is shown in Figure 4. In the thermal image, the brightness reflects the CO_2 radiation intensity. The strongest emission occurred near the spark plug well after the top dead centre from the burnt gas. It then diminished in the subsequent expansion and exhaust strokes and disappeared in the intake and compression strokes. However, it is the boundary between the burnt and unburnt gas mixture in the power stroke, indicating the arrival of the flame front, that we are interested in. The other regions are of no concern to the present study.

Data Acquisition and Processing— In order to study the effect of flame propagation on the development of cylinder pressure and the mass fraction burned rate the cylinder pressure and the flame propagation measurements were made simultaneously.

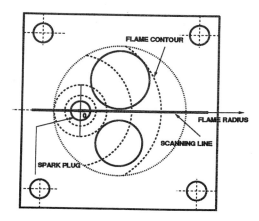

Figure 3: Schematic drawing of the Hydra engine cylinder head and the Position of the Scanning line

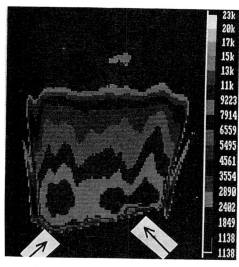

spark plug   flame front

Figure 4: A sample thermal image used to measure flame speeds.

A general purpose counter generated triggering pulses at a preset crank angle. An analogue switch enabled triggering pulses, which synchronized recordings of the pressure and flame propagation, 40° BTDC in the present study. The pressure trace was marked at 360° BTDC of each cycle. The ignition timing was identified on the thermal image by the sudden increase of the I.R radiation detected.

The thermal images were stored as ASCII files on the PC computer initially. They were transferred to an Apollo minicomputer afterwards for processing.

In general, these thermal images show sharp boundaries between burnt and unburnt mixtures, which are identified as the flame front. By converting the ASCII values into integers of each pixel in each image a computer program determined the flame position according to a preset threshold. Flame propagation speeds, by which we mean the rate

206

© IMechE 1992 C448/029

of development of the flame across the cylinder, were measured from the adjacent lines of known spacing (0.4ms)

Engine Operation Conditions— The operation conditions are listed in Table 1 and Table 2. IMEP was calculated from the pressure traces over many cycles. The relative air/fuel ratio λ was measured by a NTK Air Fuel Ratio Meter.

The engine was on WOT(wide-open-throttle full load condition) and the spark timing was 20° BTDC in all cases. This timing was at MBT for the 1200rpm and $\lambda=1$ test condition and was not reset to MBT for the other engine speeds and air/fuel ratios conditions in order to simplify the interpretation of our data.

Table 1: The Hydra Engine Operation Conditions

Bore	84.4mm
Compression Ratio	9:1
Inlet Valve Closes	41 deg. ABDC
Exhaust Valve Opens	61 deg. BBDC
Window Material	Silicon
Air Inlet Temperature	20°C
Type of Cooling	Water
Coolant Temperature	80°C
Oil Temperature	70°C
Fuel Type	Isooctane
Spark Timing	20 Degree BTDC

Table 2: The Experiment Data Sets

Speed	λ	Load	Brake Torque	IMEP
900rpm	0.9	WOT	23 N.m	1085 KPa
900rpm	1.0	WOT	22 N.m	1036 KPa
900rpm	1.2	WOT	16 N.m	837 KPa
1200rpm	0.9	WOT	27 N.m	997 KPa
1200rpm	1.0	WOT	26 N.m	962 KPa
1200rpm	1.2	WOT	20 N.m	771 KPa
1500rpm	0.9	WOT	25 N.m	833 KPa
1500rpm	1.0	WOT	22 N.m	785 KPa
1500rpm	1.2	WOT	12 N.m	473 KPa

Results and Discussion

The objective of this paper is to demonstrate the measurement of flame development using an infra-red camera. The data presented here serve to illustrate how such measurements can be interpreted to yield flame propagation data which are1 similar to those obtained by other methods, for example, high speed photography or ionization detection. The advantage of the IR method lies in the clear distinction of the flame boundary and the line scanning techniques lends easily to automated data analysis by a computer.

Figure 5-10 show the measured flame speeds averaged over many cycles plotted against the average flame radius under various engine speeds and air/fuel ratios conditions. The negative flame radii in the figures indicate the distance between the spark plug and the flame front moving toward the near

side of the cylinder wall, opposite to the main flame propagation direction. They show a consistent picture of the stages of flame development in a spark ignition engine similar to that described in reference [8]. The flame accelerates to a maximum value when the flame radius reaches approximately 30mm and then slows down as the flame moves towards the edge of the combustion chamber. Because of the off-centre location of the spark plug, the flame is slowed down more rapidly in the direction towards the near side of the wall. The maximum flame speed is approximately 17m/s and remains unchanged for the higher engine speeds and for the stoichiometric and rich mixtures. At engine speed of 900rpm, the maximum flame speed is lower because of reduced turbulence. At $\lambda=1.2$, the maximum flame speed is also lower because of lean combustion.

To study the correlation between the flame propagation speed and the mass fraction burnt in the engine, the mass fraction burnt was calculated from the recorded pressure traces based on an one-zone model developed at MIT[9]. Figure 11-16 show the flame speed and mass fraction burnt parameters plotted against engine crank angle relative to TDC under various engine speeds and air/fuel ratios conditions. They show a consistent picture where the crank angle position of the 50% mass fraction burnt is a fair indication of the deviation from MBT spark timing described in reference [10]. From our test results, the MBT condition at 1200rpm and $\lambda = 1$ corresponds to the 50% mass fraction burnt occurring at 17° ATDC. The maximum flame speed position when the the flame radius reaches approximately 30mm from the the spark plug corresponds to less than 10% mass fraction burnt and occurs at 5° BTDC. For other engine speeds and air/fuel ratios conditions, the crank angle positions of the maximum flame speed and the maximum mass fraction burn rate both change in a similar manner as that of the 50% mass fraction burnt crank angle position, thus reflecting a consistent effect of spark timing on the different stages of flame development. The slower burning test conditions would require an increase in spark timing advance to bring the flame development in line with that of the MBT conditions.

From the above data, it is possible to derive according to reference [8] additional parameters which further describe the flame development and propagation process, for example, the flame front expansion speed, the burning speed and the gas speed ahead of the flame front. The infra-red technique described in this paper makes the flame propagation measurement and the derivation of these flame development parameters easier and more adapted to automated method by a computer.

Conclusion

A new technique using an infra-red camera for the study of flame propagation in a spark ignition engine has been described. The technique is capable of measuring flame speed with good temporal resolution over many cycles. In combination with heat release analysis from in-cylinder pressure measurements, it provides a useful way to study the combustion processes under various engine operating conditions.

Acknowledgement

Support for this work was provided by Ford Motor Company and is gratefully acknowledged.

Reference

1. Withrow L. and Boyd, T. A. Photographic Flame Study in the Gasoline Engine. Ind. Eng. Chem., 23, pp 539-547, 1931.

2. Beretta, G. P., Rashidi, M. and Keck, J. C. Turbulent Flame Propagation and Combustion in Spark Ignition Engines. Combust. Flame 52, pp 217-245, 1983.

3. Namazian, M., Hansen, S., Lyford-pike, E., Sanchez-Barsse, J. and Rife, J. Schlieren Visualization of the Flow and Density Fields in the Cylinder of a Spark Ignition Engine. SAE Paper 800044, 1980.

4. Witze, P. O. and Vilchis, F. R. Stroboscopic Laser Shadowgraph of the Effect of Swirl on Homogeneous Combustion in Spark-Ignited Engine. SAE Trans. 90, pp 979-992, 1981.

5. Ziegler, G. F. W., Zettlitz, A., Meinhardt, P., Herweg, R., Maly, R. and Pfister, W. Cycle-Resolved Two-Dimensional Flame Visualization in a Spark-Ignition Engine. SAE Paper No. 881634, 1988.

6. Zhao, H., Collings, N. and Ma, T. The Cylinder Head Temperature Measurement by Thermal Imaging Technique. SAE Paper 912404, 1991.

7. Zhao, H., Collings, N. and Ma, T. Warmup Characteristics of Surface Temperatures in an I.C Engine Measured by Thermal Imaging Technique. SAE Paper 920187, 1992.

8. Heywood, J. B. Internal Combustion Engine Fundamentals, McGraw-Hill, 1988.

9. Chun, K. M. and Heywood, J. B. Estimating Heat-Release and Mass-of-Mixture Burned from Spark-Ignition Engine Pressure Data. Combust. Sci. and Tech., pp. 133-143, Vol.54, 1987.

10. Wiseman, M. W., Spark Ignition Engine Combustion Process Analysis, Ph.D Thesis, Department of Mechanical Engineering, Nottingham University, 1989.

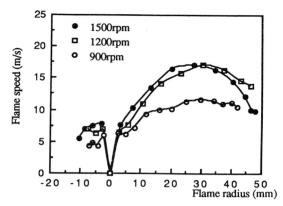

Fig.5 Average flame speed against the flame radius at λ=0.9 WOT

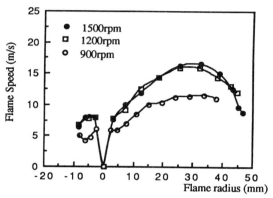

Fig.6 Average flame speed againt the flame radius at λ=1.0 WOT

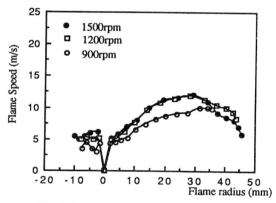

Fig.7 Average flame speed against the flame radius at λ=1.2 WOT

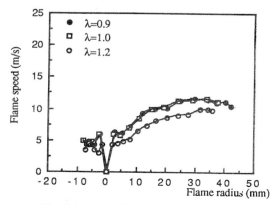

Fig.8 Average flame speed against the flame radius at 900rpm WOT

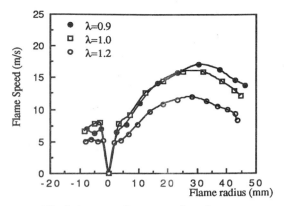

Fig.9 Average flame speed against
the flame radius at 1200rpm WOT

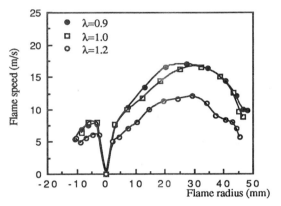

Fig.10 Average flame speed against
the flame radius at 1500rpm WOT

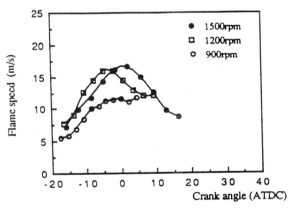

Fig.11 Average flame speed againt crank
angle with λ= 1.0 WOT

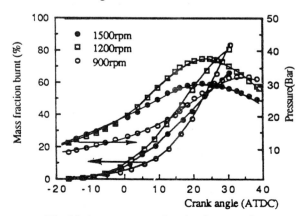

Fig.12 Average mass fraction burnt and
pressure v.s crank angle λ =1.0 WOT

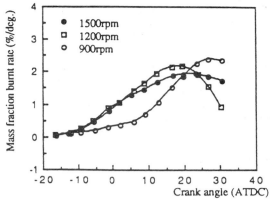

Fig.13 Average mass fraction burnt
rate against crank angle λ=1.0 WOT

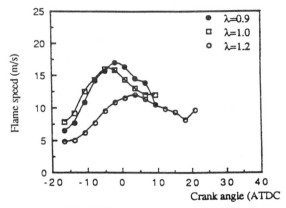

Fig.14 Flame speed againt crank
angle at 1200rpm WOT

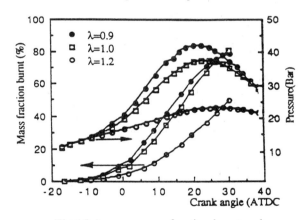

Fig.15 Average mass fraction burnt and
pressure v.s crank angle at 1200rpm WOT

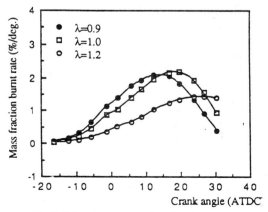

Fig.16 Average mass fraction burnt rate
against crank angle at 1200rpm WOT

C448/042

On the electric wind formation in spark plugs for SI engines

G PINHASI, MSc and E SHER, PhD, CEng, MIMechE
Department of Mechanical Engineering, Ben Gurion University of the Negev, Israel
A POKRYVAILO
Spectronix Limited, Tel-Aviv, Israel

SYNOPSIS It has been shown earlier that it is possible to improve the ignition process in SI engines by introducing a wind of a certain velocity which carries away the ignited mixture from the plug electrodes thus reducing heat losses to the electrodes. A convenient way of producing this wind is by electrical means, namely by a corona discharge.

Electric (corona) wind in gaps relevant to the spark plugs has been studied both theoretically and experimentally. Short (up to several mm) point to-plane gaps were employed. Data on electric wind in such gaps are very scarce in the literature. It has been shown that a fundamental difference exists between the mechanisms of the electric wind formation at positive and negative polarities. In particular, at negative polarity electrode the wind velocity strongly depends on the gap distance, and increasing the distance results in an increase in the wind velocity. At positive polarity there is no such strong dependence and the wind velocity is considerably higher than at negative polarity for the same current.

A theoretical model of mass transfer in corona discharges proposed by Sigmond has been modified to the case of negative polarity in electronegative gases (air and air-fuel mixtures are included). The modified theory readily explains the anomalies in wind behavior at negative polarity by electron attachment mechanism. A fair agreement between theoretical and experimental results has been obtained.

1 INTRODUCTION

It has been suggested by Pischinger & Heywood[1] that it is possible to improve considerably the performance of a SI engine by introducing a wind of a certain velocity, typically 3-6 m/s, in the vicinity of the spark gap. This wind carries away the ignited mixture from the electrodes thus reducing heat losses. As a result, cycle variability and misfire occurrence reduce significantly. Attempts to build a wind producer by conventional means (blowers, modification of the cylinder and piston geometry, etc.) meet obstacles which are hard to overcome. A promising device for producing the required gas flow is a corona discharge where direct conversion of electrical energy to mechanical energy without moving parts takes place. Sher et al.[2] have explored the possibility of employing the corona discharge as a wind source in SI engines and have obtained promising results by using a negative corona device.

Corona discharges are involved with mass transfer phenomenon, an effect which is often called "electric", or "corona" wind. An extensive collection of experimental data may be found in Levitov et al.[3], while a plausible theoretical treatment is given by Sigmond[4]. These and other works lack however information on the mechanism of formation and behavior of electric wind in short gaps at negative polarity in electronegative gases, which is the case in SI engines. The point of the present communication is to fill in this blank.

2 THEORETICAL

When corona occurs, the space between the electrodes is characterized by two different regions; the ionization region in which ionization takes place in a very thin layer in the vicinity of the sharp electrode, (typically less than 1 millimeter) and the drift region in which charged particles are accelerated toward the blunt electrode.

In negative corona the charge carrying particles in the drift zone are mainly free electrons. However, when the medium is an electronegative gas (air or SF_6) the charge carrying particles include also negative ions which are created when electrons are caught by neutral molecules having electron affinity (oxygen molecules in air). In this case the electrons travel some distance before being caught. The travelling distance of the free electrons is characterized by the attachment coefficient which depends on a number of parameters such as he field strength, humidity and pressure. For atmospheric pressure in air for the field strength 20 kV/cm, the travelling distance is in the order of 2 mm. In the drift region the negative ions exchange their momentum with the neutral molecules by collisions thus producing electric wind. In positive corona, the positive ions are created in the ionization region and are accelerated toward the negative (blunt) electrode while exchanging their momentum with neutral molecules in the drift region. The rate of momentum exchange depends very much on the mobility of the charge carriers (either positive or negative). As the mobility of the positive ions is lower, the maximum velocity of an electric wind is somewhat higher for the positive polarity[4] (at the same current).

We consider here unipolar DC corona discharge which is to be distinguished from the case of streamer coronae. It is assumed that the domain enclosed by the gap is mainly a drift region, i.e., a region in which ions are accelerated toward the opposite electrode. This is true for air gaps exceeding fractions of a millimeter at atmospheric pressure. Point-to-plane gaps are considered though the results obtained below are basically applicable for other kinds of corona gaps.

Sigmond[4] has suggested that during transition the momentum transferred by a single charge q to the gas through the gap along a field line of length L and field strength E is:

$$p_1 = \int_0^L \frac{qEdx}{E\mu} = q \int_0^L \frac{dx}{\mu} \tag{1}$$

where μ is charge mobility. Note that the mobility of the electrons is three orders of magnitude higher than that of the ions and therefore their contribution to the momentum exchange with the neutral gas is negligible.

Consequently, for positive polarity equation (1) becomes:

$$p_1 = q \int_0^L \frac{dx}{\mu_i} \tag{2}$$

where μ_i is the ion mobility. Equation (2) suggests that the momentum transfer does not depend strongly on the field strength and for simplicity is assumed constant.

The mean distance passed by an electron before its capture, is characterized by the attachment coefficient η. For the simplicity sake let assume that η does not change along the trajectory path and all electrons are stable ions. Then the number of negative ions increases with the coordinate x exponentially as follows:

$$N = N_o(1-\exp(-\eta x))$$

where N_o is the number of electrons starting at x = 0. The increment of the ion number along the path dx is therefore:

$$dN = N_o\, \eta\, \exp(-\eta x)dx \tag{4}$$

When an electron is caught by a neutral molecule at a distance x, the gas momentum transfer is:

$$p_1 = q\left(\int_0^x \frac{dx}{\mu e} + \int_x^L \frac{dx}{\mu_i} \right) = q\frac{L-x}{\mu_i} \tag{5}$$

where μ_e is the electron mobility. The momentum increment is:

$$dP = p_1 dN \tag{6}$$

therefore the total momentum is:

$$P = \int_0^L q \frac{L-x}{\mu_i} N_o \eta e^{-\eta x} dx = \tag{7}$$

$$= L\frac{1}{\eta}(1-e^{-\eta L}) + \frac{1}{\eta^2}\left[e^{-\eta L}(\eta L + 1)-1\right]$$

The total force applied on a cylindrical domain by a corona with current I:

$$F = \frac{I}{\mu_i}\left\{ L(1-e^{-\eta L}) + \frac{1}{\eta}[e^{-\eta L}(\eta L + 1)-1]\right\} \tag{8}$$

where B is the cylinder basis area and L its height. The gas velocity is then derived to yield:

$$v_g = \sqrt{\frac{I}{\mu_i \rho B}} \sqrt{L(1-e^{-\eta L}) + \frac{1}{\eta}[e^{-\eta L}(\eta L + 1)-1]} \tag{9}$$

where ρ is the gas density.

Analysis of (9) shows that v_g tends to zero when $\eta \to 0$, which means that all electrons traverse the gap without being caught. When $\eta \to \infty$, all electrons are caught at x = 0 and

$$v_g = \sqrt{\frac{IL}{\mu_i \rho B}} \tag{10}$$

Equation (10) which is the result obtained by Sigmond, is virtually valid only for positive polarity.

In the case when the product $\mu_i\rho$ does not depend on pressure and η is proportional to it[5] we find that for the same current v_g increases with pressure. Thus in electronegative gas at a certain pressure a considerable electric wind may be developed in short gaps where it is negligible at lower pressures. This has been observed experimentally by Sher et al.[2]. Corona wind velocities of over 7 m/s were observed at 1 mm air gap for a pressure of 0.5 MPa while at 0.2 MPa no corona wind velocity was observed for the entire voltage range up to the breakdown, though the electric current was quite considerable. That means that the measured current was a current of a streamer corona, and unipolar corona did not occur.

It is of interest to derive an expression for the maximum corona wind velocity in a given gap. The limit is obviously postulated by the maximum possible current in an unipolar corona. Following Sigmond and Goldman[6] the corona saturation ion current is given by

$$I_s \sim V^2/L, \tag{11}$$

where V is voltage across the gap. According to Pashen's Law at high pressure breakdown voltage depends linearly on the electrode distance (a reasonable approximation for air gaps up to several centimeter). Therefore,

$$I_m \sim L \tag{12}$$

If we further assume that the volume occupied by corona in space is L^3, i.e., $B = L^2$, and disregard the difference between positive and negative coronas then the maximum velocity of the electric wind becomes:

$$v_{g\,max} \sim \sqrt{\frac{1}{\mu_i \rho L}} \sqrt{L(1-e^{-\eta L}) + \frac{1}{\eta}[e^{-\eta L}(\eta L + 1)-1]} \tag{13}$$

and for positive corona

$$v_{g\,max} \sim \sqrt{\frac{1}{\mu_i \rho}} \tag{14}$$

which means that the maximum corona wind velocity for a positive corona is independent on the gap size and pressure.

It should be stressed here again that the trends shown by the above derivations are valid only within the boundaries of unipolar coronae.

3 EXPERIMENTAL AND CONCLUSIONS

The experimental part was conducted with a sewing steel needle as the sharp electrode. A copper gauze having a mesh of 0.5 x 0.5 mm and a wire diameter of 0.1 mm was situated at a distance L as shown in Fig. 1. The wind velocity along the axis of symmetry was measured by using a miniature probe situated in the proximity to the gauze, while the pressure difference was measured by a workshop apparatus (Sher, et al.[2]). Some results are plotted in Fig. 2.

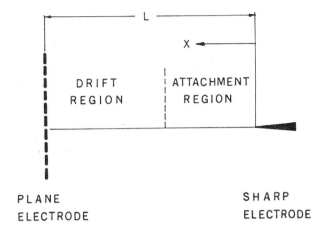

Figure 1   A schematic layout of the experimental set-up.

It seems that while the positive corona (positive sharp electrode) produces a constant wind velocity of about 12 m/s independent of the gap size, the wind velocity of the negative corona increases from 3 m/s for 0.5mm gap width to 12 m/s for 15 mm. This is attributed to the travelling distance of the electrons (in negative coronae) before being captured by the oxygen molecules. Also are shown in Fig. 2, the predicted results as estimated by Eq. 9. The theoretical curves were produced by using the data as follows:

			negative	positive
η	attachment coef.	1/m	100	→ ∞
μ$_i$	ion mobility	m²/V·s	2.2x10⁻⁴	2.2x10⁻⁴
ρ	density	kg/m³	1.2	1.2
I/L	current/gap width	A/m	.07	.04

Taking into account numerous assumptions such as a simplified relation between current voltage and gap size as well as ignoring the effect of the curvature of the sharp electrode, predictions are found in good agreement with the experimental observations. This qualitative agreement supports our proposal for the mechanism of corona wind formation. It should be mentioned here that for short gaps (L < 1 mm) corona is very sensitive to the sharp electrode curvature and for dull electrodes, spark breakdown may occur before corona inception.

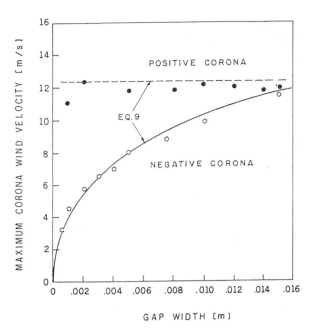

Figure 2   Maximum corona wind velocity at atmospheric pressure vs. gap width.

In order to implement the above findings to practical ignition systems, information about the maximum voltage allowed before breakdown and the associated electric current is of an importance (note that the corona wind velocity is proportional to the square root of the electric current). Figs. 3 and 4 show some results obtained with an experimental set-up which includes a sharp electrode situated at the center of a ring shape electrode. It appears that for a gap of 1 mm, a corona occurs already at 0.15 MPa and that above a pressure of 0.6 MPa, a maximum current of 80 μA can be achieved with negative corona. It appears that the maximum electric current and therefore the resulting corona wind velocity, are much lower for the positive coronae and the superiority of the negative corona over its counterpart is prominent. The current-voltage characteristics (Fig. 4) suggest that the electric current depends very strongly (for a negative corona) on the voltage. This means that for real engine applications, where the cylinder pressure varies with time, a constant current source is needed in order to maintain a corona wind.

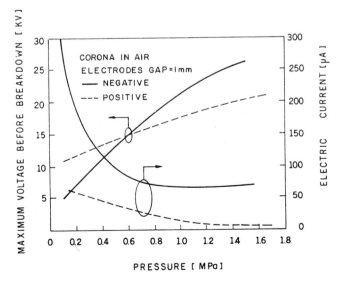

Figure 3   Maximum voltage before breakdown and the associated electric current vs. pressure.

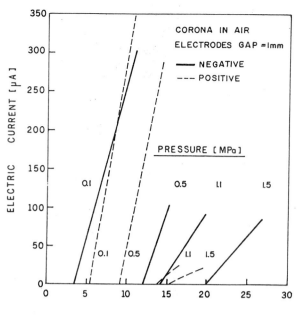

Fig. 4.   Electric current vs. pressure.

REFERENCES

(1)   PISCHINGER S. and HEYWOOD, J.B. "A Model for Flame Kenel Developed in a Spark-Ignition Engine," 23rd Symposium (International) on Combustion, pp. 1033-1040, The Combustion Institute, 1990.

(2)   SHER, E., BEN-YAISH, J., POKRYVAILO, A. and SPECTOR, Y., A corona spark plug system for spark ignition engines, SAE Paper No. 920810, 1992.

(3)   LEVITOV, V.I., *Electrostatic Precipitators*, Moscow, 1980.

(4)   SIGMOND, R.S., Mass transfer in corona discharges, *Rev. Int. Hautes Temper. Refrac.*, Fr., 25, 201-206, 1989.

(5)   MANFRED, B., WOLFRAM, B., KLAUS, M. and WALTER, Z., *Hochschpannungstechnik*, Springer-Verlag, 1986.

(6)   SIGMOND, R.S. and GOLDMAN, M., Corona discharge physics and applications, In *Electrical breakdown and discharges in gases*, Plenum Press, 1983.

Acknowledgements

This research was supported by grant no. 88-00158 from the United States-Israel Binational Science Foundation (BSF), Jerusalem, Israel, and by the Basic Research Foundation administered by the Israel Academy of Sciences and Humanities.

Life on earth at the point of inflection

A K OPPENHEIM, Dipl-Ing, PhD, DSc, DIC
Department of Mechanical Engineering, University of California, Berkeley, United States of America

SYNOPSIS Involvement with pollutant emissions manifests our concern over the negative influence the rapidly decaying quality of the environment may have upon the growth rate of life on earth. For engines, this brought about an over-reliance upon principally peripheral means, such as the catalytic converter associated with alternative and reformulated fuels, whereas the actual genesis of pollutants, the mode in which the exothermic process of combustion is executed in the cylinder, has been so far neglected. It is in catering to this aspect that, I hope, our conference will be of particular assistance.

1 DEMAND

The fact that life on earth is at the point of inflection, its growth rate increment turning over from positive to negative, is manifested by the menacing signs of the decaying quality of the environment (1-12). Today the protection against this impending disaster is a hot political issue. Just to get an impression of the relevant activities undertaken on the international scale over the last few years, consider the following events. The World Meteorological Organization (WMO) established the World Climate Research Program that, in turn, inaugurated the International Geosphere-Biosphere Program, adopted in 1986 by the International Council of Scientific Unions. In 1988 WMO sponsored, in conjunction with the United Nations Environmental Programme, the Intergovernmental Panel on Climate Change which in 1980 convened the Second World Climate Conference. Associated with these activities are the Montreal (1987) and London (1990) Protocols on protection of the ozone layer. All that culminated this year in the United Nations World Conference on Environment and Development, the so-called Earth Summit in Rio de Janeiro.

The universally accepted measures of atmospheric quality are the concentrations of ozone and carbon dioxide with the concomitant rise in surface temperature. The evolution of these parameters is presented, respectively, in Figs. 1, 2 and 3. Causally, the latter is the outcome of the greenhouse effect: the reduction of radiation from earth to space due to the long-wave filtering interference of the atmosphere. The particular culprits in this respect are the so-called greenhouse gases, i.e., quoting from the OECD report (1), "carbon dioxide (CO_2), methane (CH_4),

chlorofluorocarbons (CFCs), nitrous oxide (N_2O) and tropospheric ozone (O_3)".

Ultimately, their escalation is the consequence of the population explosion, Fig. 4 (13), and the concomitant human activities. Among them, of particular concern here are those associated with the rapidly growing use of internal combustion engines for transportation, as illustrated by Fig. 5 whose future trend is easily predictable in view of its amazingly linear character. Depicted in Fig. 6 is the contribution of the transportation sector in comparison to other sources of major pollutants (14). The concomitant effects on public health were, of course, a subject of extensive studies, as exemplified prominently by the compilation of comprehensive articles on various aspects of the subject edited by Watson, et al. (15).

2 RESPONSE

Interestingly enough, mankind's concern over the environmental menace has been first manifested, following the cataclysm of the Second World War, by the space program - an activity tantamount to admitting that life on earth may soon become unbearable so that exploring the universe for new homelands is a worthy pursuit. To begin with, thanks to the governmental support in the Soviet Union and United States, the program has been spectacularly successful.

Then, in response to the premise that, perhaps something could be done to deter the menace by protecting the quality of the atmosphere, a legislature was introduced in the U.S.A. by the Muskie Committee in terms of the 1970 amendments to the Clean Air Act, restricting significantly pollutant emissions from combustion systems with particular attention paid to road vehicles - an act giving rise to the Environmental Protection Agency. In recognition of the especially acute local air quality, in California this was associated with the establishment of the State's Air Resources Board.

Consequently, impressive advances have been made in reducing drastically the allowable pollutant emissions, reflected in the progressively tightening standards displayed in Figs. 7 and 8. As the time went on, the industry became complacent and the society satisfied that adequate measures are taken to alleviate the menace of atmospheric contamination.

3 COMMENTS

The space program was earmarked by clearly defined goals, such as the famous command of J. F. Kennedy to reach the moon, and it was enforced by a coordinated effort involving effective cooperation between government agencies with their research laboratories and the aerospace industry enhanced by close association with distinguished scientists.

Nothing of the sort has been done, as yet, for the protection of the environment - a cause evidently devoid of the appeal that the adventures of space exploration and warfare have to offer. Instead, the full brunt of responsibility for the development of preventive measures was left entirely to the industry in the belief that it is affluent enough to provide a satisfactory solution on its own, while the government provided not much more than legislative constraints and law enforcing agencies.

As a result, overwhelming reliance was put upon peripheral devices, such as the use of catalytic converters - chemical processing plants in the exhaust system - on one side, and, on the other, the improvement in fuel quality - an aim achievable by the introduction of alternative (in particular, natural gas, liquid petroleum gas, and methanol) and reformulated (16) fuels. Comparatively little has been accomplished, however, in catering to the essential origin of pollutants: the exothermic process of combustion taking place in the engine cylinder.

Less I be misunderstood, let me stress that this does not mean at all that there were no attempts made with this aim in mind. On the contrary, it is the notable failure of such well known concepts, conceived in this connection, as the *stratified charge* and *lean burn*, that is at the root of the major resistance of the industry to progressive techniques, especially those born out of the rapidly advancing frontiers of science: the fluid mechanics of turbulent flow (17) and the chemical dynamics of elementary steps in exothermic reactions (18). Paradoxically enough, the primary reason for this failure was the lack of awareness of the advances

made concurrently in these fields of scientific endeavor. To top it off, there have been a number of studies carried out on the future of engine technology [vid. e.g. (19)], where the possibility of a significant progress achievable by ameliorating the combustion process in the cylinder has been conspicuously neglected. This omission became, in fact, characteristic of practically all the investigations concerning automotive pollution, as prominently manifested by the *Auto/Oil Air Quality Improvement Research Program* (20, 21) currently under way in the U.S.A.

The preventive legislature introduced in the U.S.A. has been imitated by all the industrial nations of the world, quite soon in Japan, and, with a considerable delay, in Europe. In spite of the time made thereby available, all the programs followed the same hasty pattern: urgently imposed legislative restrictions or incentives issued by the government, based primarily on what can be achieved by the use of catalytic converters, and unimaginative response of the industry, earmarked by notable omission of what can be accomplished in the engine cylinder.

4 RECOMMENDATIONS

In view of the above, the recommendations I have to offer are self-evident:

- let us concentrate upon not only what is going on in the engine cylinder, but also what can be done to optimize the execution of the exothermic process of combustion.

- let us then launch an attack upon the conversion of the cylinder-piston enclosure from solely a source of power, as it is treated today, to a full-fledged chemical reactor where the formation of pollutants is appreciably curtailed.

- recognizing that the combustion process takes place in a turbulent field, let us exploit for this purpose the salient properties of turbulence: the large scale vortex structure created by shear with the concomitant entrainment and spiral mixing.

- realizing that our principal task in this respect is to affect the chemistry of the process, let us take full advantage of recent advances made in chemical dynamics of exothermic reactions.

In one phrase:
- let us get on with the development of a *clean combustion engine* by bridging the gap between the progress of science and its engineering implementation.

5. REFERENCES

(1) ANON., The state of the environment, *Organization for Economic Cooperation and Development*, Paris, 1991, 297 pp.

(2) SOLOMON, S., The mystery of the Antarctic ozone 'hole', *Reviews of Geophysics,* 1988, 26, 131-148.

(3) WATSON, R. T., PRATHER, M. J., and KURYLO, M. J. (Chairs), Present state of knowledge of the upper atmosphere: An Assessment Report of Ozone-Trend Panel, Ad Hoc Theory Panel and NASA Panel for Data Evaluation, NASA Reference Publication 1208, 1988, VI + 200 pp.

(4) MACCRACKEN, M. C., BUDYKO, M. I., HECHT, A. D., and IZRAEL, Y. A. (Editors), Prospects for future climate, Prepared under the auspices of the US/USSR Agreement on Protection of the Environment, Lewis Publishers, Chelsea, Michigan, 1990, XIII + 270 pp.

(5) MACCRACKEN, M. (Chair), Energy and climate change, Report of the DOE Multi-Laboratory Climate Change Committee, Lewis Publishers, Chelsea, Michigan, 1990, XVI + 161 pp.

(6) KEELING, C. D., The carbon dioxide cycle: Reservoir models to depict the exchange of atmospheric carbon dioxide with the oceans and land plants, in Chemistry of the Lower Atmosphere (S. I. Rasool, Ed.), Plenum Press, New York, 1973, XII + 335 pp.

(7) BODEN, T. A., KANCIRUK, P., and FARRELL, M. P., Trends '90: A compendium of data on global change, ORNL/CDIAC-36, Oak Ridge National Laboratory, Oak Ridge, Tennessee, 1990, XV + 257 pp.

(8) JONES, P. D., RAPER, S. C. B., BRADLEY, R. S., DIAZ, H. F., KELLY, P. M., and WIGLEY, T. M. L., Northern hemisphere surface air temperature variations: 1851-1984, J. Clim. Appl. Meteor, 1986, 25, 161-179.

(9) IPCC (Intergovernmental Panel on Climate Change), Climate change: the IPCC scientific assessment, Cambridge University Press, Cambridge, Mass., 1990, XXIX + 365 pp.

(10) EVANS, D. J. (Chair), Policy implications of greenhouse warming, Synthesis Panel of the Committee on Science, Engineering and Public Policy, U. S. National Research Council, National Academy Press, 1991, XIII + 127 pp.

(11) WUEBBLES, D. J., and EDMONDS, J., Primer on greenhouse gases, Lewis Publishers, Chelsea, Michigan, 1991, XX + 230 pp.

(12) BROWN, L. R., FLAVIN, C., and POSTEL, S. (Directors), State of the world 1992, a world watch institute report on progress toward a sustainable society, W. W. Norton & Co., 1992, XV + 256 pp.

(13) MCEVEDY, C., and JONES, R., Atlas of world population history, Penguin, U. K., 1978, 368 pp.

(14) ANON., National air quality and emissions trends report, U. S. Environmental Protection Agency, 1991, XI + 129 pp.

(15) WATSON, A. Y., BATES, R. R., and KENNEDY, D. (Editors), Air pollution, the automobile, and public health, National Academy Press, Washington, D. C., 1988, VIII + 692 pp.

(16) SAWYER, R. F., Reformulated gasoline for automotive emissions reduction, Twenty-Fourth Symposium (International) on Combustion, Sydney, Australia, 1992.

(17) CHORIN, A. J., Computational fluid mechanics, Academic Press, New York, 1989, XV + 223 pp.

(18) LEE, Y. T., Molecular beam studies of elementary chemical processes, Nobel Lecture, Les Prix Nobel en 1986, Nobel Foundation, 1987, 168 pp.; *Chem. Scripta*, 1987, 27, 2489; *Science,* 1987, 236, 793; *Angew Chem.,* 1987, 99, 967.

(19) STEPHENSON, R. R. (Principal Investigator), Should we have a new engine? An automobile power systems evaluation, Vol. 1. Summary, X + 109 pp., Vol. 2. Technical Reports, VII + 540 pp., Jet Propulsion Laboratory, (SAE SP399 and SP400), 1975.

(20) KOEHL, W. J., BENSON, J. D., BURNS, V. R., GORSE, R. A., HOCHHAUSER, A. M., and REUTER, R. M., Effects of gasoline composition and properties on vehicle emissions: a review of prior studies -- auto/oil air quality improvement research program, SAE Paper 912321, 1991.

(21) ANON., Auto/oil air quality improvement research program, SAE Special Publication SP-920, Warrendale, PA, 1992.

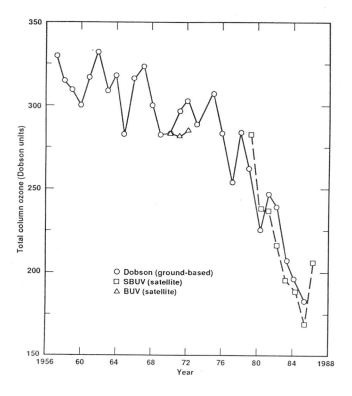

Fig. 1 Evidence of the ozone hole
Observed decrease in spring-time total column ozone amount (in Dobson units) over the Halley Bay Station in Antarctica based on results from a ground-based Dobson spectrophotometer and satellite-based solar back-scattered ultraviolet (SBUV) and ultraviolet (BUV) radiator.

[1DU = 10^{-5}m thickness of ozone layer @ STP, whereas global mean amount of ozone (8×10^{22} molecules m^2) = 300 DU] (4,1).

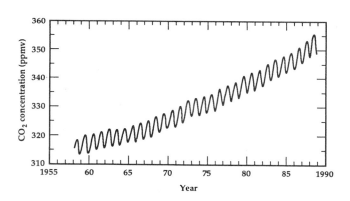

Fig. 2 Monthly average CO_2 concentration at Mauna Loa, Hawaii l (6,7,1).

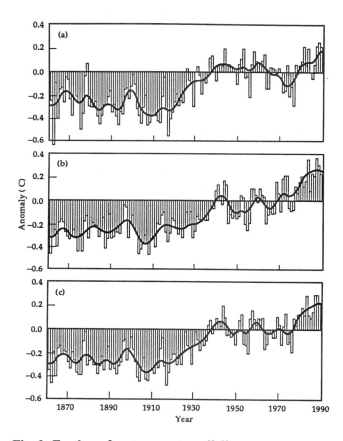

Fig. 3 Earth surface temperature (8,9)
(a) Northern Hemisphere; (b) Southern Hemisphere; (c) Global Mean

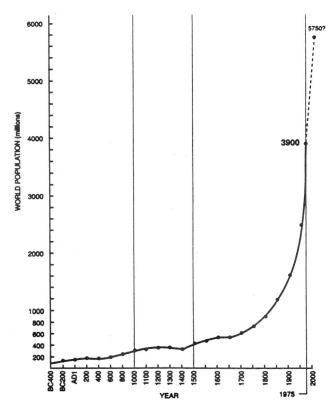

Fig. 4 World population (13,10)

218

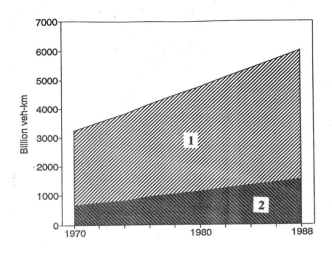

Fig. 5 Trends in road traffic (1)
1-passenger cars; 2-commercial vehicles

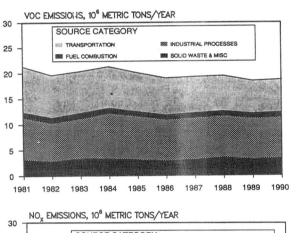

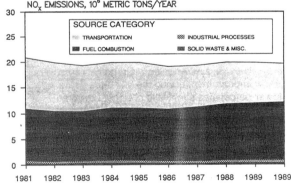

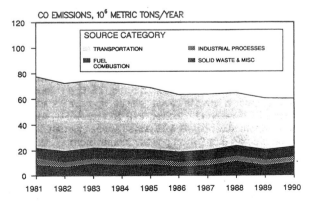

Fig. 6 National trends in pollutant emissions (14)
VOC - Volatile Organic Compounds

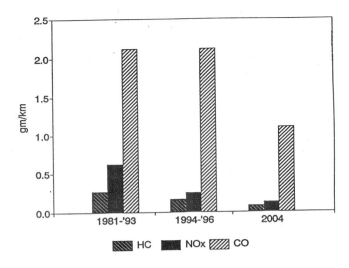

Fig. 7 United States passenger car emission standards
1960 pre-control emissions: HC:6.6; NO_x:2.6;
CO:52.5 gm/km

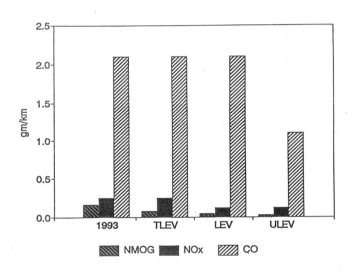

Fig. 8 California Air Resources Board passenger car low emission standards
TLEV: Transition Low Emission Vehicle
LEV: Low Emission Vehicle
ULEV: Ultra Low Emission Vehicle
NMOG: Non Methane Organic Gases

C448/019

The development of a dry low No$_x$ combustor for the 6.2MW Tornado industrial gas turbine engine

M F CANNON, BSc, CEng, MIMechE and D J ECOB
European Gas Turbines Limited, Lincoln
F W STRINGER, BSc and S M DE PIETRO, BEng
Aero and Industrial Technology Limited, Burnley, Lancashire

SYNOPSIS The development of a lean burn natural gas fired combustor for the 6.2MW Tornado Industrial Gas Turbine Engine to meet European emissions standards is described. The development effort involved testing a full scale combustor at pressures up to 8 bar. The performance in terms of combustion efficiency, NO$_x$, CO, UHC, smoke, exhaust temperature pattern factor and metal temperatures is presented. Tests at atmospheric pressure were carried out to determine the ability of the combustor to light-up and generate sufficient power for engine acceleration. Where possible comparisons were made with the performance of the conventional combustor and predictions of engine performance are given. The impact of the new combustor design on the fuel control system is also discussed briefly.

1 INTRODUCTION

Current and future emissions legislation, resulting from public and political concern over the environment, has driven the intensive search for dry solutions to NO$_x$ control. The requirement for a Dry Low NO$_x$ solution provides technical challenges in both combustion and fuel control. These challenges can be met but require a change in the level of technology and system complexity.

2 NO$_x$ CONTROL MECHANISMS

In a natural gas fired combustor, with little fuel bound nitrogen, NO$_x$ reduction essentially means control of thermal NO$_x$ i.e. 'time at temperature'. A mechanism that controls flame temperatures and the time the combustion gas spends at these temperatures also controls the thermal NO$_x$ emissions. The options available for NO$_x$ control fall into three categories; diluent addition, catalytic and combustor design.

This paper describes the development of a low NO$_x$ combustor design capable of meeting the NO$_x$ emissions target without diluent addition or catalytic assistance.

Possible approaches considered were:-

i) Rich burn/rapid quench - where the fuel is initially mixed with insufficient air to burn fully, thus not attaining high primary zone temperatures, and is then rapidly quenched and diluted to limit the residence time at high temperatures.

ii) Lean burn/well mixed - which relies on maintaining relatively uniform combustion gas temperature below the level at which high levels of NO$_x$ are formed.

Of these two fundamental design options it was felt that the latter offered the more immediate solution with a greater depth of background experience.

3 MODULE SYSTEM

The lean burn system adopted was based on a module concept developed for low emissions aero engine combustors in the 70's. Reports by Bunn and Winter (1) and Winter (2), Antos, Mumford and Winter (3), Stringer (4) illustrate the principle that low NO$_x$ emissions are achievable by a fuel/air injector module with rapid mixing of air and fuel jets.

The basic module configuration is shown in Fig. 1. This jet mixing element forms the fundamental mixing mechanism of the module. Each element consists of a radial fuel jet and a co-axial radial air plume jet which provides an element of primary mixing. An axial backplate air jet is configured such that it hits and rapidly mixes (secondary mixing) with the combined radial fuel and air jet as it leaves the module plume orifice. Each module injector has six of these mixing elements positioned symmetrically around its periphery. The six elements then combine to form a mixing re-circulation stabilised on the bluff face of the module.

One of the significant advantages of the module system, relying as it does on internal flame tube mixing, is that the major concern of premix flashback is eliminated.

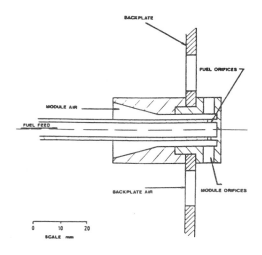

Fig 1 Combined air and fuel injector module

4   DESIGN SPECIFICATION

The design brief for the combustor was to provision a combustion system capable of satisfactory operation over the range of engine operation and in doing so meet the emissions and performance targets specified below.

It was necessary for the combustor to have retrofit capability.

110% load was selected as the combustor design point in order to leave sufficient NO_x margin for future engine uprating.

The defined targets were:

NO_x <   60 ppmv (corrected to 15% O_2)

CO <   15 ppmv at 110% load
     <   60 ppmv at   25% load

Exhaust Gas Temperature Distribution at 110% load:

OTDF less than 20%
RTDF between   5 & 7%

Smoke <   Bacharach No. 2 at continuous operation.
     <   Bacharach No. 3 at start-up.

Metal Temperatures less than 800°C.

5   DESIGN APPROACH

The Tornado engine operates over a wide air/fuel ratio range from 222/1 at idle to 58,5/1 at 110% load. This wide range dictated the need for a conventional diffusion flame pilot combustor for low load stability and a fuel staged module system (inward and outward jets fuelled independently) for both good part load efficiency and full load NO_x performance.

From the emissions requirements, the operational constraints of the engine and the operating margins of the modules, a fuel schedule across the load range was established (Fig. 2). Between idle and 40% load the pilot alone provides the heat input for engine operation. Between 40% and 85% load, only the inward firing jets of the modules are fuelled (the inner stage). Above 85% load all the module jets are fuelled (the inner and outer stages).

Fuelling inner jets only at part load enables the inner stage of the module to burn at a richer local air/fuel ratio than would have been possible if the whole of the module was fuelled. This fuel staging, therefore, enables the module air/fuel ratio to be maintained within the identified stable operating boundaries.

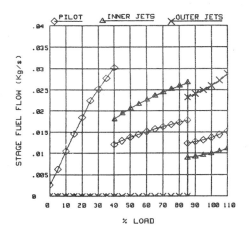

Fig 2 Fuel Schedule

6   DEVELOPMENT PROGRAMME

The test rig and combustor arrangement is shown in Fig. 3 and the combustor itself is shown in Figs. 4 and 5. The rig was configured to simulate the engine compressor delivery flow by admitting the air at the discharge end of the combustor and allowing it to flow in a reverse flow manner towards the head.

A standard Tornado combustor was tested across the load range to validate the test rig.

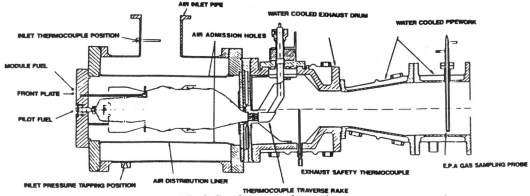

Fig 3 Combustor and test rig

Fig 4 G60 Combustor

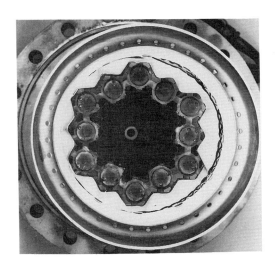

Fig 5 G60 Combustor

6.1 Light-up operation

The Tornado engine light-up fuel schedule follows conventional practice of fixing a start fuel flow and then maintaining that flow over initial engine acceleration. After initial acceleration of the starter the combustor is lit at an air/fuel ratio of around 50/1. However, by the time of disengaging the starter the air/fuel ratio is weaker at 160/1 due to the increased air mass flow. At this weak end of the light-up curve only the combustor provides the energy input to continue the acceleration to idle. If the drop off in combustion efficient at this point, due to weakening of the combustor, is severe enough the engine will not reach self-sustain speed.

Combustion efficiency data supplied by EGT is shown in Fig. 6 for a standard Tornado combustor which is known to accelerate the engine. In Fig. 6 this efficiency data is compared with the calibration data of the standard combustor tested on the AIT rig and the G60 low NO_x combustor performance data. The good correlation between EGT supplied, and AIT measured data can be seen by the match of the two data sets. It can be seen that the G60 combustor efficiency is notably better at the critical higher air mass flows where the starter has disengaged. This is a direct consequence of operating the pilot combustor, which takes 25% of the total air mass flow, at a near ideal local air/fuel ratio of 40/1 (i.e. 160/1 overall air/fuel ratio). To achieve this nominally constant efficiency curve, however, necessitates altering the light-up fuel schedule. The objective is to enable the G60 combustor to be lit at an overall air/fuel ratio of 160/1 and then to be maintained at that constant air/fuel ratio for the initial engine acceleration.

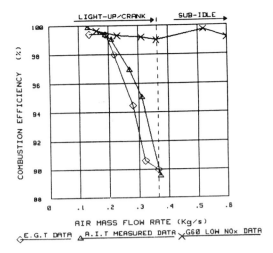

Fig 6 Light-up/crank and sub-idle performance

6.2 Sub-idle acceleration

After initial light-up and starter crank acceleration the acceleration up to idle is further complicated by the Tornado sub-idle bleed valve operation. In order to prevent compressor surge, after initial light-up, a compressor air bleed opens and diverts air from the combustion chamber. The effect of this is to produce a combustor air/fuel ratio of 59/1 (i.e. richer than the 100% load air/fuel ratio). As the engine then continues to accelerate toward idle the air mass flow increases as both the engine speed increases and the bleed valve closes until at idle the engine is at full speed with bleed valve closed.

The implication of this rich sub-idle operation, for a fuel staged combustor designed with a small diffusion flame pilot passing around 25% of total air mass flow, is that in order to achieve the overall combustor air/fuel ratio of 59/1 the pilot combustor has to operate at near stoichiometric conditions. The further implication of this, therefore, is that the module stages with their narrow stability range would have to operate at low compressor outlet temperatures and pressures in order to take the excess fuel from the pilot.

The combustion efficiency of the low NOx G60 combustor was measured at the reduced air mass flow operating points. The measured efficiency with module stages fuelled in rich operation proved to be adequate to continue the engine acceleration to idle as illustrated in Fig. 6.

The general conclusion drawn for the low NO_x combustor light-up and acceleration performance was that the combustor could be expected to perform at least as well as the standard combustor.

6.3 Load operation

The low NO_x combustor development programme centred around the air distribution to the modules, film cooling, and the dilution system. This distribution has a direct effect on the performance of the low NO_x combustor as shown in Table I.

An additional and very significant variable on a fuel staged combustor is the capability to transfer fuel between stages to influence emissions performance and exhaust gas temperature distribution. This is particularly significant in that specific fuel distribution can be used to control temperature and hot gas velocity interactions between stages.

A comprehensive programme of testing was completed which determined the optimum build configuration i.e. air mass flow distribution, and the optimum fuel distribution between stages across the load range. The optimised fuel schedule based on the stage fuel ratio's (i.e. fuel split pilot/inner/outer) and their variation across the load range, has a very significant impact on the engine fuel system requirement and is discussed later.

The low NO_x combustor was tested across the load range from idle through to simulated 110% load at the test facility limiting pressure of 8 bar. Data were recorded for exhaust gas emissions and temperature distribution, combustor wall metal temperatures, pressure loss, combustion vibration, NO_x, CO and UHC.

Table 1

	Direct Effect	Performance Effect
Module Aeration Level	Local Stoichiometry	NO_x/CO
Module Type	Ratio of Module to Backplate Air and Fuel/Air Mixing	NO_x/CO
Film Cooling	Air Admission Control	Metal Temperatures NO_x/CO
Dilution	Air Admission Control	Exhaust Temperature Distribution NO_x/CO

Corrected NO_x values (Dry, corrected to engine pressure and to 15% O_2) for the standard and low NO_x combustors are shown in Fig. 7. The estimated engine values for the standard combustor, based on rig measured values corrected by the application of the conventional square root dependence on the pressure of thermally generate NO_x, agreed with the EGT engine measured values thereby validating the methods of estimating engine NO_x from rig measurements.

The measured carbon monoxide levels for the G60 combustor are shown in Fig. 8 with the rig measured values for the standard combustor.

Both Figs. 7 and 8 illustrate the typical saw tooth performance pattern associated with all fuel staged combustion systems without air schedule control.

The exhaust gas temperature distribution of the low NO_x combustor met the defined specification limits. Actual values achieved were 18.2% OTDF and 6.8% RTDF.

Rig measurements of smoke indicated that engine smoke levels across the full operating range would be negligible.

Since the close interaction of fuel and combustion systems is an essential feature of the low emissions programme the early introduction of the combustor into the engine environment enables the fuel control system to be developed in parallel with continued combustor improvement.

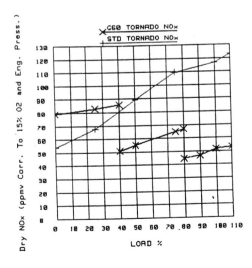

Fig 7 NO_x Performance

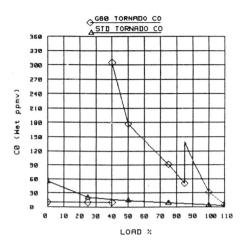

Fig 8 CO Performance

7 FUTURE PROGRAMME AND OBJECTIVES

7.1 Performance targets

Tighter emissions specification for the low NO_x engine, particularly with regard to the CO levels acceptable at off design points, are currently being targeted for inclusion into the future Tornado low NO_x engines.

There are three main performance solutions currently under investigation:

i) Minimising cooling films in sensitive areas by use of thermal barrier coating and increased use of impingement cooling.

ii) Utilising the production engine improved casing design to allow increased combustor length for improved CO burnout.

iii) Combustor airflow control.

7.2 Serviceability and reliability

The penalty for emission reduction is complexity. All low NO_x combustors by their nature are more complex than conventional systems. This means that in addition to the emissions performance challenge, one of the principal engineering objectives is making the system live successfully in an industrial gas turbine environment. The production design objective now is to simplify the combustor, to reduce initial cost and to ensure that serviceability and reliability levels match those of a conventional system.

8   FUEL CONTROL

Staged combustion chambers inevitably make considerable demands on control system flexibility and the close linkage of fuel control and combustor air and fuel schedule development is a necessary development of the G60 combustion programme. See Fig. 9.

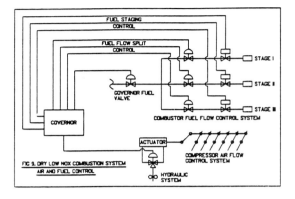

Fig 9 Dry low NO$_x$ combustion system air and fuel control

The G60 Combustor makes a number of demands of the control system. It requires:-

1)   Three independently controlled fuel feeds.

2)   Modification of the light-up/crank fuel schedule.

3)   Modulation of the compressor air flow to enhance the low NO$_x$ performance range.

These demands are further complicated by the operational requirements of the combustor and engine i.e.:-

i)   The variation of fuel split between fuel staging points.

ii)   The 'load shed' requirement means that response times of the fuel control valves and their consequent ability to re-distribute the fuel from module to pilot stage becomes a factor.

iii)   Change-up/change-down hysteresis must be designed into the control system logic to prevent the engine 'cycling' if the stage change over point becomes a running point, rather than a transient.

The option has been retained, however, to override this for development purposes by 'Soft Key' control.

Engine trials are planned at Lincoln having a number of objectives relating to fuel and air control systems:-

1)   proving of the fuel system.

2)   identification of interaction effects between engine, combustor and fuel system.

3)   demonstrating critical areas of fuel system design to aid simplification of production engine fuel control.

4)   demonstrating the effectiveness of compressor air flow control in enhancing the low NO$_x$ performance range over each operating stage.

9   CONCLUSION

The general conclusion of the rig programme was that the low NO$_x$ combustor met all the original specified targets. Measured metal temperatures, structural vibrational testing and stress analysis of the development combustor showed it to be acceptable to offer for trial on a Tornado development engine at the EGT site in Lincoln.

ACKNOWLEDGEMENTS

The authors wish to thank European Gas Turbines Plc and Aero & Industrial Technology Ltd., for permission to publish the paper and the dedicated staffs participating in the work both at EGT, Lincoln and AIT Ltd, in Burnley, in particular Dr. A. F. Al-Shaikly for his contribution on fuel scheduling.

REFERENCES

(1)   BUNN, G. E. and WINTER, J. The Evolution of a Low Emission Combustion System for Aircraft Engine Application. Oral presentation at the 24th ASME Gas Turbines Conference, San Diego, California USA. March 1979.

(2)   WINTER, J. The performance of a Low NO$_x$ Combustor on a Range of Fuels. Paper T3.3 15th Cimec Conference, Paris. June 1983.

(3)   ANTOS, R. J., MUMFORD, S. E. and WINTER, J. A Dual Fuel Dry Low NO$_x$ Combustion System for Industrial Combustion Turbines. American Power Conference 53rd Annual Meeting, Chicago, Illinois April 29 - May 1 1991.

(4)   STRINGER, F. W. A Lean Burn Approach to Dry Low NO$_x$ Combustor Design. Paper presented at 1991 Yokohama International Gas Turbines Congress held in Yokohama, Japan October 7 - November 1 1991.

C448/065

New developments in catalytic combustion for industrial gas turbines

S BLUMRICH, Dipl-Ing and B ENGLER, Dr-Ing
Degussa AG, Hanau, Germany

SYNOPSIS Existing combustion systems in industrial gas turbines produce unacceptable levels of NO_x due to the high temperatures necessary for the complete combustion of the fuels. Legislation exists in many countries and is being considered in many others to limit NO_x emissions to very low values, e.g. below 10 ppm. In some gas turbine systems these limits cannot be reached with existing combustion techniques. The installation of secondary flue-gas cleaning systems, such as SCR systems using ammonia injection for example, is considered not to be very economical. Therefore it is necessary to find new combustion systems which generate ultra low NO_x emissions. A possible solution could be the development of a catalytic combustion system which converts hydrocarbons into carbon dioxide (CO_2) and water (H_2O) at temperatures far below the temperature at which thermal NO_x is formed. These systems are under development now.

CONTENTS

1 Introduction

2 Actual emission situation

3 Existing technologies for reducing pollutant emissions

4 Catalytic combustion systems

5 Conclusion

1   INTRODUCTION

In the early 1970s exhaust emission control legislation to alleviate the effects of the environmental pollution caused by combustion exhaust products was discussed for the first time. Typical pollutants from coal combustion, for example, are dust, sulphur dioxide and nitrogen oxides. The two latter pollutants are held responsable for the incidence of acid rain. Combustion of hydrocarbon containing fuels such as petrol or natural gas produces nitrogen oxides, unburnt hydrocarbons and carbon monoxide.

Initially, coal fired power plants were equipped with dust removal units, which were followed by desulphurisation and some years later by the DENOX abatement units. Some countries, such as the U.S.A. and the EEC, have made compulsary the use of catalytic converters in automobiles. Nowadays stationary combustion units, like industrial gas turbines or gas engines in co-generation plants, are also subject to emissons restrictions.

2   ACTUAL EMISSION SITUATION

Today, worldwide environmental awareness has led to more and more stringent legislation restricting the discharge of pollutants. Further reduction of pollutant emissions is permanently under review in many countries. The most stringent values known are those of California, where, for example, the NO_x limit for gas turbines is set to 9 volume parts per million (vppm). European emission limits vary from 220 ppm to 60 ppm (15% O_2, dry) for natural gas fired engines and from 293 to 80 ppm (15% O_2, dry) for industrial gas turbines fired with liquid fuels.

The CO emission limit in most countries is 100 mg/mN (15% O_2, dry), and in one country only 243.75 mg/mN^3 (15% O_2, dry) is allowed.

Calculations have shown that worldwide 1.37 x 10^6 t/year of NOx are emitted by gas turbines, 30 % of this (0.41 x 10^6 t/year) in the countries of the European Community. In table 1 some interesting values of NO_x emissions are listed.

3   EXISTING TECHNOLOGIES FOR REDUCING POLLUTANT EMISSIONS

The main pollutants from combustion of fuels containing hydrocarbons are nitrogen oxides (NO_x), carbon monoxide (CO), unburnt hydrocarbons (UHC) and, if sulphur containing fuels (like oil) are burned, sulphur dioxide (SO_2). NO_x is mainly formed due to high temperatures (thermal NO_x), while CO and UHC are formed by an incomplete combustion. Thermal NO_x emissions can be suppressed by reducing the flame temperature, but this, in most cases, increases the formation of CO and UHCs. Furthermore, there are limits beyond which the flame temperature cannot be reduced within the limits of stable combustion.

Conventional approaches to combustion in use or being developed for the reduction of NO_x emissions of stationary/industrial gas turbines are:

3.1   Methods which modify the conditions of combustion

(a) Steam or water injection

This method, in spite of some disadvantages, became very popular. The addition of water or steam in the combustion zone lowers the flame and gas temperature and therefore suppresses the formation of NO_x. Published sources state that, by these means, up to 70 % of NO_x suppression can be achieved [5].

The disadvantages of this method are as follows:

- Requirement for large quantities of high quality water and therefore additional capital and operating costs.
- Increase of CO and UHC emissions.
- More complex engine control and monitoring.

- Shortening of the life of major engine components.
- Increase of fuel consumption by up to 5 %, especially at low loads and therefore an increase of CO_2 emissions (only for water injection).

(b) Premix, Lean Burn Combustion

This method is based on the same theory as the above injection techniques - reducing the flame temperatures. For a stable com-combustion, this method requires fuel/air mixtures to be tightly regulated in each combustion stage, which adds complexity to the engine control and monitoring. NO_x levels of 25 vppm are predicted, but the possibility of remarkable increases of CO and UHCs concentrations in the exhaust gases cannot be excluded.

(c) Emulsified fuel oil

Some manufacturers of heavy industrial gas turbines are testing liquid fuels emulsified with water. It seems, that under special conditions significant suppression of NO_x can be achieved. At this moment nothing is known about any increase of CO and UHCs.

3.2   Exhaust gas clean-up techniques

(a) Selective Catalytic Reduction (SCR) for NO_x reduction

This technique is often used for the purification of exhaust gases from all types of combustion processes which contain an excess of oxygen (power plants, dualfuel engines, gas turbines etc.). The good performance of this system is well proven on base load systems. But it has the disadvantage of requiring a supply of reduction means (ammonia/urea), necessitating storage, dosage, monitoring and security equipment. This incurs heavy capital and operating costs.

(b) Catalytic exhaust-gas cleaning with oxidation catalysts

This method only reduces the concentrations of CO and UHC by oxidizing them to CO_2 and H_2O. Due to the excess of oxygen in the exhaust a reduction of NO_x is not possible. This system has a few disadvantages, such as increase in pressure drop and investment costs for the catalyst.

4 CATALYTIC COMBUSTION SYSTEMS

As described at the beginning of section 3, the control of the temperature during combustion is the most important factor in suppressing NO_x formation. The main rule is: the lower the temperature, the lower the amount of formed NO_x. On the other hand, the lower the temperature, the higher the concentration of the products of incomplete combustion, such as CO and UHCs. The consequence is a decrease in efficiency, resulting in an increase of CO_2 emissions. CO_2 is not directly regarded as a pollutant, but its influence on the greenhouse effect necessitates the best possible reduction.

Because of the above discussed reasons a future combustion system needs to achieve "cold and complete" conversion of the fuel.

Such a system which fulfills all these requirements is catalytic combustion.

The heart of a catalytic combustor is the catalyst. A catalyst is a material which increases the rate of a chemical reaction and is not consumed in the process.

From the chemical point of view, combustion is the oxidation of the hydrocarbons in the fuel. The aim of combustion is to produce hot gases. Combustion efficiency depends only on the completeness of the oxidation process. Therefore, it makes no difference whether a fuel is combusted in an open flame at temperatures between 1500 °C and 2500 °C or in a catalyst below 1500 °C. The difference is the NO_x production: at temperatures higher than 1500 °C thermal NOx is formed in significant amounts, whereas below this temperature only fuel-bonded nitrogen may oxidize into NO_x, which normally is a very small amount.

Many studies [1],[2],[3] have shown that catalytic combustion of gaseous fuels and vaporized liquid fuels is in principle possible. But it was also shown that in most cases very special conditions are necessary to have a complete combustion. In the following overview some important aspects of catalytic combustion will be discussed:

4.1 Fuels

All types of hydrocarbon containing gases, such as methane, propane, methanol etc. as well as vaporized liquid fuels, can be used as fuels in a catalytic combustor.

The chemical compounds of the fuel are of great influence on the "ignition" temperature. The hydrocarbons can be divided into long-chained hydrocarbons or short-chained hydrocarbons. The shortest "chain" is the very stable methane molecule, which is very difficult to oxidize (i.e. to combust). The sequence for an easier combustion is: methane, ethane, propane, butane etc. The general rule is: the longer the chain, the easier the oxidation, and consequently the lower the "ignition" temperature.

4.2 Catalyst systems

Catalyst systems are normally made of two components:

- the catalyst (active material) and
- the support system (substrate and intermediate layer, the so-called "washcoat").

Both components greatly influence the combustion efficiency. The requirements for a succesful catalyst system are:

- Large geometric surface area in combination with a low pressure drop.
- High surface of the "washcoat".
- Excellent catalyst distribution on the "washcoat".
- Low "ignition" temperatures.
- High catalyst activity to maintain complete combustion at the lowest levels of air preheat and the highest value of mass throughput.
- Excellent thermal shock resistance.
- High durability of "washcoat" and catalyst.

4.2.1 Substrates

Due to the necessity of large geometric surface area combined with low pressure drop the honeycomb structure is the most suitable substrate for these applications. Honeycombs and monoliths can be made of ceramic (cordierite, mullite etc.) or thin metal foils. The major differences between the two materials are:

- heat capacity
- heat conductivity
- wall thickness

The coating technologies for these materials are different but lead to the same result.

4.2.2   Intermediate Layers (Washcoats)

The intermediate layer has different functions: It offers an extremely high surface for the fine distribution of the active materials, it has storing functions (for example: oxygen), it influences the selectivity of the catalyst system and it stabilises the active materials. As intermediate layers the following are in use: alumina oxide (Al_2O_3), titania oxide (TiO_2), silica oxide (SiO_2) etc. The addition of secondary materials into the washcoat, such as rare-earth components, ensure for example the ability to store oxygen.

4.2.3   Catalyst (Active Material)

Many materials can be used as active materials to accelerate chemical reactions. In the field of catalytic combustion, precious metal and metal oxide catalysts are successfully used.

4.2.3.1 Precious metal catalysts

A considerable amount of literature and data is available regarding the use of platinum and palladium as oxidation catalysts, and it has been found that these materials are among the most active catalysts [4]. Furthermore, for catalytic combustion applications the use of precious metals other than Pt and Pd, such as Ru, Rh, Os and Ir, appears to be limited due to their high volatility and ease of oxidation [1].

The reason for the high activity of Pd and Pt is their ability to activate H_2, O_2, C-H and O-H compounds. For the oxidation of carbon monoxide, olefins and methane Pd is more active than Pt. To oxidize paraffinic hydrocarbons, C_3 and greater, Pt is more active than Pd. For the oxidation of aromatic compounds both precious metals work similarly.

Platinum and palladium can be impregnated in a highly dispersed form on several washcoats. Due to the high specific activity only small amounts of them are necessary for good catalytic activity.

At temperatures higher than 700 °C the sintering effect of the precious metals has an influence on catalytic activity. The result of the sintering effect is a reduction of the precious metals surface area which leads to a reduction of the catalyst activity. In addition to this it is known, that precious metals may be oxidized under these conditions, which also reduces the activity.

The aims of today's development programmes are to stabilize the catalyst on the intermediate layer and to avoid the above mentioned sintering and oxidizing processes.

4.2.3.2 Base metal oxide catalysts

For the oxidation of carbon monoxide and hydrocarbons also metal oxides, such as Cr_2O_3, Co_3O_4, NiO, CuO and La_2O_3 can be used. Mixtures of these materials have greater stability and activity than single oxides. Compared with precious metal catalysts, they have a lower activity and higher light-off temperatures.

4.3   "Ignition" of the catalytic combustion

Depending on the fuel and the catalyst used, the minimum temperature for igniting the reaction may be as high as 500 °C. Most of the systems described in literature use additional heating systems, e. g. preburners, which are installed upstream of the catalyst.

4.3.1   Preburner systems

A very simple way of heating up the gases in front of the catalyst is the installation of a preburner. This precombustor, which may be a homogeneous diffusion burner, is fed with part of total fuel and air mass flow. The hot exhaust gases are mixed with the main air and fuel mass flow. By varying these two flows the temperature in front of the catalyst can be held easily in a narrow range. The disadvantage of this system is an open flame with hot zones, in which thermal NO_x is formed.

4.3.2 Pre-catalyst systems

A more elegant system is a catalytic precombustor which is fed with a fuel that oxidizes very easily at low temperatures, such as propane, methanol or, under extreme conditions, hydrogen. These "support fuels" deliver the necessary heat so that no additional burner is needed. It is suggested that the injection of the supporting fuels operates only in the start phase and at low load.

4.4 Test results of laboratory catalytic combustion

A series of catalysts for catalytic combustion have been prepared in recent months. They have been tested in laboratory test equipment under realistic conditions. The results of two catalysts will be presented. Both catalysts are coated on ceramic (cordierite) monoliths with a cell density of 300 cpsi. The geometrical data of the samples are listed in table 2.

Sample A is a medium loaded palladium catalyst, sample B is a high loaded one. The intermediate layer is based on Al_2O_3.

4.4.1 Laboratory test conditions

The samples were tested in laboratory test installations with synthethic gas mixtures:

Component	Gas composition [Vol.-%]	
	rich combustion	lean combustion
Methane (CH_4)	3	3
Oxygen (O_2)	5.88	10
Nitrogen (N_2)	balance	balance

Gas flow: 2.0 m^3/h
Temperature: variable, by electric heating of the gas upstream of the reactor.

4.4.2 Analyses

Methane concentrations were measured as total hydrocarbons with a flame ionisation detector, oxygen concentrations with the paramagnetic sensor, carbon monoxide and carbon dioxide concentrations with infrared and nitrogen oxides with chemoluminescence. The methane conversion rate was not only determined by the different methane concentrations upstream and downstream of the catalyst, it also had to be confirmed by the corresponding carbon dioxide and oxygen concentrations in the "exhaust" gas mixture.

4.4.3 Results

The methane conversion rates of sample A are shown in the Fig. 1 for both rich and lean conditions. The influence of the residence time (space velocity) on the methane conversion can be clearly seen.

Fig. 2 shows the results from the testing of sample B in the same way. The higher Pd-loading results in better conversion rates. In Fig. 3 and Fig. 4 conversion rates and measured temperatures up- and downstream of the catalyst can be seen.

Fig. 5 shows the influence of the velocities. The conspicuous bend in the lines at a space velocity of 500,000 1/h could not yet be explained and needs further investigation. Fig. 6 shows the temperature dependence of the methane conversion of sample B under lean conditions with a space velocity of one million. From the fact that no significant increase of conversion with an increasing temperature occurs, it could be concluded, that the reaction, under these conditions, is controlled due to mass transfer limitations.

In all experiments the methane conversion rates under rich conditions were significantly higher than those under lean conditions (excess of oxygen). This could be explained by the partial "poisoning" of the catalytic materials with oxygen. Comparison samples with low precious metals loadings were also tested, but lost their activity after only a few test hours under lean conditions due to the above mentioned effect. Further influence factors are the amount and the composition of the washcoat as well as a proper selection of the raw materials.

The measured concentrations of nitrogen oxides were below 1 ppm, which is equivalent to the detection limit. Under lean conditions no carbon monoxide could be measured.

5. CONCLUSION

In order to achieve reduction of exhaust pollutant emissions from combustion processes the so called "cold and complete" conversion of hydrocarbons into carbon dioxide and water has to be applied. Catalytic combustor systems offer the possibility of a selective combustion which avoids the formation of undesired combustion by-products such as nitrogen oxides.

Furthermore, with modern catalyst technologies a combustion efficiency of > 99,5 % can be achieved. For industrial gas turbines, replacement of conventional combustion chambers by catalytic combustors seems to be possible in the future.

Literature:

[1] J.P.Kesselring, W.V.Krill, H.L.Atkins, R.M.Kendall, J.T.Kelly: Design Criteria for Stationary Source Catalytic Combustion Systems, EPA-Report No. 600/7-79-181, 1979

[2] H. Fukuzawa, Y. Ishihara, Y. Hasegawa: Catalytic Combustion for Gas Turbines, Presentation on the 5th workshop on Catalytic Combustion, San Antonio, Texas, 15.-16.9.1981

[3] Y. Ozawa, J. Hirano, M. Sato, M. Saiga, S. Watanabe, M. Okahata: Preliminary Test Results of Catalytic Combustor for Gas Turbine, Presentation on JECAT '91, December 2-4, 1991, Tokyo, Japan

[4] A. Schwartz, L.L. Holbrook, H. Wise: Catalytic Oxidation Studies with Platinum and Palladium, J. Catal., 21, 199 (1971).

[5] CITEPA-Report Nr. B 661-90-007872: The limitation of pollutant emissions into the air from gas turbines (second interim report), 01.07.1991

[6] SRI International: Report No. 200: NOx-Removal, May 1989

[7] OECD Environmental Data 1991

	USA	European Community	Japan	Worldwide
Total	19.8	9	3.6	n.a.
Stationary Sources	11.7	4.5	1.2	n.a.
Industry	4.5	1	n.a.	n.a.
Mobile Sources	8.1	6.4	0.5	n.a.
Passenger Cars	n.a.	n.a.	n.a.	14
Gas Turbines	n.a.	0.41	n.a.	1.37

n.a. = not available

Table 1: NO_x- emissions in 10^6 t/year   Sources: [6,7]

Sample	Space velocity [1/h]	Length [mm]	Width [mm]	Volume [litre]	Nr. of channels
A	100,000	50	22	0.024	196 (14x14)
A	200,000	25	22	0.012	196 (14x14)
A	400,000	12.5	22	0.006	196 (14x14)
B	200,000	42.5	15.3	0.010	100 (10x10)
B	250,000	35	15.3	0.008	100 (10x10)
B	500,000	17	15.3	0.004	100 (10x10)
B	1,000,000	17	10.7	0.002	49 (7x7)
B	2,000,000	17	7.65	0.001	25 (5x5)

Table 2: Geometrical Data of Catalyst Samples

Catalytic Combustion
Long Time Behaviour (Sample A)

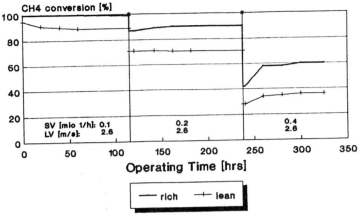

Fig. 1 [CC53]

Catalytic Combustion
Long Time Behaviour (Sample B)

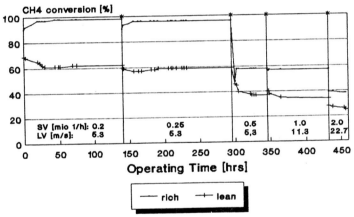

Fig. 2 [CC50]

Catalytic Combustion
Long Time Behaviour (Sample B)

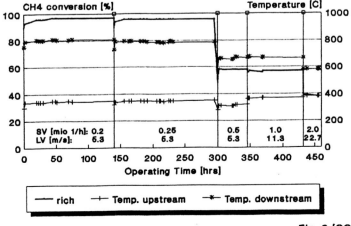

Fig. 3 [CC51]

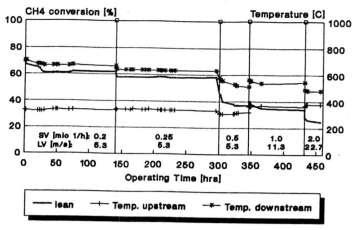

Catalytic Combustion
Long Time Behaviour (Sample B)

Fig. 4 [CC52]

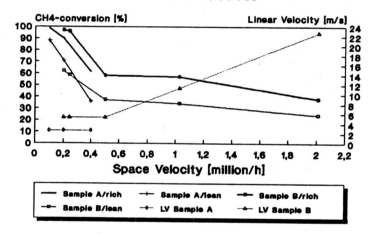

Catalytic Combustion
Influence of Velocities

Fig. 5 [CC54]

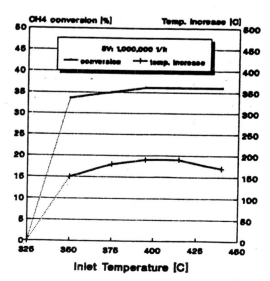

Catalytic Combustion
Conversion versus temperature
(Sample B)

234

Fig. 6 [CC55]

C448/018

Warm-up characteristics of a catalyst heated by exhaust gas ignition

N COLLINGS, BSc, PhD, CEng, MIMechE, MSAE
Department of Engineering, University of Cambridge
T MA, BSc, PhD, CEng, MIMechE
Ford Motor Company Limited, Basildon, Essex
T HANDS
Cambustion Limited, Cambridge

ABSTRACT In a recent paper, it was shown how extremely rapid light-off of automotive catalysts was possible by igniting a mixture of exhaust gases and air just in front of the catalyst. In this way, working catalyst temperatures are reached in a few seconds, rather than minutes. The present paper discusses how the catalyst behaves following this brief light-up phase. In particular, it is shown how the high temperature 'slice', formed on the front section of the heated catalyst during EGI, moves rearwards. Unless special measures are taken, the moving hot section reduces in temperature to the point of inactivity, before the exhaust system between the engine and catalyst has become hot enough to cause 'normal' light-off conditions.

1 INTRODUCTION

A great deal of effort is presently being made to create catalyst systems which reach working temperature much more quickly than current production systems. Some estimates put the contribution to the total emissions output during the warm up period (which is typically 2 minutes of a 20 minute drive cycle) at about 80% of the total (1). Various schemes are under study to reduce the warm up period. Placing a catalyst very close to the exhaust port is very effective since, even at a cold start, the cylinder exit temperatures are about 500°C - the problem is that such a catalyst may easily overheat at other operating conditions. (This problem may be overcome by using by-pass systems, but these are necessarily rather complex.)

As the engine out exhaust gas is hot, even at a cold start, reducing the heat transfer to the exhaust pipe work is another option - for example making the exhaust pipe with a double skin construction, in which the inner skin is of very low thermal inertia. Once again, there are overheat problems, and light-off is still not as fast as would be wished.

Heat storage systems are another method for rapid warm-up, in which a super-insulated thermal storage battery holds heat stored from the previous engine operation, and then transfers this heat into the engine at the next cold start. This has significant advantages other than in rapid light-off of the catalyst, such as reduced cold start enrichment requirements (and consequently reduced emissions), reduced friction and rapid passenger compartment warm-up, but the improvements in light-off time are smaller than is desirable.

Electrically heated catalysts offer a route to very significant reduction in light-off times, by pre-heating the catalyst prior to the start, together with additional heating immediately after the start (2). This technology is still in its development stage, and there are some open issues. These particularly relate to the very high electrical powers required, equivalent to 300-500 amps for a 12 volt system. This requires a special battery, and associated control, charging and wiring systems. The need for a pre-heating period does not fit well with the established mode in which vehicles are used, and the post start period is problematic off idle, as the electrical power required to maintain catalyst temperature rapidly becomes unachievable.

The present paper relates to a competitive technology to those above, which while at a very preliminary stage of development, offers the possibility of using the catalyst in a mode which is well established, while offering the shortest time to light-off of any of the systems discussed above.

The Exhaust Gas Ignition (EGI) system (3) is extremely simple in concept - fig 1. At start, the engine is calibrated to produce rich exhaust products (the combustible components consisting mainly of carbon monoxide, hydrogen and unburnt hydrocarbons). These are mixed with air from a pump in the exhaust, and the mixture is ignited just in front of the catalyst. Operation of EGI is for a period of about 5 seconds, after which time the front face, or more accurately a front slice of the catalyst is hot enough to be reactive, and in a sense, the catalyst has lit-off.

The normal light-off sequence, is 'safe' as there is no danger of the catalyst re-extinguishing (in the short term). This is definitely a danger following EGI - the exhaust system is still cold, and the gases at catalyst inlet can, and do, rapidly cool the catalyst, unless measures are taken to prevent it. Catalyst Temperature Management (CTM) is the topic of the present paper.

2 CATALYST TEMPERATURE MANAGEMENT

Fig. 2 shows the changing temperature profiles at stations along the catalyst brick following a 4 second EGI flame, during a cold start of a conventional 4 cylinder 1.8l engine. Immediately following EGI, the secondary air flow was cut off, and the normal calibrated fueling trajectory was operated. This figure therefore relates to the situation when no special measures are taken to keep it warm following EGI. In this situation there will be virtually no exotherm in the catalyst as there is virtually no oxygen available. The resulting behaviour is perhaps predictable; the region of peak temperature moves rearwards, and with reducing temperature, until at some time, short with respect to the warm up time of the exhaust system, the peak temperature reduces to the point when reaction (effectively) ceases.

Although a detailed model of the transient behaviour has been developed, which will be published elsewhere, some important features of the observed behaviour can be explained at a very simple level. The most important issues are a) when will the temperature wave exit the back of the catalyst, and b) how long will the catalyst stay reactive i.e. how quickly does the peak temperature drop?

The heat transfer between the gas and the catalyst is very high (as it must be by analogy with the mass transfer mechanism), and consequently the gas and brick temperature do not differ significantly. On this basis, the velocity with which the region of peak temperature moves rearwards can be inferred by making an energy balance. In time δt, let the temperature wave (assumed to maintain a top hat profile) move rearwards by a distance δx.

Then

$$\delta t.\dot{m}.c_{pg}.delT = \delta x.\sigma.c_{pc}.delT$$

where c_{pg}, c_{pc} are the gas and catalyst specific heats, $\dot{m}$ is the gas mass flow, σ is the mass of the catalyst per unit length, and delT is the height of the top hat. The wave velocity is thus $\delta x/\delta t$.

For the (idle) conditions of fig 2, the mass flow of engine exhaust was about 0.006 kg/s, and the gas specific heat is close to 1.0 kJ/kgK; the catalyst specific heat was about 0.8kJ/kgK, and σ was about 8.75 kg/m. This gives a velocity for the temperature wave of about 1mm/s - close to the observed value.

It is more difficult to model the drop in peak temperature by such simple ideas. It would be valid to assume that the total internal energy of the catalyst, post-EGI, remained constant up to the point that either a significant temperature rise occurred at the exit, or the inlet temperature rose significantly. The broadening of the temperature profile could be modeled using the ideas developed by Taylor for dispersion in capillaries (4). A major difficulty in applying the concepts directly is that there is an axial temperature profile in both the catalyst and the gas, (which causes the asymmetry in the temperature profile) but the basic result, that the initially sharp profile should broaden in proportion to the square root of distance downstream should still hold, since the width of the catalyst and gas temperature profiles will always be similar. Combined with the energy conservation idea, this suggests that the peak temperature should also fall as the root of the distance from the inlet.

The finite response time of the thermocouples (c. 3 seconds) makes it very difficult to assess the validity of this concept. In any event, as catalysts are typically of the order of 100mm long, the temperature wave will exit long before the catalyst is heated to reaction temperatures by the rising exhaust gas temperature. (In fact for the (race track) catalyst that we have used (length 60mm), it appears that the temperature wave reaches the rear of the can at roughly the time that peak temperature has reduced to one half of its initial value. For a typical EGI start the catalyst extinguishes at almost exactly the same time that the wave reaches the exit.)

What can be done to avoid these problems? An obvious way of addressing the problem of the reducing maximum temperature is to continue with some (moderate) engine enrichment and air pump flow after the EGI period. In this way it can be arranged for an exothermic reaction to continue in the catalyst while the exhaust system is warming up. It turns out that this procedure also slows down the rate at which the temperature wave moves rearwards, as the exotherm takes place on the leading edge of the temperature profile. This effect is dominant, and with an appropriate enrichment, the temperature profile at the front of the catalyst becomes stable, while the trailing edge of the temperature profile moves continuously towards the rear of the catalyst. Fig 3 shows this behaviour in practice. (Note that in this figure, the thermocouple positions are the same as in fig 2. Also thermocouple 3 has been replaced by a smaller diameter probe, with an improved time constant.)

It appears that heat can actually conduct/radiate in the upstream direction, judging by the behaviour of thermocouple 4. The temperature at position 4 drops when the throttle is opened, and recovers at idle. Note also that the front face of the catalyst becomes hot at idle, as thermocouple 3 seems to be radiatively, (or possibly convectively if there is any reverse flow) heated when 4's temperature rises.

Using the simple ideas discussed above, we can write for this 'steady state' condition, and for a constant inlet temperature to the catalyst,

$$\delta t.\dot{m}.cv = \delta x.\sigma.c_{pc}.delT = \delta t.\dot{m}.c_{pg}.delt$$

where cv is the calorific value of the gases. $\delta x/\delta t$ is now the velocity with which the trailing edge of the temperature profile travels towards the back of the catalyst. In other words, the rate of heat release is equal to the rate of increase of the catalyst's internal energy (up to the point that the exit gas temperature rises significantly).

In reality the catalyst inlet temperature rises during the warm up period, and so the required calorific value reduces (i.e. the engine is run with ever reducing enrichment).

The fact that an air pump is already necessary for EGI itself makes CTM very attractive, as only a calibration issue is then involved. In addition, it is worth emphasising that this technique can maintain an exotherm in the catalyst off idle, up to a limit imposed by the available air flow. The effect on fuel economy has to be addressed, but it will certainly be small, and clearly a lot less than going via an electrical route.

In using this technique, it is important to neither over-heat the catalyst, nor to allow it to drop below light-off temperature. In essence then, we need to know the target catalyst temperature, the gas inlet temperature to the catalyst, and the gas mass flow at any instant. With this information the exotherm required in the catalyst may be deduced. The relationship between the engine AFR and the exhaust gas heating value is straightforward, though prediction of the actual composition is a little less easy. The composition information is useful for making estimates of the flammability limits.

3 EXHAUST GAS HEATING VALUE

The calorific value of rich exhaust gases depends to some extent on the detailed chemistry, since frozen reactions, for example the water shift reaction, effect the quantities of the combustible components. Following conventional wisdom about exhaust gas composition (as used in the Spint AFR calculation method for example), the calorific value (CV) is given approximately by

$$CV = constant.(AFR_s - AFR_e)/(AFR_e + 1)$$

(Note that this is the calorific value of the exhaust gases - 'cv' in the relationships developed previously referred to the calorific value per unit mass of the exhaust plus secondary air.)

Here AFR is the air fuel ratio, and the subscripts s and e refer to stoichiometric and engine respectively. The constant depends on the fuel C/H ratio - for normal unleaded pump fuel it is equal to about 4000 kJ/kg. (One reason this is approximate is that the value of the constant changes slightly with air fuel ratio.)

Various approaches can be taken to the prediction of the catalyst inlet temperature. Lumped heat capacity models are popular and appropriate in view of the uncertainties in the values of the inner and outer heat transfer coefficients. Another source of significant error is the input condition, i.e. the gas temperature at exit from the engine cylinders - this is a function of the mass flow rate, and engine speed, even at fixed air fuel ratio and coolant temperature. We have used a simple 4 element lumped heat capacity model. The details of the model are not relevant to the present paper, and are not fundamentally different from the previous recently published work in this area e.g. (5).

With the model, it is possible to estimate, during the warm-up, how much exotherm is required to keep the catalyst hot following EGI. Clearly the required temperature rise due to reaction in the catalyst, plus the inlet temperature of the gas into the catalyst inlet should be approximately equal to the desired catalyst brick temperature, say about 500°C.

Fig 4 shows a typical record, illustrating the actual and predicted catalyst inlet temperatures, and the catalyst temperature at mid-brick, change during the warm up period. This is a particularly severe test, as it consists of periods of idle and wide open throttle (at 2500 rev/min). In this test, the air flow was constant, and the air fuel ratio of the engine was adjusted to generate the required heat release in the catalyst.

4 DISCUSSION AND CONCLUSIONS

The basic technique described for catalyst temperature management appears to be appropriate, but the actual choice of hardware, and the finesse with which the overall air fuel ratio can be controlled during the warm up period are still open issues. Until detailed calibration work is performed, it is difficult to decide whether it is worth trying to run the system as a three-way or oxidation catalyst during CTM. As the engine is running slightly rich during CTM, the feed gas NO_x levels are lower than normal anyway, so it is possible that it will prove safer (from the overall emissions point of view) and cheaper to stay slightly in the lean region overall. It is also not entirely clear what level of sophistication will be required for the control of the pump air flow.

The maintenance of catalyst temperature in the period between exhaust gas ignition and normal light-off conditions is clearly of great importance. This paper has shown how the catalyst temperature may be controlled during this period, by a controlled exotherm originating from a slightly rich calibration and added air.

5 REFERENCES

1. Gottberg,I., Rydquist,J.E., Backlund,O., Wallman,S., Maus,W., Bruck,R., Swars,H., "New Potential Exhaust Gas After-treatment Technologies for 'Clean Car' Legislation", SAE paper 910840 (1991)

2. Kubsh,J.E., Lissiuk,P.W., "Vehicle Emission Performance with an Electrically Heated Converter System", SAE paper 912385 (1991)

3. Ma, T., Collings, N., Hands, T., "Exhaust Gas Ignition (EGI), a Novel Technique for Rapid Light-off of Automotive Catalysts", SAE paper 920400, (1992)

4. Taylor,G.I., "Dispersion of Soluble Matter in Solvent Flowing Slowly Through a Tube", Proc. Roy. Soc., 219, pp 186-203, (1953)

5. Zhang, Y., Phaneuf, K., Hanson, R., Showalter, N., "Computer Modeling on Exhaust System Heat Transfer", SAE paper 920262 (1992).

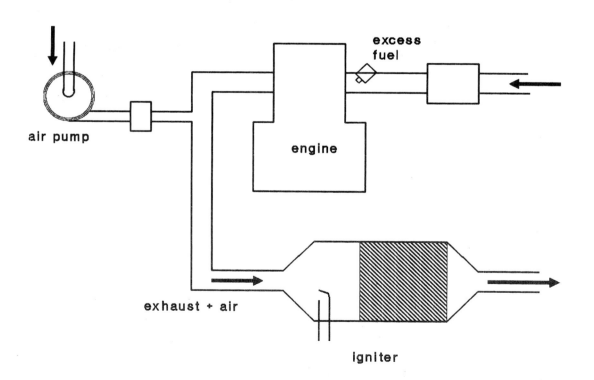

Fig. 1. General experimental arrangement.

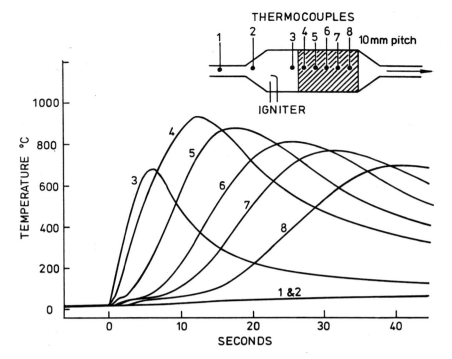

Fig. 2. Temperatures along the catalyst during EGI, but without CTM. Idle throughout.

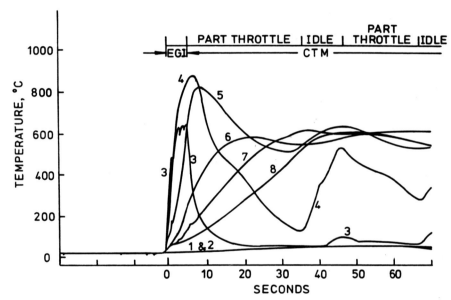

Fig. 3. Temperatures along the catalyst during EGI and CTM. Idle and 2000 rev/min part throttle. Thermocouple postions as Fig. 2.

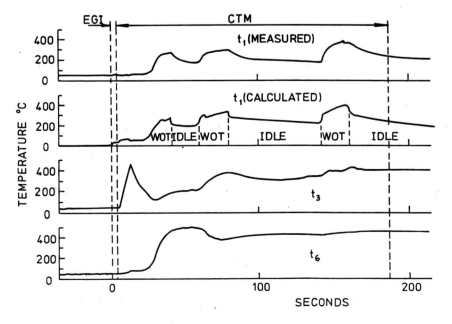

Fig. 4. Measured and predicted temperatures at various positions (as fig. 2). Idle and 2000 rev/min wide open throttle.

Effect of certain catalysts in the combustion chamber of a two-stroke SI engine on exhaust emissions

P RAMESH BABU, B NAGALINGAM and K V GOPALAKRISHNAN
Indian Institute of Technology, Madras, India

ABSTRACT

The concept of catalytic combustion where the oxidation of the fuel is brought about with the aid of a catalyst offers the advantages of higher brake thermal efficiency and lower exhaust emissions. This paper discusses the performance, combustion and emission characteristics of a two-stroke spark ignition engine in which the combustion occurs in a catalytically activated combustion chamber. Catalysts such as Copper, Nickel and Chromium in the form of coatings on the piston top and cylinder head have been investigated. Among the various catalysts used, copper shows the maximum percentage improvement in the brake thermal efficiency, about 10.8 percent. Hence the effects of lean fuel-air mixtures and higher compression ratios have been studied for the copper coated combustion chamber. Results indicate that copper as catalyst is very effective at higher compression ratios with lean fuel-air mixtures. The knock limit is extended and a significant reduction in HC and CO emissions is obtained with copper catalyst when using lean fuel-air mixtures. A hemispherical catalytic prechamber has been developed and tested with platinum metal deposited on its surface. There is a marked difference in the exhaust emissions in the presence of platinum catalyst.

1 INTRODUCTION

The conventional two-stroke spark ignition (SI) engine is a favoured prime mover for light vehicles and portable equipment applications due to its high specific power output, compactness, low production and maintenance costs. However, this type of engine has two serious drawbacks : (i) high specific fuel consumption and (ii) high carbon monoxide and unburned hydrocarbon emissions resulting from loss of fresh charge during scavenging and large exhaust gas dilution. The high amount of residual gases inside the cylinder reduces the speed of flame propagation leading to poor combustion [1]. In India, two-stroke SI engines play an important role in the transportation sector, consuming about 60 percent of the gasoline. Improvement in the combustion of two-stroke SI engines is therefore extremely desirable to obtain better fuel economy and lower emissions.

The present work aims to improve the combustion of two-stroke SI engines through the presence of catalysts in the combustion chamber and there by to obtain lower exhaust emissions and lower fuel consumption.

2 CATALYTIC COMBUSTION

The phenomenon of heterogeneous oxidation of hydrocarbons at a catalyst surface is well known and widely reported [2,3] in a variety of combustion systems. Catalytic surface reaction can be hypothesized as partial or total , liberation of reactive intermediate species as well as heat. Both these can lead to activation of the adjacent combustible mixture. As a result of catalytic prereaction, the required ignition energy is reduced and the flame velocity is increased.

Thring [4] suspended a platinum coated wire mesh in both direct and indirect injection diesel engines, which improved the performance over a broad range of operating conditions and permitted lower quality fuels to be utilized. Pfefferle[5] investigated catalytic combustors for gas turbines. The advantages of high volumetric heat release rates and exceptional combustion stability were demonstrated. Karim and Kibrya [6] investigated combustion of a homogeneous lean methane-air stream in the presence of different catalytic materials. The sequence of these materials in terms of the improvement of lean limit of combustion is in the following order: Platinum-Copper-Brass - Chromium - Nickel-Cadmium-Stainless steel. Rychter[7] proposed a catalytic prechamber, as a means of supporting lean combustion in an SI engine. Detailed engine tests revealed that this concept extends the lean burn limit, minimizes cycle-to-cycle variations and increases brake thermal efficiency. Catalytic combustion study deserves serious consideration because it does not need any modifications which complicate fuel delivery, ignition or exhaust systems.

3 PRESENT WORK

The present work deals with an experimental study involving the coating of various catalysts such as Copper, Chromium and Nickel on the combustion chamber wall and determining its effects on the engine performance, combustion and emission characteristics. The effect of lean fuel-air mixtures and higher compression ratio were studied with the best catalyst of these, namely copper. These coatings were applied using electroplating techniques. Copper was coated using cyanide copper bath

[8], since it produces a coating of porous nature and a fine deposition of metal on the surface.

Platinum metal could not be deposited over the piston top and cylinder head unlike the coatings listed above, because of its high cost and absence of a suitable method to deposit on the aluminium alloy materials (piston and cylinder head). Hence, a hemispherical prechamber made of stainless steel was fabricated and platinum metal was deposited on its surface. The spark plug employed with the prechamber was of surface discharge type located centrally in the chamber and thus eliminates the further clearance volume surrounding the spark plug. The original compression ratio was retained by reducing the clearance volume between the piston and cylinder head. The prechamber volume occupies 4.7 percent of the total clearance volume. A platinum paste consisting of 83 percent platinum powder, a flux containing lead (5 percent), butyl carbitol acetate 10 percent and resins was used. After application of the platinum paste the sample is required to be maintained at 650 deg C and above for nearly 2 hours so as to get a smooth platinum coating on the sample.

All the above catalyst materials were not tested for durability. Since the catalysts are present in the form of thin coatings (about 20-60 microns), their durability as well as the effect of carbon deposit on the catalyst surface are likely to be questioned. Nevertheless, the present work mainly aims to find the relative effectiveness of various catalysts for engine applications.

4 EXPERIMENTATION

A small capacity (150 cc, 4.3 kW at 5200 r/min), single cylinder, air-cooled, loop scavenged, two-stroke SI engine was employed in the present investigations. It was coupled to an eddy current dynamometer (VIBROMETER type CEB 104) for torque and speed measurements. Calibrated standard instrumentation was used to measure fuel and air flow rates. An infrared analyser (HORIBA, Mexa 324 FB type) was employed to measure Hydrocarbons (HC) and carbon monoxide (CO) levels in the exhaust gas. The cylinder pressure, measured by a flush mounted piezo-electric Kistler transducer was recorded and analysed in a data acquisition system (IWATSU Signal Analyser). A computer program based on instantaneous cylinder pressure and volume was developed to study the heat release rates. Cyclic variation of the peak pressure was measured by examining nearly one hundred successive pressure cycles. A schematic diagram of the experimental test setup is shown in fig 1.

Variable load tests were carried out at three different fixed speeds of 2000, 3000 and 4000 r/min respectively with corresponding MBT spark timings. Carburetor jet sizes of 0.98 to 0.80 mm diameter were available for the tests. A standard jet size of 0.84 mm was chosen for the experiments based on minimum exhaust emission levels. For some experiments, however, a jet size of 0.80 mm giving air-fuel ratio of the order of 15-17 was employed.

At each operating point, about 10-15 values of

speed and torque were continously recorded and averaging has been done to minimize the effects of variations. Similarly fuel and air consumption rates were recorded on the average basis. After calculating the performance parameters, a polynomial curve fitting was employed to draw the mean curves. The deviation of points from the mean curve based on the above method was kept minimum and the maximum deviation was less than 2 percent..

5 RESULTS AND DISCUSSION

5.1 Effect of Different Catalytic coatings

Figure 2 compares the performance and exhaust emission characteristics of different catalytic coating materials such as Copper, Nickel and Chromium, for engine operation with a carburetor jet size of 0.84 mm and engine speed of 4000 r/min. It can be seen that with all these catalysts, there is an improvement in the brake thermal efficiency and it is maximum with the copper catalyst. The percentage improvement over the base engine is about 10.8 percent in the case of the copper catalyst. The improvement in performance is mainly due to better combustion as a result of catalytic surface pre-reactions in the fuel-air charge prior to ignition.

HC emissions are lower with the Copper catalyst, but higher with Nickel and Chromium coatings. Lower HC emissions with copper catalyst indicates its superiority over other catalysts as a result of better oxidation of hydrocarbons. It can be seen that CO emissions are also significantly lower with all these catalysts when compared to the normal engine.

5.2 Effect of fuel-air mixtures with the Copper catalyst

Earlier works on lean combustion in the presence of catalyst [7] revealed that a significant reduction in cycle-to-cycle variations, extension of lean misfire limit and lower exhaust emissions could be achieved. Hence in the present work two fuel jet sizes of 0.88 mm(relatively rich mixture) and 0.80 mm(leanest) were tried. Copper catalyst was selected since it was the best among the catalysts tried. Fig 3 shows the performance and exhaust emission characteristics with and without the copper catalyst in the combustion chamber for the two fuel jet sizes of 0.88 mm and 0.80 mm at a constant speed of 4000 r/min. There is a clear difference in the brake thermal efficiency over the entire range of engine operating conditions, with and without the presence of catalysts, for both the fuel jets. The improvement in brake thermal efficiency is observed to be significant with lean fuel-air mixtures in the presence of the catalysts. The percentage improvement in the brake thermal efficiency for copper catalyst over normal engine operation is about 8.7 percent with fuel Jet size of 0.88 mm and 12.3 percent with the fuel Jet of 0.80 mm size.

HC and CO emission levels are significantly reduced for both the fuel jets in the presence of the catalyst. These improvements are mainly due to catalytic

242

activation of the charge leading to better oxidation of the hydrocarbons.

Figure 4 depicts the variation of combustion parameters such as peak pressure, ignition delay and combustion duration with the brake power at 4000 r/min. It can be observed that peak pressures are higher and ignition delay and combustion duration are generally lower in the presence of the catalyst. This effect is more significant with lean fuel-air mixtures. It has been shown [3] that heterogeneous catalysis lowers the barriers associated with gas phase reactions both in terms of higher flame velocity and reduced ignition energy. It allows more stable combustion by reducing both irregular and long duration of combustion initiation phase which result in lower ignition delay, reduced combustion duration and higher peak cylinder pressure.

5.3 Effect of Compression Ratio on Lean Mixture Catalytic Combustion

Several criteria for successful catalytic charge activation have been identified in the earlier studies[3]. These are:

(i) activation taking place in a well defined volume,

(ii) reasonably adiabatic conditions during the later stages of compression,

(iii) catalytic surface possessing as little thermal inertia as possible, and

(iv) catalytic surface having maximum contact between the catalyst and the surrounding gas.

One method of meeting the above requirements for better activation of the charge is increased compression ratio.

Figure 5 shows the performance and emission characteristics of the Copper coated combustion chamber at compression ratios (CR) of 9:1 and 7.4:1, with a Jet size of 0.80 mm and engine speed of 3000 r/min. The results are compared with normal engine operation. It is observed that the brake thermal efficiency of the engine operating with lean fuel-air mixture at the compression ratio of 9 is better compared to the performance at other conditions. It is also evident that the knock limit for maximum torque obtainable from the engine is higher at engine operation of CR 9 and at engine speed of 3000 r/min with the copper catalyst. The maximum torque increases from 8 N m to 9 N m.

HC and CO emissions are significantly lower with the catalyst at both the compression ratios over a broad range of engine operating conditions.

Figure 6 shows the variation of combustion parameters with brake power. Comparison has been made at both the compression ratios with and without the catalyst. Improved combustion is obtained as a result of charge activation process in the presence of the catalyst. This again leads to lower ignition delay and combustion duration and higher peak pressures particularly at the compression ratio of 9.

5.4 Hemispherical Catalytic Prechamber

Catalytic prechamber is an attractive concept which serves to initiate preignition reactions in the mixture adjoining the ignition source. Preignition catalysis acts to reduce the mixture ignition energy and improves the conditions for flame propagation. The primary reaction products which are produced at the catalytic surface are quickly removed from the surface and pushed into the space of the main chamber. Combustion initiation phase proceeds much faster and more uniformly as well as virtually synchronous with engine speed due to swirl combustion in the prechamber [9].

Figure 7 shows the schematic diagram of the catalytic prechamber and its position on the cylinder head.

Figure 8 compares the performance and emission characteristics without and with Platinum coating on the hemispherical prechamber (denotes HSP1 and HSP2 respectively) with those of the normal engine at 3000 r/min. Platinum is a very effective catalyst for oxidation of hydrocarbon mixtures. It can be seen that there is an improvement in the brake thermal efficiency and decrease in HC and CO emissions in the presence of platinum catalyst. But the prechamber without the catalyst affects the engine performance adversely due to more heat and flow losses in the swirl chamber and main combustion chamber, in particular at higher speeds and at higher outputs. Further investigations are continuing in an effort to find an optimum catalytic prechamber design with minimum losses even at higher speeds and at higher outputs.

6 CONCLUSIONS

- Coating the combustion chamber wall with catalytic materials has shown definite improvement in the fuel economy and the exhaust emissions of both HC and CO.

- Among the different catalysts investigated, copper is very effective in reducing both HC and CO emissions together with an improvement in the brake thermal efficiency of about 10.8 percent for the normal engine operating condition.

- Catalytic activation is more effective with lean-fuel-air mixtures and at higher compression ratios.

- There is an improvement in the brake thermal efficiency of about 12.3 percent at 4000 r/min with the lean jet in the presence of copper catalyst. Significant reductions in both HC and CO are also obtained.

- Ignition delay and combustion duration are lower and peak pressures are higher with the copper catalyst, with lean mixture combustion. Knock limited maximum torque increases at the compression ratio of 9 in the presence of copper catalyst.

- The presence of platinum catalyst in a hemispherical prechamber decreases CO and HC emissions, but there is a certain drop in maximum power output at higher speeds due to flow and heat losses. Further research work is necessary for developing an optimum catalytic prechamber.

ACKNOWLEDGEMENTS

Funds from the Department of Science and Technology, Government of India as well as donation of platinum paste by Johnson and Matthey, U.K. have made this research work possible and hence they are gratefully acknowledged.

REFERENCES

(1) TOMOYUKI WAKISAKA, YESHISUKE HAMAMOTO, SHUNICHI OHIGASHI, TAKASHI BANARI AND MAKOTO HOSOI
Limits of flame propagation in two-stroke cycle gasoline engines
Bulletin of the JSME
Oct.1976, Vol.19, No.136, 1204–1211

(2) PFEFFERLE, W. C. AND PFEFFERLE, L. D.
Catalytically Stabilized Combustion
Prog. Energy and Combustion Science
1986, Vol.12, 25–41

(3) STEVEN W.BEYERLEIN AND STARISLAW WOJCICKI
A lean burn catalytic engine
SAE Transactions
1988, Vol.97, Sec6, 1041–1051

(4) THRING, R. H.
The catalytic engine – Platinum improves fuel economy and reduces pollutants from a range of fuels.
Platinum Metals Review
1980, Vol.27, No.4, 126–133

(5) PFEFFERLE, W. C.
The catalytic combustor : An approach to cleaner combustion
Journal of Energy
1978, Vol.2, No.3, 142–146

(6) KARIM, G. A. AND KIBRYA, M. G.
Variations of the lean blowout limits of a homogeneous methane-air stream in the presence of a metallic wire mesh
Trans of ASME, Jl of Engg for Gas Turbine and Power
July 1986, Vol.108, 446–449

(7) RYCHTER, T. J. SARAGHI, R. LEZANSKI,T. AND WOJCICKI,S.
Catalytic activation of a charge in a prechamber of a SI lean-burn engine
The Combustion Institute
1981, Eighteenth Symposium (Int'l) on combustion, 1815–1824

(8) DHANDAPANI, S.
Experimental and Theoretical Investigations on a Single Cylinder, 4-stroke, lean burn SI engine
Ph.D Thesis
1989, IIT, Madras, India

(9) REINHARD LATSCH
The Swirl-chamber spark plug : A means of faster more uniform energy conversion in the spark ignition engine
SAE Transactions
1984, Vol.83, Sec.3, 365–377

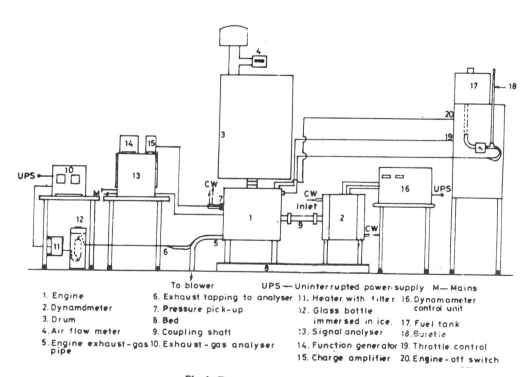

To blower   UPS — Uninterrupted power-supply   M — Mains

1. Engine
2. Dynamometer
3. Drum
4. Air flow meter
5. Engine exhaust-gas pipe
6. Exhaust tapping to analyser
7. Pressure pick-up
8. Bed
9. Coupling shaft
10. Exhaust-gas analyser
11. Heater with filter
12. Glass bottle immersed in ice.
13. Signal analyser
14. Function generator
15. Charge amplifier
16. Dynamometer control unit
17. Fuel tank
18. Burette
19. Throttle control
20. Engine-off switch

Fig.1 Experimental test set-up

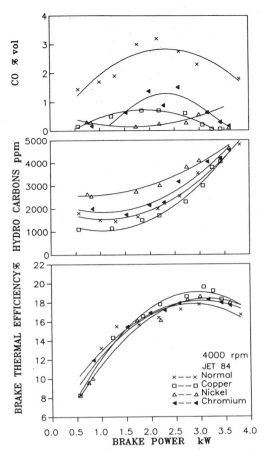

Fig.2 Variation of brake thermal efficiency, HC and CO emissions with brake power for different catalysts

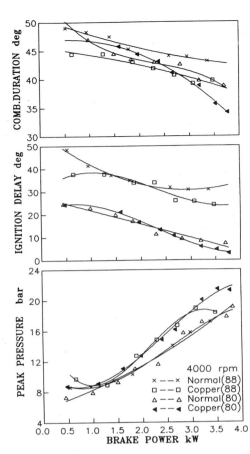

Fig.4 Variation of combustion parameters with brake power for Cu catalyst at different fuel jets

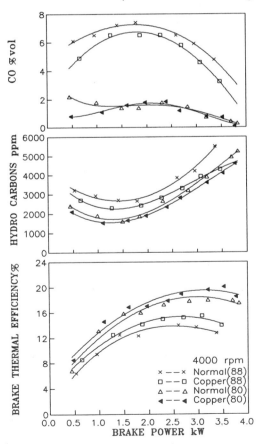

Fig.3 Variation of brake thermal efficiency, HC and CO with brake power for Cu catalyst at different fuel jets

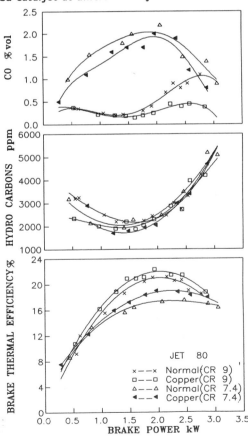

Fig.5 Variation of brake thermal efficiency, HC and CO emissions with brake power at 3000 r/min for different compression ratios

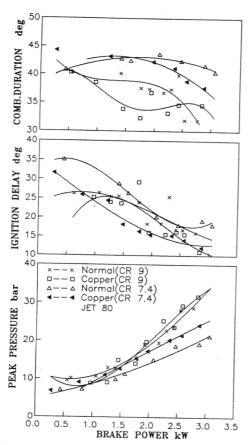

Fig.6 Effect of compression ratio on combustion parameters with Cu catalyst at 3000 r/min

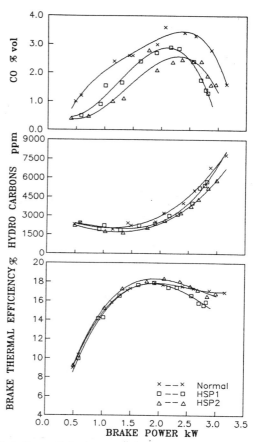

Fig.8 Variation of brake thermal efficiency, HC and CO emissions with brake power at 3000 r/min for catalytic prechamber

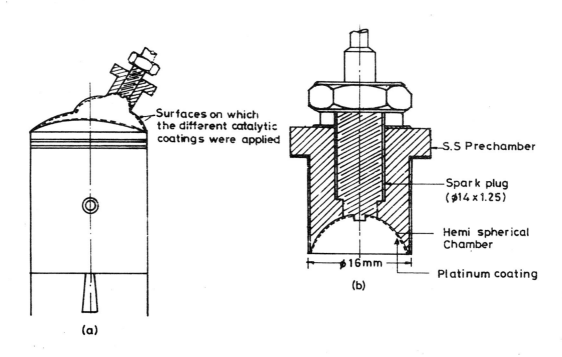

FIG. 7 .(a). SCHEMATIC VIEW OF THE MAIN CHAMBER AND PRECHAMBER
(b). ENLARGED VIEW OF THE CATALYTIC PRECHAMBER.

C448/033

Post-flame release of hydrocarbons in a spark ignition engine

G J HYMAS, PhD, CChem, MRSC and V J GRIGULIS, BSc, PhD
Department of Chemistry, Queen Mary and Westfield College, University of London
P G BROWN, BSc, MSc, PhD and W A WOODS, PhD, FEng, FIMechE
Department of Mechanical Engineering, Queen Mary and Westfield College, University of London

Synopsis   The background to the formation and discharge of hydrocarbon emissions from spark ignition engines is briefly discussed.   Recent work has revealed that hydrocarbon mole fractions measured close to the cylinder wall increase substantially during the post flame period of the expansion stroke. Results of recent experiments from an engine run on propane fuel, using rapid acting sampling valves in conjunction with an on-line gas chromatograph, have enabled further understanding of the processes involved.   Work has focussed on developing a technique to quantify the amount of fuel escaping combustion.   A new parameter has been formulated using the concept of an equivalent crevice, in which inflow and outflow are based upon the continuity equation.   This has become an additional tool in analysing the measured hydrocarbon mole fraction within an engine cylinder.   The model assumes that all storage mechanisms, crevices, deposits and oil films can be represented by a single volume acting directly on the sampling valve catchment volume.

Conclusions from the study are that the hydrocarbon species within the engine cylinder fit into two groups: fuel type and intermediate products. Both types of species undergo similar storage and release mechanisms, and the hydrocarbon mole fraction measured in the exhaust gases may not be a reliable indicator of the amount of fuel that escapes combustion. The simple mathematical description that has been derived has promising potential for helping to analyse this aspect of studying engine emissions.

1 INTRODUCTION

The need to reduce exhaust emissions from automobile engines is well known [1]. The particular problem of unburnt hydrocarbons is influenced by the chemistry of combustion reactions, the fluid mechanics of mixing and blowdown flows and the effects of thermodynamics and heat transfer. The sources of the unburnt hydrocarbons were suggested some time ago by Heywood [2] and aspects of some of these ideas have since formed the basis of many subsequent investigations [3], [4] and [5].

The work described in this paper is part of a wider study of hydrocarbon emissions from spark ignition engines which has been carried out to further the understanding of processes within the cylinder. A section of the programme of work was aimed at further understanding of the oxidation process by using gas chromatography in conjunction with rapid acting sampling valves, and taking the work of Kaiser et al [6] as the starting point.   To simplify the chromatographic analysis, these tests were carried out on an engine run on propane; all other tests have used petrol. The study has concentrated on testing at light load and slow speed as this aspect of engine operation is one about which relatively little is known.   Additionally, future emissions regulations will include all the exhaust from an engine, including cranking and cold start.   Understanding port load operation of an engine thus becomes important.   The work is related to a keynote paper [7], to be presented at the Conference.

The objects of this paper are, first, to present a different approach to the analysis of time resolved hydrocarbon measurements made in the combustion chamber of a firing engine and, secondly, to discuss the results of an experimental investigation carried out on a production 2l OHC spark ignition engine, using a rapid acting sampling valve in conjunction with an on-line gas chromatograph.

The outcome of this dual disciplinary approach is a new hydrocarbon storage parameter which is useful in helping to interpret experimental results. The paper contains a short description of the engine test bed, and the instrumentation and control systems. The hydrocarbon storage model is outlined and the test results and the analysis of them are presented. The discussion and conclusions show that the storage model interprets the result in a plausible and convincing way. It is shown that measurements from within an engine cylinder must be treated with caution before they are interpreted as indicators of the quantities of unburnt hydrocarbons which leave a spark ignition engine.

2. EXPERIMENTAL APPARATUS

The test bed consists of a Ford 2*l* SOHC engine coupled to a Heenan and Froude DPX2 water brake. The engine carburettor was modified with a commercially available conversion kit for use with propane fuel so that gaseous or liquid fuel could be used. The engine cylinder head was drilled to give access to the combustion chamber for a sampling valve and a pressure transducer. The shape and size of the sampling valve body restricted access to the combustion chamber, and it was not possible to take samples away from the wall. Thermocouples were installed to measure coolant, lubricant, and exhaust gas temperatures. Gas analysis equipment was fitted to measure exhaust pipe hydrocarbons, oxygen, carbon dioxide and carbon monoxide. The output signals from these instruments were connected to an analogue to digital converter, coupled to an IBM PC-XT computer. The signal from the hydrocarbons analyser was also sent to a chart recorder. The test bed arrangement is shown in schematic form in Fig. 1.

The engine ignition was modified to incorporate an Ignition Timing and Measurement System (ITAMS). This enabled the investigators to over-ride the standard distributor setting and use any chosen angle of advance instead. Gas analysis using Spindt's equation [8] was used to determine the air fuel ratio at which the engine was operating.

The total hydrocarbons analyser was used to measure the propane equivalent mole fraction of samples from any one of three sources: the exhaust pipe average tube, the exhaust pipe sampling valve or the combustion chamber sampling valve. For this purpose, changeover valves were installed in the sample line as shown in Fig. 1. The sample line was heated to 180 °C to avoid condensation. The experimental apparatus and sampling valves are described in detail in Ref. [9].

The gas chromatograph used was a Pye PU4550 fitted with a flame ionisation detector. This was used in conjunction with a Philips PU4810 computing integrator. A sample loop and mixing chamber were connected to the sample line from the sampling valve to the total hydrocarbon analyser; these were used to extract samples for chromatographic analysis, as shown in Fig. 1. Hydrocarbon mole-fractions were determined with reference to an external standard of propane in nitrogen. Relative response correction factors for the saturated hydrocarbons:- methane, ethane, propane, isobutane, and n-butane, were calculated from an analysis of an equimolar reference mixture in nitrogen. Unsaturated species were assigned the same response as their saturated homologues. Further details of the chromatographic analysis, can be found in Ref. [10].

3. HYDROCARBONS STORAGE MODEL

Results obtained by LoRusso et al [5] and measurements taken in the combustion chamber in the present work showed that the hydrocarbon mole fraction close to the cylinder wall increased subsequent to the flame traversing the combustion chamber. By the time the exhaust valve opened, the hydrocarbon mole fraction had increased to about one third of that of the unburned mixture. This increase was interpreted as the release of unburned and partially burned fuel that had escaped combustion and was convected into the catchment-volume of the sampling valve. This effect was present in both the standard and compact cylinder head engines and for both propane and petrol fuels. The conclusion drawn from this result was that the observed rise in hydrocarbon mole fraction was a fundamental process not confined to a specific engine geometry or type of fuel. It is important to note that the increase observed is measured as a mole fraction and not a hydrocarbon concentration.

A description of the release of unburnt hydrocarbons was made based upon the premise that the unburned fuel was stored in a hypothetical crevice which emptied into the catchment volume of the sampling valve, as illustrated in Fig. 2. A simple inflow and outflow mathematical description applied to the scheme, illustrated in Fig. 2, can be used to derive the equations given below.

Assuming the pressure and temperature in the hypothetical crevice are the same as those in the engine cylinder, that subscript o denotes the conditions at the start of the release process and subscript θ the conditions at any subsequent crank angle, then the number of moles of hydrocarbons that have left the crevice by crank angle θ is given by:-

$$\frac{V_c}{R} M_c \left[\frac{P_o}{T_o} - \frac{P_\theta}{T_\theta} \right]$$

where V_c is the crevice volume

     M_c is the hydrocarbon mole fraction of the gases stored in the crevice

     R is the Universal gas constant

     P is the pressure in the cylinder and in the crevice

and T is the temperature in the cylinder and in the crevice.

The unburnt and partially burnt air fuel mixture that leaves the crevice is assumed to displace an equivalent volume of burnt gas from the sampling valve catchment volume. If the burnt gas occupies a volume V_r, then, at the start of the release process, this volume will be equal to V_s, the volume of the sample. At any subsequent crank angle θ, V_r is less than V_s by the amount displaced by the expansion into the sample volume of the crevice gases. The sample gases are measured as the mean hydrocarbon mole fraction M_m which is given by

$$\frac{P_\theta V_s}{R T_\theta} M_m = \frac{V_c}{R} M_c \left[\frac{P_o}{T_o} - \frac{P_\theta}{T_\theta} \right] + \frac{P_\theta}{T_\theta} \frac{V_r}{R} M_r$$

(1)

where V_s is the sampling valve catchment volume

M_m is the measured mean hydrocarbon mole fraction

V_r is the volume of the burnt gases in the sample

and M_r is the hydrocarbon mole fraction of the burnt gas.

By expressing the volume of burnt gas in terms of the crevice volume, the above equation can be manipulated and simplified to give:-

$$M_m = A V_\theta + B$$

(2)

where

$$A = \frac{V_c}{V_s} \left[\frac{M_c - M_r}{V_o} \right]$$

and

$$B = \frac{V_c}{V_s} \left[M_r - M_c \right] + M_r$$

Thus, for gaseous storage in a crevice, the release process can be expressed as a linear relationship with the cylinder volume. The slope of the release line is represented by the parameter A. The derivation of this is given in Ref. [10].

4. RESULTS AND DISCUSSION

Hydrocarbon species analysis of combustion chamber samples was confined to experiments using propane fuel in order to simplify the analysis. The individual species that made up the measured hydrocarbon profile could be divided into two distinct groups according to profile shape. These two groups are described in detail below.

a) COMBUSTION PRODUCTS

The mole fraction of these species prior to ignition is primarily dependent upon the residuals from the previous cycle. This group consists of methane and ethylene, which were present in the fuel in negligible quantities compared to those arising from combustion. The mole fraction of ethylene measured in the combustion chamber is shown in Fig. 3a, as a function of crank angle. It is representative of the characteristic shape of this group of hydrocarbons.

The mole fraction, prior to ignition, is the lowest in the range 300° to 350° crank angle. It rises to a peak around TDC, as the flame front passes through the catchment

volume of the sampling valve. The mole fraction then falls to a minimum at 400° crank angle, then increases during the rest of the expansion stroke to a maximum at EVO.

b) FUEL COMPONENTS

The mole fraction of these species prior to ignition is dependent upon the composition of the fuel. This group consists of propane, propylene, ethane and the butanes. The mole fraction of propane measured in the combustion chamber is shown in Fig. 3b as a function of crank angle. This is representative of the characteristic shape of this group of hydrocarbons.

The mole fraction is greatest prior to ignition, in the range 300° to 350° crank angle. There is a rapid decrease to a minimum at around 400° crank angle, as the flame front passes through the sampling valve catchment volume. During the rest of the expansion stroke, the mole fraction increases to a maximum around EVO, then for the remainder of the cycle shows a slight decrease.

This group of species typically accounts for more than 80 percent, by mass, of the post flame hydrocarbons, i.e. the samples are primarily of unburned fuel air mixture that had escaped combustion.

The variation of total hydrocarbons in the combustion chamber has the same profile as the propane, shown in Fig. 3b. Similar results have been reported previously [5], [11]. The species analysis discussed above shows that this is because the sample is primarily made up of original fuel. It is apppropriate to point out that while the level of total hydrocarbons measured at blowdown can be as great as one third of that recorded prior to ignition, this does not mean that only two thirds of the fuel is consumed. The results shown are only applicable to the region sampled, i.e. close to the cylinder wall. The recorded levels are not representative of the bulk gas conditions. However, the recorded increase in hydrocarbon mole fraction following combustion, after 400° crank angle in Fig. 3b, shows clearly that fuel has escaped the flame front by one means or another and is being convected into the sample volume as the cylinder pressure falls. This effect has been observed for various designs of combustion chamber and for a variety of fuels.

As the species analysis showed, the sample is primarily of fuel. Therefore, the profile of the total hydrocarbon trace, which is similar to Fig. 3b, could be used for the purpose of comparing the post flame period of the expansion stroke for different engine operating conditions. The hydrocarbon mole fraction during this period of the cycle, 400° crank angle to EVO, shown in Fig. 3b, is interpreted as the release of unburned fuel into the catchment volume of the sampling valve. The measured hydrocarbon mole fraction has been found to be directly proportional to the cylinder volume, see Fig. 4. This is in agreement with equation 2, derived in the previous section. By applying a linear least squares analysis to the experimental data, values for the parameters A and B were determined.

The parameter A is proportional to the volume of stored gases, i.e. the size of the hypothetical crevice volume. It is also approximately proportional to M_c, the hydrocarbon mole fraction of the unburned air fuel mixture. For fuel storage in a crevice, changes in engine operating conditions would not be expected to change significantly the ratio of storage volume to sampled volume, V_c/V_s. Neither would changes in operating conditions be expected to change significantly the mole fraction of the residuals, M_r following flame passage, or the cylinder volume V_o, at which the release process starts. Therefore, changes in the value of A are indicative of changes in the amount of unburned and partially burned air fuel mixture that escapes combustion.

As the ratio V_c/V_s is not known, a storage parameter ψ is defined as $\psi = A/A^*$ where A^* is a reference value of A. The variation of the storage parameter is shown in Fig. 5. For this figure, a value of A^* used was the value corresponding to a fuel-air ratio of 13.5 and a spark timing of 20° BTDC.

This parameter shows that fuel storage increases with equivalence ratio and angle of advance. This trend is to be expected for a fuel storage mechanism, as increasing the equivalence ratio increases the mole fraction of the hydrocarbons in the stored gases and advancing the spark timing increases peak cylinder pressure, resulting in more air fuel mixture being stored. The storage parameter is a tool to provide a means to measure the relative extent of fuel storage for different operating conditions.

Based as it is on the magnitude of ψ the storage parameter can only show the extent to which fuel storage mechanisms have changed with engine operating conditions, and not the absolute amount of fuel that has been stored. The value of ψ of the stored material, is based on the assumption that the storage took place in a hypothetical crevice associated with the sampling valve catchment volume. However, the result of using the storage parameter, shown in Fig. 5, clearly indicates that the basic assumption of a hypothetical crevice volume is correct, as the storage parameter varies in the manner expected for a crevice storage mechanism.

The results of applying the storage parameter to the measurements for a variable engine speed are shown in Fig. 6a. In this experiment, the engine was run on petrol at an air fuel ratio of 13 and an angle of advance of 24 degrees

BTDC. The engine speed was varied between 900 and 1300 r/min by adjusting the applied brake load. It can be seen that the storage parameter, taking the reference condition A^* to be 1000 r/min at an air-fuel ratio of 13 and spark timing of 24° BTDC, indicates an increase in fuel storage as the engine speed is increased. In contrast, the exhaust average hydrocarbon mole fraction decreased with increasing engine speed, see Fig. 6b. The exhaust pipe mean gas temperature also increased with increasing engine speed, see Fig. 6c.

This shows that it would be misleading to take the result in Fig. 6b and come to the conclusion that because the exhaust average hydrocarbon mole fraction was reduced with increased engine speed, the fuel storage was reduced at the higher speed. Application of the storage parameter to measurements from the combustion chamber shows that increasing the speed by 30 percent almost doubles the fuel storage. This can be observed in Fig. 6a.

The result has other implications concerning the type of fuel storage mechanism. The increase in the storage parameter with an increase in engine speed shows that more fuel is being stored. If lubricating oil absorption and desorption of fuel vapour were a significant storage mechanism under the conditions tested, then less fuel storage should take place because the increased speed would cause a reduction in penetration depth into the oil film. At the higher engine speeds, the peak cylinder pressure is reduced because the speed change was effected by altering the applied brake load. This too is consistent with reducing the effect of lubricating oil film absorption and desorption. Note also that a reduction in peak cylinder pressure would be expected to reduce fuel storage in a crevice. This suggests that an alternative storage mechanism such as quench layers, which would be thicker with reduced peak cylinder pressure, or bulk quenching could be responsible for the increased fuel storage.

Summarizing, this parameter shows that, for an engine run on petrol at various speeds fuel storage increases when the exhaust average hydrocarbon mole fraction suggests decrease and that there may be a change in emphasis between storage mechanisms as engine conditions alter.

As the samples taken close to the cylinder wall from an engine run on propane were found to be composed primarily of fuel-type hydrocarbons, it has been shown that it is possible to use a simple inflow and outflow model to describe the measured release effect. This approach has been used by the authors in work on warm-up tests as part of the UK engine emissions consortium hydrocarbons project P4.

5. CONCLUSIONS

1. Measurements of hydrocarbon species mole fraction within the combustion chamber of a spark ignition engine run on propane have shown two distinct groups of species. These are fuel components and those arising from combustion.

2. The samples taken from the combustion chamber are comprised primarily of fuel hydrocarbons which form more than 80 percent by mass.

3. The concept of a hypothetical crevice to describe the measured post flame release effect has been introduced and shown to be consistent with the readings obtained.

4. The concept of a storage parameter, to help analyse the fuel storage processes under different engine operating conditions, has been derived.

5. The storage parameter shows that the measured hydrocarbon mole fraction release effect is consistent with the storage mechanism being that of fuel storage in a crevice.

6. The results from a test on an engine run on petrol in which the speed was varied have shown that the exhaust average hydrocarbon mole fraction does not necessarily reflect the conditions within the engine cylinder. For the work reported here, the exhaust average hydrocarbon reading showed a reduction, while the combustion chamber measurements showed a corresponding increase in fuel storage.

7. Use of the storage parameter for the case of varying engine speed suggests that lubricating oil film absorption and desorption of fuel vapour may not be a significant fuel storage mechanism under the conditions tested.

ACKNOWLEDGEMENT

The authors wish to thank the Science and Engineering Research Council for a research grant No. GR/D/40029 which supported this work. They also wish to thank the Ford Motor Company Ltd., particularly Mr T Biddulph and Dr T Ma for their help and support in connection with this research.

REFERENCES

1. Heywood, J. B. "Internal combustion engine fundamentals". McGraw Hill Inc. New York, 1988, pp. 1-930.

2. Heywood, J. B. "Engine combustion modeling - an overview". pp. 1-38. Combustion modeling in reciprocating engines, Proc. of a symposium held at General Motors Research Labs. 1978. Edited by J N Mattavi and C A Amann, Plenum Press, New York, 1980, pp. 1-605.

3. Dyer, T. M. "Sources of unburned hydrocarbon emissions from homogeneous charge spark ignition engines". Sandia report SAND 83-8241, UC 96. Sandia National Labs. Albuquerque, New Mexico 87185 and Livermore, California 94550, Sept. 1983.

4. Ramos, J. I. "Internal combustion engine modeling". Hemisphere Publishing Corp. New York, 1989, pp. 1-422. (See p. 290-344).

5. LoRusso, J. A., Lavoie, G. A., and Kaiser, E. W. "An Electrohydraulic Gas Sampling Valve with Application to Hydrocarbon Emissions Studies." SAE paper 800045 Trans. Vol. 89, 1980.

6. Kaiser, E.W., Rothschild, W.G., Lavoie, G.A., "Storage and Partial Oxidation of Unburned Hydrocarbons in Spark Ignited Engines - Effect of Compression Ratio and Spark Timing". Combustion Science and Technology, Vol. 36, p. 1771, 1984.

7. Boam, D.J. et al. "The Sources of Unburnt Hydrocarbon Emissions from Spark Ignition Engines during Cold Start and Warm-up".

8. Spindt., R., S. "Air Fuel Ratio's from Exhaust Gas Analysis" SAE paper 650507 1965.

9. Panesar, A., Woods, W. A., Brown, P.G. "Instrumentation for Determining the Time History of Hydrocarbons in Spark Ignition Engines". Proc. I.Mech.E Seminar London Dec 1991 pp 19-27.

10. Hymas, G.J. "In-Cylinder Release of Unburned Hydrocarbons in a Spark Ignition Engine". Phd. Thesis, University of London, 1991.

11. Panesar, A., Brown, P.G., Woods, W.A. "The Results of Recent Experiments on Unburnt Hydrocarbons" Proc. I.Mech.E Int. Conf. on Combustion in Engines ISBN 085298653x C50/88 1988, pp. 261-271.

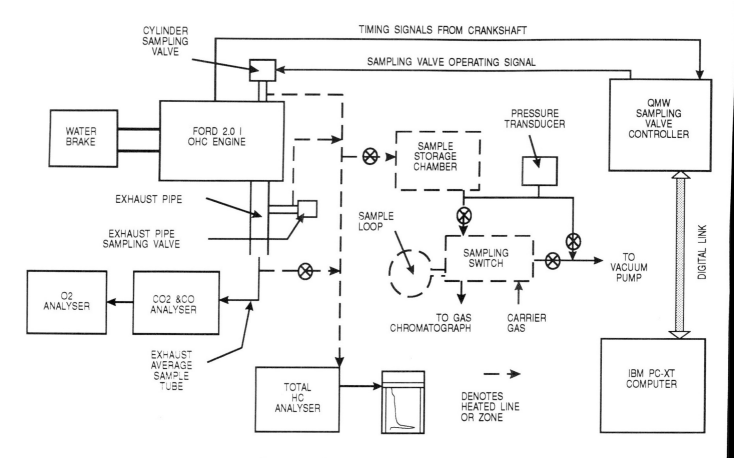

FIG. 1. ENGINE TEST BED SCHEMATIC

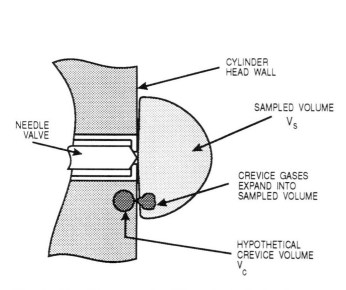

Fig. 2. The Concept of a Hypothetical Crevice used to describe the Storage and Release of Unburned and Partially Burned Air-fuel Mixture.

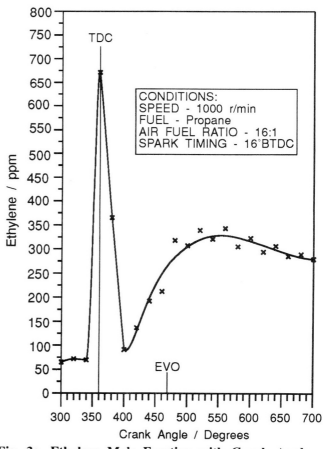

CONDITIONS:
SPEED - 1000 r/min
FUEL - Propane
AIR FUEL RATIO - 16:1
SPARK TIMING - 16°BTDC

Fig. 3a. Ethylene Mole Fraction with Crank Angle, Characteristic Profile of Combustion Products.

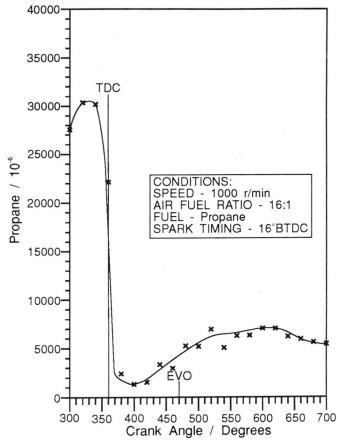

Fig. 3b. Propane Mole Fraction with Crank Angle, Characteristic Profile of Fuel Type Species.

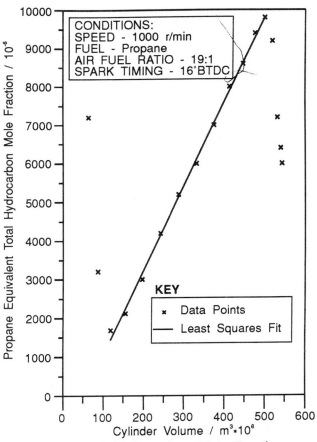

Fig. 4. Total Hydrocarbon Mole Fraction Linear with Cylinder Volume.

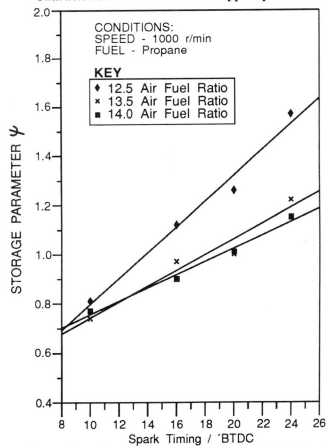

Fig. 5. Variation of Storage Parameter with Spark Timing and Air Fuel Ratio

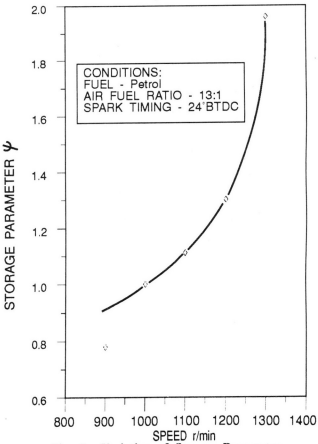

Fig. 6a. Variation of Storage Parameter with Engine Speed.

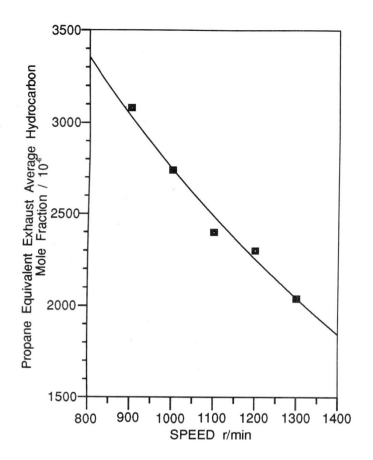

Fig. 6b. Variation of Exhaust Hydrocarbons with Engine Speed

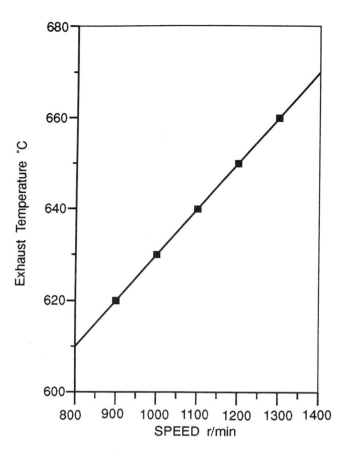

Fig. 6c. Variation of Exhaust Temperature with Engine Speed.

254